Muenscher's Keys to Woody Plants

The leaves of poison-ivy and poison-sumac and some harmless plants with which they are often confused.

1, Poison-ivy, *Toxicodendron radicans* L. Petiole bearing three leaflets; buds visible. 2, Virginia creeper, *Parthenocissus quinquefolia* (L.) Planch. Petiole bearing five leaflets. 3, Silky dogwood, *Cornus amomum* Mill. Leaves opposite, simple. 4, Fragrant sumac, *Rhus aromatica* Aiton. Petiole bearing three leaflets: buds hidden under base of petiole. 5, Poison-sumac, *Toxicodendron vernix* L. Leaves alternate: petiole bearing several leaflets with entire margins: buds visible in the axils. 6, Dwarf sumac, *Rhus copallina* L. Margin of leaflets entire or toothed, leaf axis winged. 7, Smooth sumac, *Rhus glabra* L. Margin of leaflets serrate, buds hidden under base of petiole. 8, Staghorn sumac, *Rhus typhina* L. Similar to 7, but with hairy leaves and twigs. 9, Mountain-ash, *Sorbus americana* DC. Margin of leaflets serrate: buds visible. 10, Black ash, *Fraxinus nigra* Marshall. Leaves and buds opposite. 11, Elderberry, *Sambucus canadensis* L. Leaves and buds opposite. (After Muenscher, in *Cornell University Experiment Station Extension Bulletin 191.*)

MUENSCHER'S KEYS TO WOODY PLANTS

An Expanded Guide to Native and Cultivated Species

EDWARD A. COPE

Comstock Publishing Associates

A DIVISION OF CORNELL UNIVERSITY PRESS

Ithaca and London

Original edition, *Keys to Woody Plants* by W. C. Muenscher, copyright © 1930, 1936, 1946, 1950 by Comstock Publishing Company, Inc.

First published, July, 1922
Second edition, July, 1926
Third edition, July, 1930
Fourth edition, March, 1936
Fifth edition, June, 1946
Sixth edition, May, 1950

Muenscher's Keys to Woody Plants: An Expanded Guide to Native and Cultivated Species, first published 2001 by Cornell University Press

First published 2001 by Cornell University Press
First printing, Cornell Paperbacks, 2001

Printed in the United States of America

Library of Congress Cataloging-in-Publication Data
Cope, Edward A., 1948–
Muenscher's keys to woody plants: an expanded guide to native and cultivated species / Edward A. Cope.
p.   cm.
Enl. ed. of: Keys to woody plants / W.C. Muenscher. 6th ed., rev. 1950.
Includes bibliographical references (p.   ) and index.
ISBN 0-8014-3852-7 (cloth: alk. paper)—ISBN 0-8014-8702-1 (pbk.: alk. paper)
1. Trees—United States.   2. Shrubs—United States.   I. Title: Keys to woody plants.   II. Muenscher, Walter Conrad Leopold, 1891–1963. Keys to woody plants.   III. Title.
QK115 .C69 2001
582.16′0974—dc21   00-010701

Cloth printing   10 9 8 7 6 5 4 3 2 1
Paperback printing   10 9 8 7 6 5 4 3 2 1

CONTENTS

PREFACE

It is a privilege and a source of much satisfaction to present a new version of one of the classic practical volumes in eastern North American botany, *Keys to Woody Plants* by Walter C. Muenscher. I have held this book up as the very model of a good vegetative key to plants of a region. Widely used by amateurs, students, and professionals, this book, originally published in 1922, has been available in six editions, the most recent in 1950.

Taxonomy is a summation of the current knowledge about organisms and their relationship to one other at any particular time, with any further studies typically yielding information that requires changes in the names of plants. Hence botanical nomenclature, which reflects these relationships, must constantly change, ever evolving to reflect new information affecting our understanding of plant relationships. In the fifty years since Muenscher's sixth edition was published, many such changes have occurred, including many name changes required by rules of the International Botanical Congress. It is frustrating, particularly for the student or beginner, to key a plant successfully and then to have to search the library for the current correct name. Until I began the revision, this need to update the nomenclature was my prime purpose in reworking *Keys to Woody Plants*.

While approaching this project with a hefty respect for Muenscher and his fine work, I discovered that many parts of the keys could be strengthened and made more consistent. I also realized that Muenscher had included more than a few cultivated plants, many of which were not common. This discovery suggested the possibility of widening the focus to include an extensive coverage of cultivated plants. For some time I have wanted to offer a book of vegetative keys to cultivated plants, as presented in the equally dated, out-of-print, and heretofore only such comprehensive work, *Trees and Shrubs in Eastern North America* by Benjamin Blackburn.

The result is a much lengthier work, which is essentially two books in one. The first is an extensive reworking of the keys—still with the basic Muenscher structure—to include native and cultivated plants together. The second is an abridged key, which identifies only native genera and introduced genera that have become naturalized. The user can try the second, much shorter key to identify a plant more quickly and simply when he or she is in a natural setting, whereas the other set of keys is available for an obviously cultivated or horticultural situation, or when it is not known whether a plant is native.

In expanding the 120 genera and 187 species in his first publication of *Keys to Woody Plants*, Muenscher had added 44 genera and 270 species by the time he published his sixth edition, for a total of 164 genera and 457 species. Of these species, 277 were native, 40 were naturalized, and 140 were cultivated. I have increased these totals to 335 genera (174 native, 161 cultivated or naturalized) and 1171 species (589 native, 582 cultivated, including 198 naturalized). I have made few alterations to the glossary, and I have annotated and greatly reduced the bibliography, retained all illustrations, and, of course, substantially revised the systematic list of species. With such an extensive revision, the fear arises that the book is so altered that its utility is sacrificed. I hope this is not the case, and to avoid that, I have tried to be mindful of the objectives laid out in Muenscher's preface—to help the user identify woody plants and to familiarize her or him with keys and descriptive terms.

Keys to Woody Plants has influenced many amateurs and professionals, including members of my own family. My father, James B. Cope, took a course from Muenscher at Cornell and handed his copy of the fourth edition (with notes on some excluded species in the margins) to me at an early age. I am indebted to him for sparking an interest in tree identification and teaching me to learn trees by their bark. A widely circulated Cornell bulletin, called *Know Your Trees*, by my grandfather, Joshua A. Cope, an extension forester at Cornell, also was useful in this revision. I am grateful for support from William L. Crepet, director of the Bailey Hortorium at Cornell University; Kevin Nixon, curator of the Bailey Hortorium Herbarium (who also developed the oak keys with me); and Robert Dirig and Sherry Vance, colleagues at the Bailey Hortorium. I thank Rick Uva and James B. Cope for testing some of the keys; Peter Prescott, who has been encouraging and helpful in setting realistic deadlines for completion of the manuscript; and Nancy Winemiller, whose work with the format, attention to detail, and willingness to understand my intent has produced a much better book.

In addition to Muenscher, I am indebted to the many plant systematists and foresters whose descriptions and keys I have freely consulted. I have been fortunate to have easy access to the sizeable herbarium of the Bailey Hortorium, which contains one of the country's most extensive collections of cultivated plants, as well as many collections of Muenscher's, including specimens he examined while writing *Keys to Woody Plants*.

E. A. C.

PREFACE TO THE FIRST EDITION

A woodsman or farmer from practical experience usually recognizes by general appearances or by certain "ear marks" the common trees and a few shrubs of the region with which he is familiar. But take him out of this region and place him among different trees and his familiarity with general appearances will be of little value to him in trying to identify his new tree friends either with or without the aid of a book. The reason is twofold. First, it is very difficult to describe and incorporate general appearances in descriptions of plants, and second it therefore follows that usually the best books on trees are written in the terminology of systematic botany which is often too technical for the average person who is or who should be interested in the woody plants around him. The student of systematic botany on the other hand must depend to a large extent upon the floral structure and arrangement in his descriptions of plants and is often puzzled when asked to identify a plant when it is not in flower or when it is in a leafless condition, because of a lack of knowledge of general appearances.

The purpose of this little guide is twofold: first, to aid the beginner in identifying some of the more common woody plants in the summer or winter condition, and second, to familiarize the student with descriptive terms and the use of keys and teach him how and what to look for when examining a woody plant, or a part of it, with a view to its identification. It is hoped that the beginner who has been able to use this guide with a reasonable degree of success will have acquired sufficient skill in the use of keys and the observation of diagnostic characteristics of woody plants in the summer or winter condition to enable him, with the aid of more comprehensive works, to identify practically all the woody plants desired.

In these keys the aim has been to point out, as far as possible, diagnostic field characteristics, exclusive of flowers. The summer keys are based primarily upon leaves and fruits; the winter keys are based primarily upon bud and twig characters, supplemented by those of bark, general habits, and fruits.

In the preparation of these keys the writer has made free use of every available source of information. Although he has referred to many manuals and books on trees and shrubs, the keys are based upon actual study of the woody flora about Cornell University and other regions. Every species included in the keys has been studied from living material.

W. C. M

July 1922

A NOTE ABOUT
WALTER C. MUENSCHER

Keys to Woody Plants, the basis and the inspiration for the present book, was first published privately by Walter C. Muenscher in July 1922. It has been issued in six editions and many printings. Affectionately known as the "Wizard of Weeds" among his Cornell students and colleagues, Professor Muenscher wrote over 125 articles and books during his lifetime. Having identified thousands of weeds and poisonous plants that were sent to him as a professor of botany, he continued in these endeavors until his death in 1963.

Awarded the Certificate of Merit from the Botanical Society of America, Muenscher served at various times as a botanist with the USDA and New York Biological Survey. He also was a board member of the Wild Life Preservation Society, trustee of the Wildflower Preservation Society of America, and an active member of the Torrey, the New England and the California Botanical Clubs. For decades he worked identifying hundreds of plant specimens in the swampy regions of upstate New York. Eventually his plant studies took him into every state in the country.

Professor Muenscher, a graduate of State College of Washington, earned a Master's degree in botany in 1915 from the University of Nebraska and a Ph.D. in 1921 from Cornell, where he became assistant professor in 1923 and full professor in 1937 before retiring in 1954.

Notes to Muenscher's previous editions acknowledge thanks for assistance throughout the preparation and proofreading of the original keys to Minnie Worthen Muenscher, W. E. Manning, K. M. Weigand and A. J. Eames, and A. S. Forster and members of the staff at the Arnold Arboretum. Thanks for special assistance is also given to former students and other users, including Robert T. Clausen and Babette I. Brown.

MUENSCHER'S KEYS TO WOODY PLANTS

INTRODUCTION

In 1922 Walter Conrad Muenscher published *Keys to Woody Plants*, which was to become one of the finest and most useful woody plant guides ever to adorn the botanist's and forester's bookshelf. Six editions and multiple printings of these field manuals were published, many of which were used by students in learning the identlty of woody plants of temperate eastern North America. Well-designed, effective keys and restraint in the use of technical terminology are prominent reasons the book has met with so much success.

The revised sixth edition, published in 1950, included 164 genera and 457 species. In this edition I have added 171 genera and 710 species to make a total of 335 genera and 1171 species. Some of these are additional southern native species, but about 80 percent of my additions are cultivated species, bringing the total cultivated species included to 582 (161 genera). Of these, 198 species are naturalized from horticultural or from inadvertent introductions of non-native species, although about half are only occasionally or rarely naturalized.

Although useful outside of these boundaries, this revision includes woody plants found in eastern North America, defined as that area east of the Mississippi River to the Atlantic Ocean, north to the Arctic and south to include northern Mississippi, Alabama, and Georgia, and all of South Carolina except for five southeastern counties (Hampton, Colleton, Jasper, Charleston, and Beaufort). This range includes many coastal plain trees and shrubs. The range for cultivated plants included is the same, although the southern boundary does not extend as far south. The southern limit for cultivated plants should be considered roughly the northern borders of Mississippi, Alabama, and Georgia, and perhaps the northern third of South Carolina, a line that is roughly equivalent to the northern border of USDA Hardiness Zone 7b.

My decisions about which cultivated plants to include were not always consistent but were based on deductions about commonly cultivated plants and those that are most familiar to me. A general rule was to include any tree or shrub that is grown outside an arboretum or botanical garden, although I will not deny a preference for taxa that are more familiar to me or more easily worked into the keys. Sacrificing consistency for increased information seems worthwhile, as this work includes nearly any genus and most species of woody plants that the user will encounter in the geographical area covered.

Additions of native woody plant species are mostly accounted for by a more precise and somewhat expanded southern boundary of the geographical range covered, my tendency to include more of the locally endemic species, and the inclusion of woody grasses or herbs that appear to be woody, such as *Polygonum cuspidatum*.

The appearance and structure of this revision should bring to mind the original. Two plates illustrating terms from the original publication are retained, as these remain useful for depicting the terms used. A third plate from the original publication is included showing the most important and widely distributed rash-causing poisonous plants and their harmless look-alike relatives. The decision was made to keep this volume less expensive and more useful in the learning process by using words rather than relying on illustrations for many of the plants. Illustrations can be found in some of the volumes listed in the annotated bibliography.

HOW TO USE A DICHOTOMOUS KEY

A key is a device that facilitates identification of an unknown object. The most effective key for plant identification is a dichotomous key. A dichotomous key offers two alternatives at each step. The plant should match one of the two sets of characteristics described. The choices are numbered identically and set at equal distances from the margin. Having made a choice at the first step, the user proceeds to the next step (that is, to the next number in sequence) beneath that choice. This next step is further indented. The process is repeated until the user arrives at the plant name. The two coordinate leads always bear the same index number. The subordinate leads under the main leads are numbered and indented consecutively. When no further sudivision is possible the name of the genus or species is determined. Descriptions, drawings, photographs, or herbarium specimens may be consulted to verify identification.

It is important always to examine both alternatives at each step in the key. Several distinctive characters may be offered at each step to help the reader decide which path to follow. When a decision seems especially difficult, one path should be followed, then the other. Sometimes it is more efficient to work only partway through a key to reduce the number of possibilities. Descriptions, pictures, or herbarium specimens of the remaining possibilities may then make proper identification possible. At certain difficult points in a key the same name can be reached through either choice. For example, in the key to the species of *Ilex*, *I. opaca* can be found under either the first step 16 (Leaves entire, terminating in a spine;

fruit red) or the second step 16 (Leaves with coarse, spiny teeth; fruit red or black). Leaves of *I. opaca* are normally spiny, but sometimes entire.

EXPLANATION OF THE KEYS

This book contains several categories of keys. Two principal keys are themselves broken down into smaller categories. The first key is an all-inclusive, unabridged key to nearly every genus of woody plant—whether it be native, naturalized, or cultivated—that you might find growing outside of a botanical garden or arboretum or specialized private estate in the range defined earlier. The second key is a much smaller key that includes only native and naturalized genera of woody plants. Each of these overarching keys is broken into seven smaller keys:

Plants with needlelike, scalelike, or awl-shaped leaves
Plants with opposite or whorled simple leaves
Plants with opposite or whorled compound leaves
Plants with alternate simple leaves
Plants with alternate compound leaves
Plants lacking leaves, with the leaf scars opposite or whorled
Plants lacking leaves, with the leaf scars alternate

The inclusive key to genera allows exhaustive coverage for a broad range of plants. The abridged key to genera allows faster and easier use (more like the original Muenscher key) while still being somewhat more inclusive, as many genera have been added even here. Although thorough, an inclusive key is harder to use than an abridged key, which covers far fewer genera. For this reason and because I wanted to make the book faster and easier to use for those who know that they have a native or naturalized plant in hand, I shortened the inclusive key by eliminating all genera that have only cultivated (non-native and non-naturalized) species to create the abridged key. The genera, including species and common names, are arranged alphabetically in the keys to species, so that a genus may be easily found after completing the keys to genera.

One or sometimes two keys to species are offered for each genus, unless, of course, a genus contains only one species. If the genus has several species that have deciduous leaves, two keys are often provided, one for summer (leaves present) and one for winter (leaves lacking).

The keys can be used by the novice, but technical terms are incorporated when precision is needed. The keys are strictly dichotomous. I have coded the plant names with abbreviations in the key to species and the systematic list of species to add useful information without using more space. These abbreviations indicate whether a plant is naturalized, occasionally or infrequently naturalized, restricted to the south, or occasionally or rarely cultivated. Additional abbreviations in the systematic list denote the place of origin for cultivated and naturalized species. Boldface type is used in these sections to distinguish native and natural-ized species from cultivated species.

Readers should keep in mind that the keys are based on the woody plants as they grow in the field. To be sure, it is not always possible or necessary to have access to the whole plant, but the specimens selected for tracing through the key should represent normal structures from well-grown individuals. Small seedlings, sucker shoots, and coppice growth frequently produce twigs, leaves, and buds quite un-like those developed by normal growth of more mature individuals. Such speci-mens, shade leaves, and leaves near the top of the tree in conifers may cause some difficulty in using the keys. The keys are most useful if several specimens or dif-ferent parts of the shrub or tree are examined for each character.

Bark color on young stems is subject to considerable variation. When color of twigs is mentioned, it refers to winter color of the shoots of the last season's growth. Bark on the older shoots (branchlets) may be quite different in color and texture. The color stated for fruit refers to that of the ripe fruit. On twigs that lack a terminal bud, the uppermost lateral bud (pseudoterminal bud) frequently ap-pears terminal, although it actually is just back from the end of the twig on one side. Such twigs have a leaf scar on one side and an abscission scar from the grow-ing point on the opposite side, usually somewhat raised on a short stub. Recogni-tion of pseudoterminal buds has caused considerable difficulty for users in earlier editions. I have considerably reduced the reliance on this characteristic by remov-ing it from early steps in the keys. The persistent or "evergreen" nature of leaves may be determined in summer by the presence of leaves on the growth of the pre-vious season, that is, on the axis below or behind the bud scars. It should be em-phasized here, as it is in the glossary, that in these keys "twig" refers to the shoot of the current year, and the term "branchlet" describes those of earlier years.

Winter keys are more difficult to use and their success is less certain compared to summer keys because the number of vegetative characters that are possible to use are fewer. I have often included information about fruit since, in many plants, it will be available for at least some of the winter. These keys are by necessity shorter than the summer keys, where I sacrifice brevity for more information by often including several characters for each choice or key step.

The keys to genera section contains fourteen keys to plant groups. The first seven keys make up the inclusive key and contain all woody plants, native, natu-ralized, and cultivated. The other seven keys make up the abridged key and include only native and naturalized woody plants. The abridged key is shorter and will more quickly be worked through to obtain an identification. Rather than refer-encing page numbers for each genus within the keys (as in the sixth edition), bold-face page numbers in the index clearly show the page on which the genus can be found in the keys to species. Page numbers are placed in parentheses after the first step of a couplet where the second step is more than a page away. I have inserted the fruit type where possible in the keys, as this information can be useful in identifying a plant.

The keys to species section lists the genera in alphabetical order, each with its key to species or a simple listing of the single species if there is only one. In most cases common names are provided. Some genera have both a summer and winter key. Others have winter and summer characters combined in the same key. In a few cases, such as the genus *Rosa*, the winter characters that are used are not sufficient to guarantee an identification. In even fewer cases a winter key has not been attempted. Because plant relationships and hence plant names are based primarily on reproductive characters, it is more challenging to produce a useful key that is based on a different set of characters such as leaves, twigs, and other vegetative features.

The systematic list of species is a completely revised list arranged in accordance with Arthur Cronquist's classification, perhaps now the most universally recognized system. The genera and species are positioned alphabetically within phylogenetically arranged families. The entries include standardized common names and some synonyms, including names from Muenscher's latest edition.

I have included a list of annotated references that are useful companions to this book. To construct the keys and to assign proper nomenclature I have consulted some of these extensively as a supplement to herbarium material and field observations. Any treatment that consists entirely of keys should be supplemented by manuals or floras that include complete descriptions or illustrations of each species.

KEYS TO GENERA

INCLUSIVE KEY: CULTIVATED, NATURALIZED, AND NATIVE WOODY PLANTS

KEY I. PLANTS WITH NEEDLELIKE, SCALELIKE, OR AWL-SHAPED LEAVES

1. Low, prostrate, or creeping shrubs or mats or parasitic and growing on trees, 10–100 cm tall (Dwarf conifers, which are in most cases derived from tree species, are keyed in alternate step 1, p. 8)

2. Leaves alternate; stems forming low mats (or creeping and
  horizontal in *Taxus*)
   3. Leaves covered with downy hairs; southern . . . .*Hudsonia*
   3. Leaves glabrous or nearly so
      4. Leaves revolute, falsely whorled, 4–7 mm long; leaf scars
      decurrent; fruit a drupe
         5. Leaves denticulate . . . .*Corema*
         5. Leaves entire . . . .*Empetrum*
      4. Leaves flattened, not revolute, not appearing whorled,
      2–20 mm long; leaf scars decurrent or not; fruit a capsule
      or drupe-like
         6. Leaves imbricated, subulate, 2–8 mm long; leaf scars not
         decurrent; fruit a capsule
            7. Leaves spreading, mostly 4–8 mm long, about
            1 mm wide . . . .*Pyxidanthera*
            7. Leaves crowded, appressed, mostly 2–4 mm long, less than
            1 mm wide . . . .*Harrimanella*
         6. Leaves spaced apart on the branchlet, not imbricated,
         linear, 5–20 mm long; leaf scars decurrent; fruit a capsule
         or drupe-like
            8. Leaves acuminate; reproductive structure drupe-like,
            fleshy, red . . . .*Taxus*
            8. Leaves obtuse; fruit a dry, brown capsule . . . .*Phyllodoce*
2. Leaves opposite or whorled, scalelike or minute; stems erect
  or creeping
   9. Plants growing on trees, parasitic; leaves barely visible . . . .*Arceuthobium*
   9. Plants erect or creeping, growing from the ground; leaves small,
   usually obvious
      10. Stems creeping or horizontal; leaves subulate or broader
      or scalelike
         11. Branchlets terete or 4-angled, groups of branchlets (sprays) in
         different planes . . . .*Juniperus*
         11. Branchlets flattened, groups of branchlets (sprays) mostly in
         horizontal planes . . . .*Microbiota*
      10. Stems erect
         12. Leaves dry, brown, the whorls distant; branchlets and
         stems green . . . .*Ephedra*
         12. Leaves green, many, crowded; branchlets and stems, if visible,
         brown, gray, or black
             13. Leaves scalelike or subulate . . . .*Juniperus*
            13. Leaves sagittate, oblong or linear
               14. Leaves opposite, sagittate . . . .*Calluna*
               14. Leaves whorled, oblong or linear . . . .*Erica*

1. Trees or erect or ascending shrubs, greater than 100 cm tall
   (All genera with dwarf conifers except *Microbiota* are
   included here.)
   15. Leaves mostly scalelike (linear or juvenile leaves often
       also present), no greater than twice as long as wide;
       twigs and branchlets usually hidden by the brown or
       green leaves
       16. Leaves alternate
           17. Leaves linear, horizontal or divergent; branchlets exposed, rough,
               gray; southern . . . .*Ceratiola*
           17. Leaves subulate, awl-shaped, or scalelike, parallel or
               ascending; branchlets exposed or not
               18. Leaves directed forward toward the tip of the branchlet
                   but not pressed against it
                   19. Secondary branchlets unbranched; leaves of similar sizes on the
                       same branchlet; southern . . . . *Araucaria*
                   19. Secondary branchlets branching; leaves often of markedly
                       different sizes on the same branchlet . . . .*Cryptomeria*
               18. Leaves pressed against and completely clothing
                   the branchlet
                   20. Leaves greater than 3 mm long; large trees; female reproductive
                       structure a cone . . . .*Sequoiadendron*
                   20. Leaves 1–2 mm long; shrubs or small trees; female reproductive
                       structure a capsule . . . .*Tamarix*
       16. Leaves opposite or whorled
           21. Leaves in whorls of 4 . . . .*Calocedrus*
           21. Leaves opposite or in whorls of 3
               22. Branchlets terete or 4-angled, the groups of branchlets
                   (sprays) in several different planes, spreading at various
                   angles from trunk or main stems
                   23. Leaves with white longitudinal band on lower surface; female
                       reproductive structure with woody, peltate scales . . . .*Cupressus*
                   23. Leaves with white longitudinal band on upper surface;
                       female reproductive structure with fleshy peltate scales fused,
                       resembling a berry . . . .*Juniperus*
               22. Branchlets flattened, the groups of branchlets (sprays) in
                   several different planes or mostly horizontal planes
                   24. Seam where margins of lateral leaves meet on upper
                       surface of twig usually visible
                       25. Facial and lateral leaves with prominent groove; cone longer
                           than wide, with imbricated scales . . . .*Platycladus*
                       25. Facial and lateral leaves smooth or facial leaves only with
                           inconspicuous groove or circular pit; cone globose, with
                           peltate scales . . . .*Chamaecyparis*

24. Seam where margins of lateral leaves meet on upper surface of
twig usually hidden
   26. Visible portion of leaves greater than 3 mm long, greater than
2 mm wide; shrubs . . . .*Thujopsis*
   26. Visible portion of leaves less than 3 mm long, less than 2 mm
wide; trees or shrubs
      27. Leaves sharp-pointed . . . .*Chamaecyparis* and ×*Cupressocyparis*
      27. Leaves blunt
         28. Sprays mostly in horizontal planes; lateral leaves as long
as wide . . . .*Thuja*
         28. Sprays in various planes, tending to be parallel to trunk or main
stems; lateral leaves longer than wide . . . .*Platycladus*
15. Leaves needlelike, at least 4 times as long as wide; twigs visible,
not hidden or clothed by the leaves
   29. Leaves white-tomentose on lower surface; twigs and branchlets square or
4-edged, with white cobwebby hairs; shrubs . . . .*Rosmarinus*
   29. Leaves glabrous; twigs and branchlets terete, glabrous or
pubescent, not cobwebby; trees or shrubs
      30. Apex of leaves extended into long spines, the longest 2–3 cm long;
female reproductive structure a legume . . . .*Ulex*
      30. Leaves lacking spines at apex, rounded or blunt to pungent;
female reproductive structure a woody cone
         31. Leaves in clusters or fascicles
(solitary only on young shoots)
            32. Leaves in fascicles of 2–5 . . . .*Pinus*
            32. Leaves in clusters or whorls of 10 or more
               33. Leaves 8–16 cm long . . . .*Sciadopitys*
               33. Leaves less than 8 cm long
                  34. Branches from trunk arranged irregularly; spurs
roughened from persistent leaf sheaths; mature leaves
stiff, sharp . . . .*Cedrus*
                  34. Branches mostly in regular whorls from trunk;
spurs smooth, with only rings of leaf scars
remaining; mature leaves soft, blunt
                     35. Leaves mostly greater than 4 cm long, greater than
1 mm wide . . . .*Pseudolarix*
                     35. Leaves mostly less than 4 cm long, less than
1 mm wide . . . .*Larix*
         31. Leaves all solitary
           36. Leaves greater than 8 cm long . . . .*Sciadopitys*
           36. Leaves less than 7 cm long
              37. Leaves curved, awl-shaped, keeled, with thick ridge
on the keel
                 38. Leaves alternate

39. Secondary branchlets unbranched; southern . . . . ***Araucaria***
39. Secondary branchlets nearly always branching . . . .***Cryptomeria***
38. Leaves opposite or whorled
    40. Leaves whorled or, if opposite, usually greater than
        10 mm long . . . . ***Juniperus***
    40. Leaves opposite, less than 10 mm long . . . .***Chamaecyparis***
37. Leaves straight, flat, or angled
    41. Leaves mostly greater than 4 cm long, covered with silver scales,
        deciduous; spiny shrubs; reproductive structure berrylike . . . . .***Hippophae***
    41. Leaves mostly less than 4 cm long (except in *Podocarpus*),
        lacking silver scales, evergreen (except in *Taxodium* and
        Metasequoia); trees or shrubs, lacking spines; reproductive
        structure a woody cone or fleshy and berrylike
        42. Buds lance-ovoid, at least $2^1/_2$ times as long as wide, sharp-pointed, the
            terminal buds usually greater than 5 mm long . . . .***Pseudotsuga***
        42. Buds globose or ovoid, less than twice as long as wide,
            usually less than 5 mm long
            43. Leaves 4-sided or angular in cross section, easily rolled between the
                fingers; branchlets very rough because of the persistent woody
                projections at the base of each leaf . . . .***Picea***
            43. Leaves flattened, not easily rolled between the fingers;
                branchlets smooth or lacking prominent woody projections
                44. Leaves less than 1 mm wide; ultimate twigs less than
                    0.5 mm wide
                    45. Leaves alternate . . . .***Taxodium***
                    45. Leaves opposite . . . .***Metasequoia***
                44. Leaves greater than 1 mm wide; twigs greater than
                  1 mm wide
                  46. Leaves opposite or whorled (appearing so in *Torreya*);
                    female reproductive structure fleshy, berrylike or
                    drupe-like
                    47. Leaves with a single white or gray longitudinal band on upper
                      surface, the margins rolled or curled upward . . . . . .***Juniperus***
                    47. Leaves with 2 white or pale longitudinal bands on
                      lower surface, margins, if rolled, rolled under
                      48. Leaves with prominent midrib on upper surface, each
                        band on lower surface 1 mm wide . . . .***Cephalotaxus***
                      48. Leaves lacking an obvious midrib, each band on lower
                        surface usually less than 1 mm wide . . . .***Torreya***
                46. Leaves alternate; female reproductive structure a
                  woody cone or drupe-like
                  49. Leaves of conspicuously different lengths and
                    orientations on the same twig

50. Leaves pulling away from branchlet cleanly, leaving indented orbicular leaf scars; branches and shoots stiff . . . . ***Abies***

50. Leaves pulling away from branchlet with bark attached; leading shoots flexible, often pendulous

    51. Leaves acute; awl-shaped leaves present on older branches . . . . . ***Sequoia***

    51. Leaves obtuse or emarginate; awl-shaped leaves absent . . . . ***Tsuga***

49. Leaves of similar lengths and orientations on the same twig

    52. Leaves mostly less than 25 mm long

        53. Leaves pulling away from branchlet cleanly, leaving indented orbicular leaf scars; cones erect, disintegrating on the tree . . . . ***Abies***

        53. Leaves pulling away from branchlet with bark attached; cones pendulous and falling intact or fleshy and drupe-like

            54. Leaves on woody projections to 2 mm long on all sides of the branchlet . . . . ***Picea***

            54. Leaves decurrent, in mostly horizontal planes, appearing to be opposite, giving branchlets a flattened appearance

                55. Leaves with raised midrib or longitudinal groove on upper surface . . . . ***Taxus***

                55. Leaves smooth, even, lacking raised midrib or groove on upper surface . . . . ***Sequoia***

    52. Leaves mostly greater than 25 mm long

        56. Leaves mostly greater than 4 cm long . . . . ***Podocarpus***

        56. Leaves mostly less than 4 cm long

            57. Leaves greater than 6 mm wide; southern . . . . ***Araucaria***

            57. Leaves less than 6 mm wide

                58. Leaves pulling away from branchlet cleanly, leaving indented orbicular leaf scars; cones erect, disintegrating on the tree . . ***Abies***

                58. Leaves pulling away from branchlet with bark attached; cones pendulous and falling intact or fleshy and drupe-like

                    59. Leaves borne on woody projections with white longitudinal bands on all surfaces or upper surface only . . . . ***Picea***

                    59. Leaves decurrent, with white or pale longitudinal bands on lower surface only

                        60. Branchlets and twigs alternate, at most subopposite . . . . ***Taxus***

                        60. Branchlets and twigs opposite

                            61. Leaves alternate, lanceolate, blue . . . . ***Cunninghamia***

                            61. Leaves opposite or appearing so, linear, green

                                62. Leaves with prominent midrib on upper surface, each band on lower surface 1 mm wide . . . . . . ***Cephalotaxus***

                                62. Leaves lacking an obvious midrib, each band on lower surface usually less than 1 mm wide . . . . ***Torreya***

KEY II. PLANTS WITH OPPOSITE OR WHORLED SIMPLE LEAVES

1. Leaves subopposite
    2. Leaves toothed . . . .*Rhamnus*
    2. Leaves entire; southern
        3. Leaves greater than 5 cm long . . . .*Lagerstroemia*
        3. Leaves less than 5 cm long . . . .*Fontanesia*
1. Leaves distinctly opposite or whorled
    4. Leaves lobed
        5. Leaves mostly pinnately lobed
            6. Margin of lobes entire; sap clear; shrubs; fruit a capsule . . . . .*Syringa*
            6. Margin of lobes serrate; sap milky or clear; trees or tall
                shrubs; fruit a capsule or head of achenes
                7. Trees; sap milky; fruit a head of achenes . . . .*Broussonetia*
                7. Shrubs; sap clear or milky; fruit a capsule . . . .*Hydrangea*
        5. Leaves palmately lobed
            8. Leaf blades less than 20 cm long
                9. Petioles with stipules and glands, or if lacking glands, the lower
                    surface of leaf densely pubescent; fruit a drupe . . . .*Viburnum*
                9. Petioles lacking stipules and glands, or if stipules present,
                    the lower surface of leaf glabrous to pubescent, not densely so;
                    fruit a samara . . . .*Acer*
            8. Leaf blades greater than 20 cm long
                10. Leaves with long tapering tip, glabrous or softly pubescent,
                    usually in whorls of 3; pith continuous; fruit a long cylindrical
                    capsule, 20–50 cm long . . . .*Catalpa*
                10. Leaves with obtuse or rounded tip, densely pubescent, like velvet,
                    always opposite; pith chambered or hollow; fruit an ovoid
                    capsule, 3–4 cm long . . . .*Paulownia*
    4. Leaves entire or toothed, lacking lobes
        11. Stems climbing, creeping, prostrate, or forming low mats, or
            plants parasitic and growing on trees (alternate step 11, p. 14)
            12. Stems climbing (vines); leaves deciduous or persistent
                13. Aerial rootlets absent; fruit a berry or capsule
                    14. Leaves lanceolate, long acuminate; fruit
                        a capsule . . . .*Gelsemium*
                    14. Leaves ovate, acute to cuspidate; fruit a berry
                        or capsule
                        15. Twigs and branchlets light red-brown; bark of stems
                            and branches gray, shredding or peeling;
                            fruit a berry . . . .*Lonicera*
                        15. Twigs and branchlets dark purple-brown or brown;
                            bark of stems and branches tight, not peeling; fruit
                            a capsule; southern

16. Leaves with 10–20 conspicuous, straight
secondary veins . . . .*Periploca*
16. Leaves with fewer than 10, inconspicuous, curved or wavy
secondary veins . . . .*Trachelospermum*
13. Aerial rootlets present on stems; fruit a capsule
17. Leaves nearly as wide or wider than long, mostly greater than
10 cm long, dull or at least not lustrous on upper surface,
dentate, rounded to truncate or cordate at the base
18. Leaves less than 6 cm long, coarsely dentate, the teeth greater than
2 mm long . . . .*Schizophragma*
18. Leaves mostly greater than 8 cm long, finely serrate, the teeth less
than 1 mm long . . . .*Hydrangea*
17. Leaves mostly at least twice as long as wide, mostly less than
10 cm long, often lustrous on upper surface, entire or toothed,
cuneate to truncate at the base
19. Leaves deciduous or persistent; branchlets
brown; southern . . . .*Decumaria*
19. Leaves persistent; branchlets green . . . .*Euonymus*
12. Stems prostrate, creeping, or forming low mats, or plants parasitic
and growing on trees; leaves persistent
20. Plants growing on trees, parasitic; fruit a berry
21. Leaves about 1 mm long, scalelike; stems 1–3 cm long . . . .*Arceuthobium*
21. Leaves 20–40 mm long; stems to 40 cm long . . . .*Phoradendron*
20. Plants prostrate, creeping, not parasitic or growing on trees;
fruit a berry, follicle, or capsule
22. Leaves crenate or serrate; plants creeping or with some stems
erect; fruit a capsule
23. Leaves rounded-ovate, crenate; branches
brown-pubescent . . . .*Linnaea*
23. Leaves oblanceolate, sharply serrate; branches glabrous
24. Leaves opposite; stems with longitudinal ridges;
plants stoloniferous . . . .*Euonymus*
24. Leaves in false whorls; stems erect, smooth; plants from
creeping rootstocks . . . .*Chimaphila*
22. Leaves entire or slightly undulate; plants creeping or prostrate;
fruit a berry, capsule, or follicle
25. Leaves with translucent or pellucid dots (visible when held up
to light) . . . .*Hypericum*
25. Leaves opaque, lacking translucent dots
26. Leaves and young stems strongly pubescent;
fruit a berry . . . .*Lonicera*
26. Leaves and stems glabrous or nearly so; fruit a berry,
follicle, or capsule
27. Leaf base cordate or rounded; fruit a red berry . . . . . .*Mitchella*

27. Leaf base narrow or cuneate; fruit a follicle or capsule
  28. Leaves greater than 2 cm long, margins flat, not revolute; stems
      slightly woody, creeping; fruit a follicle . . . .***Vinca***
  28. Leaves less than 1 cm long, margins revolute; stems woody, forming
      dense, low mats; fruit a capsule . . . .***Loiseleuria***
11. Stems erect; trees or shrubs
  29. Leaves entire (alternate step 29, p. 18)
    30. Leaves with blade mostly greater than 20 cm long, ovate, with
        cordate, subcordate, or truncate base; large trees
      31. Leaves with long-tapering tip, glabrous or softly pubescent, usually in
          whorls; pith continuous; fruit a long cylindrical capsule,
          20–50 cm long . . . .***Catalpa***
      31. Leaves with obtuse or rounded tip, densely pubescent like velvet,
          opposite; pith chambered or hollow; fruit an ovoid capsule,
          3–4 cm long . . . .***Paulownia***
    30. Leaf blade less than 15 cm long, if longer the shape not ovate,
        the base various; shrubs or small trees
      32. Leaves mostly in whorls of 3, linear to oblong-lanceolate, the lateral
          veins more than 100 on each side and perpendicular to the
          midrib; southern . . . .***Nerium***
      32. Leaves mostly opposite, variously shaped, the lateral veins
          fewer than 30 on each side of the midrib, usually ascending
        33. Leaves with yellow glandular dots, often with 1–3 remote, sharp
            teeth, ill-scented; twigs striate . . . .***Iva***
        33. Leaves green, pale, or brown, lacking yellow glandular
            dots (except *Skimmia* with black glandular dots), entire,
            aromatic or lacking odor; twigs smooth or rough, mostly
            lacking ridges or lines
          34. Leaves with fetid odor when crushed, 1–2 cm long; shrubs to
              50 cm tall; stipules persistent, with long stiff awns; twigs and
              branchlets with 2 longitudinal bands of pubescence; fruit a
              capsule; southern . . . .***Serissa***
          34. Leaves spicy and aromatic or lacking odor, of various
              lengths; trees or shrubs usually greater than 50 cm tall;
              twigs and branchlets, if pubescent, the hairs not
              arranged in 2 longitudinal bands (except in *Weigela*
              and *Diervilla*); fruit various
            35. Leaves and twigs spicy-aromatic
              36. Leaves linear-lanceolate to linear-obovate, 5–10 times
                  longer than wide; low shrub
                  (less than 1 m tall) . . . .***Lavandula***
              36. Leaves ovate, elliptic to oblong-lanceolate, mostly
                  less than 5 times longer than wide; shrub to
                  3 m tall . . . .***Calycanthus***

35. Leaves and twigs odorless or at least not spicy-aromatic
  37. Leaves and twigs covered with silvery or rusty scales . . . .*Shepherdia*
  37. Leaves and twigs lacking silvery or rusty scales
    38. Leaves ovate with cordate, subcordate, or truncate base and
      long-tapering tip
      39. Leaves with petioles pubescent on the thickened or winged margins,
        connected to the opposite leaf by a pubescent line; branchlets with
        hollow pith between the nodes; fruit a berry; southern . . .*Leycesteria*
      39. Leaves smooth, with glabrous petioles, lacking connecting
        line to opposite leaf; branchlets with solid pith; fruit
        a capsule . . . .*Syringa*
    38. Leaves of various shapes, the base acute, or if cordate, the
      apex not long-tapering
      40. Leaves mostly less than 1 cm long . . . .*Leiophyllum*
      40. Leaves greater than 2 cm long
        41. Leaves with translucent or pellucid dots (evident when held up to
          light); fruit a capsule . . . .*Hypericum*
        41. Leaves opaque, lacking translucent dots; fruit various
          42. Leaves gray-green or pale green, cuneate-attenuate or
            long attenuate at the base, acute to acuminate or
            often mucronate, minutely undulate or occasionally
            with 2–6 remotely spaced linear teeth
            43. Leaves pale green, often glossy, cuneate-attenuate at the base,
              widest below the middle, acute to acuminate, often
              minutely undulate . . . .*Fontanesia*
            43. Leaves gray-green, dull, long attenuate at the base, widest
              above the middle, mucronate, occasionally with 2–6 remotely
              spaced linear teeth; southern . . . .*Borrichia*
          42. Leaves green to brown, variously shaped at the base,
            not long attenuate, apex various, not usually
            mucronate, always entire
            44. Leaves sessile or subsessile (petiole less than 1 mm
             long); branchlets terete, quadrangular, flattened,
             or winged
              45. Leaves shiny, thick; stipules fused to the stipules of the
               opposite leaf; southern . . . .*Gardenia*
             45. Leaves dull or at least not shiny, thick or thin;
              stipules or stipules scars, if present, not connected
               46. Branchlets terete; fruit a drupe; southern . . . .*Buckleya*
              46. Branchlets quadrangular, flattened, or winged;
                fruit a capsule
                47. Leaves with lower surface glaucous, margin strongly
                  revolute; twigs flattened, brown; fruit a
                  5-valved capsule . . . .*Kalmia*

47. Leaves with lower surface light green with median white band, margin only slightly revolute; twigs quadrangular or winged, often green; fruit a 3-valved capsule . . . .*Buxus*

44. Leaves with a distinct petiole; branchlets terete, neither quadrangular, flattened, nor winged

    48. Leaves often crowded toward the end of the season's growth, often alternate or in whorls of 3, leathery, persistent; fruit a capsule or drupe

        49. Leaves with glandular dots; fruit a drupe . . . .*Skimmia*

        49. Leaves lacking glandular dots; fruit a capsule . . . .*Kalmia*

    48. Leaves scattered in pairs along the season's growth, thin, deciduous or persistent; fruit various

        50. Leaves with 3 main veins from the base, the other veins on each side of the midrib not reaching the margin . . . .*Heptacodium*

        50. Leaves with a single main vein from the base, the midrib, the lateral veins often reaching the margin

            51. Lateral buds imbedded in the bark, supra-axillary (located some distance above the axil); leaves often in whorls of 3; fruit a head of nutlets . . . .*Cephalanthus*

            51. Lateral buds not imbedded in the bark, axillary; leaves opposite; fruit various

                52. Bark of stems and branches loose, peeling off in long strips

                    53. Branchlets crowded on the slender twigs; bundle scar 1; bud scales deciduous . . . .*Symphoricarpos*

                    53. Branchlets widely spaced on the twig of medium thickness; bundle scars 3; bud scales often persisting at base of twigs . . . .*Lonicera*

                52. Bark of stems and older branches smooth, not peeling off in long shreds

                    54. Leaves with lateral veins running somewhat parallel with the margin and meeting near the apex, the veins when torn dangling silky strands; fruit a drupe . . . .*Cornus*

                    54. Leaves with lateral veins ending near the margin and not running to the apex, the veins tearing clean (sometimes leaving dangling strands in *Viburnum*); fruit various

                        55. Leaves persistent (deciduous in *Pinckneya*), leathery, the petioles often 2 mm thick

                            56. Leaves with at least a few large (at least 2 mm long) spiny teeth; southern . . . .*Osmanthus*

                            56. Leaves entire, undulate, or serrulate; if serrulate, the teeth less than 1 mm long

57. Leaves mostly less than 6 cm long
   58. Leaves entire . . . .*Phillyrea*
   58. Leaves undulate or remotely serrulate . . . .*Viburnum*
57. Leave mostly greater than 6 cm long
   59. Leaves serrulate to nearly undulate . . . .*Viburnum*
   59. Leaves entire
      60. Leaves mostly greater than 14 cm long, pubescent
         61. Leaves oblong, wrinkled, the veins impressed, less than
            4 cm wide . . . .*Viburnum*
         61. Leaves ovate, flat, the veins not conspicuously impressed, greater
            than 7 cm wide; southern . . . .*Pinckneya*
      60. Leaves mostly less than 12 cm long, glabrous
         62. Leaves obtuse to cuspidate, ovate, greater than
            4 cm wide . . . .*Viburnum*
         62. Leaves acute to acuminate, elliptic, oblanceolate, or
            lanceolate, less than 3 cm wide
            63. Leaves acuminate, the margins flat, the petiole mostly greater
               than 5 mm long . . . .*Phillyrea*
            63. Leaves acute, the margins revolute, the petiole about 2 mm
               long; southern . . . .*Osmanthus*
55. Leaves deciduous (except *Ligustrum japonicum* and some species of
   Viburnum), thin, the petioles less than 1 mm thick
   64. Lower leaf surface, especially along the midrib, scurfy, rusty brown,
      woolly or glandular (margin of leaf often finely serrate); fruit
      a drupe . . . .*Viburnum*
   64. Lower leaf surface glabrous, lightly pubescent, lacking glands,
      not scurfy or woolly; fruit a drupe, berry, samara, or achenes
      65. Leaves mostly greater than 8 cm long; fruit a drupe or achenes
         66. Leaves usually pubescent on lower surface; twigs often pubescent;
            fruit a drupe . . . .*Chionanthus*
         66. Leaves glabrous; twigs glabrous; fruit a cluster
            of achenes . . . .*Chimonanthus*
      65. Leaves mostly 1–7 cm long; fruit a berry or samara
         67. Leaves acuminate, ovate; fruit a samara . . . .*Abeliophyllum*
         67. Leaves acute, obtuse, rounded, sometimes apiculate,
            mostly oblong or elliptic, sometimes ovate; fruit a berry
            68. Branchlets usually terminating in sharp thornlike spurs;
               leaves often grooved at the midrib on upper
               surface; southern . . . .*Punica*
            68. Branchlets lacking spurs or thorns; leaves with midrib
              raised or even on upper surface, not grooved
              69. Twigs red, red-brown, or purple-brown, usually
                glaucous; southern . . . .*Nestronia*
              69. Twigs brown to gray, lacking bloom . . . .*Ligustrum*

29. Leaves serrate to serrulate, dentate to crenulate or nearly undulate
  70. Leaves mostly 10–25 cm long, cordate at the base, crenate to
      undulate, the petiole usually as long or longer than the blade;
      fruit a drupe . . . .*Clerodendrum*
  70. Leaves mostly less than 10 cm long, cordate to cuneate at the
      base, serrate to crenate, the petiole mostly shorter than the
      blade (except in *Hydrangea*); fruit various
    71. Leaves with the bases of the petioles meeting or joined by a
        distinct transverse line
      72. Buds imbedded under the petiole base, not axillary . . .*Philadelphus*
      72. Buds axillary, not imbedded
        73. Leaves with yellow, glandular dots; twigs strongly striated . . . . .*Iva*
        73. Leaves uniformly green, lacking yellow, glandular dots;
            twigs uniform, not strongly striated
          74. Twigs and branchlets with decurrent ciliate ridges or
              densely pubescent longitudinal bands between the
              nodes; leaves with main lateral veins fading before
              reaching the margin; fruit a capsule
            75. Branchlets with 2 decurrent ridges and 2 faint longitudinal
                bands lightly pubescent just below the node; petioles lightly
                pubescent; buds with outer 2 bud scales half the length of
                the bud . . . .*Diervilla*
            75. Branchlets usually with 2 densely pubescent lines between the
                nodes; petioles densely pubescent; buds with outer 2 bud
                scales greater than half the length of the bud . . . . . .*Weigela*
          74. Twigs and branchlets glabrous or pubescent, lacking
              pubescent longitudinal bands or ridges between the
              nodes; leaves with main lateral veins often extending
              to the margin; fruit a capsule, drupe, samara, berry,
              or nutlets
            76. Buds naked or covered by 2 valvate scales . . . .*Viburnum*
            76. Buds covered by several imbricated scales
              77. Twigs with horizontal pubescent lines connecting the
                  leaves of each node, the pith hollow; petioles pubescent
                  on thickened, nearly winged margins; fruit a
                  berry; southern . . . .*Leycesteria*
              77. Twigs lacking connecting horizontal lines of
                  pubescence between the leaves, the pith solid;
                  petioles glabrous or pubescent at the margin;
                  fruit various
                78. Leaves crenulate, undulate, or subentire,
                    mostly less than 4 cm long
                  79. Bark of branchlets red-brown to light brown; leaves
                      usually less than 3 cm long, the teeth close or

irregularly spaced, scalloped, never sharp; fruit glabrous
with persistent sepals . . . . ***Abelia***
79. Bark of branchlets gray to gray-brown; leaves greater than 3 cm long,
the teeth remotely denticulate but sharp; fruit bristly with
deciduous sepals . . . .***Kolkwitzia***
78. Leaves dentate to serrate, mostly greater than 4 cm long
80. Leaves coarsely doubly or triply (or more) toothed, the lateral veins
more than 20 on each side of the midrib . . . .***Acer***
80. Leaves simply toothed or finely and closely doubly toothed,
the lateral veins fewer than 20 on each side of the midrib
81. Leaves closely and finely doubly serrate; stipules present; leaf scars
ciliate; fruit a cluster of shiny black nutlets surrounded by large,
persistent sepals . . . .***Rhodotypos***
81. Leaves simply dentate or serrate; stipules absent or
apparently so; leaf scars glabrous, not ciliate; fruit a capsule
or drupe
82. Bud scales from buds of previous year persistent; leaves with
petiole usually longer than the blade; fruit a capsule . .***Hydrangea***
82. Bud scales from buds of previous year deciduous; leaves
with petiole shorter than the blade; fruit a capsule or drupe
83. Leaves and twigs stellate-pubescent; fruit a capsule . . . .***Deutzia***
83. Leaves and twigs glabrous or, if pubescent, with simple hairs;
fruit a drupe . . . .***Viburnum***
71. Leaves with petiole bases of opposite leaves distinct, not meeting
or joined by a transverse line
84. Leaves woolly on lower surface; branchlets gray-brown, dark
brown, or black; fruit a berry or drupe-like
85. Twigs and branchlets gray-brown, lacking spines; inner bark brown;
petioles 3–15 mm long . . . .***Callicarpa***
85. Twigs and branchlets dark brown or black, often terminating in sharp
black spines; inner bark yellow and bitter; petioles
7–25 mm long . . . .***Rhamnus***
84. Leaves glabrous or pubescent on lower surface, not woolly;
branchlets gray-brown, green, or red-purple; fruit various
86. Pith hollow or chambered, not continuous between
the nodes . . . .***Forsythia***
86. Pith continuous or spongy between the nodes (occasionally
chambered in *Euonymus*)
87. Leaves thick, persistent
88. Leaves serrate to crenulate most of the length; branchlets green,
the pith green or white; fruit a capsule . . . .***Euonymus***
88. Leaves serrulate, dentate, or with a few sharp teeth
toward the apex (or sometimes the entire length in
Osmanthus heterophyllus), entire toward the base;

branchlets gray-brown to dark brown or black, the pith brown;
fruit a drupe or capsule

  89. Leaves mostly less than 3 cm long, sessile or subsessile; fruit
a capsule . . . .*Paxistima*

  89. Leaves mostly much greater than 7 cm long, the petioles at least
2 mm long; fruit a drupe

    90. Leaves with a few sharp teeth toward the apex or sometimes also
toward the base; southern . . . .*Osmanthus*

    90. Leaves dentate or coarsely toothed toward the apex . . . .*Aucuba*

87. Leaves thin, deciduous

  91. Leaves woolly on lower surface; branchlets light brown to
gray-brown; fruit a capsule or drupe

    92. Buds positioned to the side of the petiole base; leaves narrowly
elliptic, long-acuminate, serrate the entire length of the margin; fruit
a capsule . . . .*Buddleia*

    92. Buds axillary to the petiole base; leaves ovate, abruptly acuminate,
serrate mostly only in apical half; fruit a drupe . . . .*Callicarpa*

  91. Leaves glabrous or pubescent on lower surface, not woolly;
branchlets gray, green, red, or purple-red; fruit a capsule,
drupe, achene, or samara

    93. Leaves orbicular, cordate, crenate . . . .*Cercidiphyllum*

    93. Leaves ovate, elliptic, or lanceolate, acute or cuneate at the
base, serrate, often remotely and obscurely so

      94. Leaves coarsely dentate, the petiole 4 cm long, nearly as long as the
blade; buds dark brown to black, mostly globose;
branchlets stout . . . .*Fraxinus*

      94. Leaves mostly finely serrate, serrulate, crenate, or
crenulate, the petiole less than 4 cm long, much shorter
than the blade; buds brown, ovoid; branchlets slender

        95. Bark of branchlets shredding or peeling; leaves ovate,
rounded, or acute at the base, the teeth remote and
obscure; fruit an achene

          96. Bark of branchlets gray to gray-brown; leaves usually greater
than 3 cm long, the teeth remote, sharp,
or denticulate . . . .*Kolkwitzia*

          96. Bark of branchlets red-brown to light brown; leaves usually
less than 3 cm long, the teeth close or irregularly spaced,
scalloped, never sharp . . . . *Abelia*

        95. Bark of branchlets tight, not peeling; leaves lanceolate,
oblong, or elliptic to ovate, rounded to cuneate at the
base, the teeth obvious, close together; fruit a capsule,
drupe, or samara

          97. Twigs mostly light gray, nearly horizontal, usually terminating
in sharp points; fruit a drupe . . . .*Forestiera*

97. Twigs light brown to gray or dark brown, ascending, flexible,
blunt, not terminating in sharp points; fruit a samara or capsule
    98. Leaves irregularly, doubly serrate; fruit a samara . . . .*Acer*
    98. Leaves singly serrate or serrulate; fruit a capsule
        99. Bud scales several . . . .*Euonymus*
        99. Bud scales 1 . . . .*Salix*

KEY III. PLANTS WITH OPPOSITE COMPOUND LEAVES

1. Leaves with 2–3 leaflets
    2. Leaflets 2; vine with branched tendrils; fruit a capsule . . . .*Bignonia*
    2. Leaflets 3; trees, shrubs, or vines; fruit a capsule, achene, berry,
    or samara
        3. Stems climbing or twining; fruit a berry or hairy achene
            4. Vines; leaflets coarsely toothed; fruit a hairy achene . . . .*Clematis*
            4. Stems weakly climbing; leaflets entire; fruit a
            berry; southern . . . .*Jasminum*
        3. Stems erect or sometimes pendulous, not climbing or twining;
        trees or shrubs; fruit a samara or a capsule
            5. Twigs bright green or purple and glaucous or brown to black and
            pubescent; leaflets 3–5, lobed, coarsely dentate, or undulate, the teeth
            large and few; fruit a samara . . . .*Acer*
            5. Twigs brown or gray, lacking bloom, usually glabrous;
            leaflets 3, serrate, the teeth crowded; fruit a glabrous capsule
                6. Petioles 3–7 cm long; leaflets of nearly equal size; older bark
                white-striped; fruit a bladderlike inflated capsule . . . .*Staphylea*
                6. Petioles less than 2 cm long; lateral leaflets reduced or mostly
                absent; bark solid brown, yellow, or black, not white-striped;
                capsule not inflated . . . .*Forsythia*
1. Leaves with more than 3 leaflets
    7. Leaves palmately compound
        8. Leaflets 5–7, mostly obovate or oblanceolate, greater than 2 cm wide,
        serrate; fruit a 3-valved capsule with large nut-like seeds . . . .*Aesculus*
        8. Leaflets 5, lanceolate or narrowly elliptic, mostly less than 2 cm wide,
        entire; fruit a drupe . . . .*Vitex*
    7. Leaves pinnately compound
        9. Stems climbing, with or without aerial rootlets
            10. Stems weakly climbing; leaflets acute or acuminate, entire; fruit a
            berry; southern . . . .*Jasminum*
            10. Vines; leaflets long-acuminate, coarsely toothed or dentate;
            fruit a fusiform, 2-celled capsule or a head of hairy achenes
                11. Leaflets with 1 main vein from the base; fruit a capsule . . .*Campsis*
                11. Leaflets with 3–5 main veins from the base or near the base; fruit a
                head of achenes . . . .*Clematis*

9. Stems erect, not climbing; erect trees or shrubs; leaflets usually acute or short acuminate and finely serrate (lobed or coarsely toothed in *Acer*)
  12. Twigs and branchlets with lenticels large, raised, and corky; older branchlets with a very large pith; shrubs; fruit a berrylike drupe . . . . *Sambucus*
  12. Twigs and branchlets with lenticels small, inconspicuous, or lacking, not corky; older branchlets with small pith; trees or shrubs; fruit various
    13. Leaflets less than 5 cm long, entire; southern . . . . *Jasminum*
      14. Leaflets 9 or more . . . . *Syringa*
      14. Leaflets 3–7
    13. Leaflets greater than 5 cm long, lobed, serrate, serrulate, or entire
      15. Leaflets 3–5, lobed or coarsely serrate; twigs often glaucous, green or purple; fruit a pair of samaras . . . . *Acer*
      15. Leaflets 5–13, entire or finely serrate; twigs glabrous or pubescent, not glaucous, gray to brown; fruit various
        16. Leaves lacking strong odor when crushed; fruit a samara or capsule
          17. Leaflets 7 or more; buds nearly globose, blunt; trees; fruit a samara . . . . *Fraxinus*
          17. Leaflets 5–7; buds narrowly ovoid, acuminate; shrubs or small trees; fruit an inflated capsule . . . . *Staphylea*
        16. Leaves and twigs pungent, aromatic when crushed; fruit a drupe or separating into sections
          18. Leaves ovate, about twice as long as wide; buds obvious in the axils; fruit separating into sections . . . . *Tetradium*
          18. Leaves ovate-lanceolate to elliptic; usually much greater than twice as long as wide; buds covered by the base of the petiole; fruit a drupe . . . . *Phellodendron*

KEY IV. PLANTS WITH ALTERNATE SIMPLE LEAVES

(Sometimes some of the leaves are clustered on short spurs.)

1. Leaves greater than 70 cm wide; southern . . . . *Serenoa*
1. Leaves less than 70 cm wide
  2. Leaves crowded and overlapping at the base, linear to linear-lanceolate, at least 10 times longer than wide; woody grasses
    3. Leaves with inconspicuous midvein, lacking strong white midrib, 1–5 cm wide, the margins smooth . . . . *Phragmites*
    3. Leaves with conspicuous white midrib projecting from the lower surface, less than 2 cm wide, margins sharp, serrulate
      4. Leaves less than 1 cm wide, the midrib on lower surface rough, scabrous, or cutting to the touch; southern . . . . *Cortaderia*
      4. Leaves 1–2 cm wide, the midrib smooth to the touch

5. Leaves mostly less than 25 cm long, about 10 times longer
   than wide . . . .*Arundinaria*
5. Leaves mostly greater than 50 cm long, at least 50 times longer
   than wide . . . .*Miscanthus*
2. Leaves mostly separated, not densely overlapping at the base,
   shaped variously, mostly less than 10 times longer than wide
   6. Leaves lobed (alternate step 6, p. 26)
      7. Stems climbing or twining; vines
         8. Leaves with 1 or more lateral lobes on each side of midrib; leaves and
            stems with rank odor when crushed; fruit a red berry . . . . . .*Solanum*
         8. Leaves palmately lobed; leaves and stems lacking rank odor
            when crushed; fruit a red, black, or blue berry or drupe
            9. Aerial rootlets present on twigs and branchlets; leaf lobes mostly
               entire; veins on upper surface of leaves white or pale;
               fruit a berry . . . .*Hedera*
            9. Aerial rootlets absent; leaf lobes mostly serrate; veins
               green, not white or pale; fruit a berry or drupe
               10. Tendrils present; fruit a berry
                   11. Tendrils ending in expanded tips or
                       disklike appendages . . . .*Parthenocissus*
                   11. Tendrils narrowed or blunt at tips, not ending in
                       disklike appendages, coiled
                       12. Bark shredding . . . .*Vitis*
                       12. Bark close, not peeling . . . .*Ampelopsis*
               10. Tendrils absent; fruit a drupe
                   13. Petiole 1–4 cm long; leaves with middle lobe longer than the
                       lateral ones, pubescent; fruit red; southern . . . .*Cocculus*
                   13. Petiole 5–15 cm long; leaves with lobes nearly equal
                       in length, glabrous or nearly so; fruit blue or black
                       14. Leaves peltate, the lobes rounded or acute;
                           fruit blue . . . .*Menispermum*
                       14. Leaves cordate, the lobes acuminate; fruit
                           black; southern . . . .*Calycocarpum*
      7. Stems erect or prostrate; trees or shrubs
         15. Leaves broadly fan-shaped with more or less dichotomously
             branched veins running to the margin, bluntly 2-lobed, mostly
             on short spur-like branches on older growth; fruit
             drupe-like; gymnosperm . . . .*Ginkgo*
         15. Leaves variously shaped, not fan-shaped, the veins branched
             or not; entire, toothed, or more than 2-lobed; short spur-like
             branches usually lacking; fruit various; angiosperm
             16. Leaves distinctly palmately lobed (alternate step 16, p. 25)
                 17. Leaves orbicular or wider than long, lacking spines,
                     or the petiole and both leaf surfaces densely covered
                     with spines

18. Petiole and both leaf surfaces densely covered with
spines; southern . . . .*Oplopanax*
18. Petiole and leaf surfaces glabrous; southern
   19. Leaves with 3–5 lobes . . . . ×*Fatshedera*
   19. Leaves with 8 or more lobes . . . .*Fatsia*
17. Leaf blades ovate to suborbicular, longer than wide, lacking spines
or spines small and scattered (Spines may be present on branches
and branchlets.)
   20. Plants with milky sap; southern . . . .*Ficus*
   20. Plants lacking milky sap
      21. Leaves with 8 lobes, these beginning at least $^2/_3$ of the way
toward the base; southern . . . .*Fatsia*
      21. Leaves with fewer than 8 lobes, shallowly or deeply lobed
         22. Leaves star-shaped, 5–7 lobed, margins evenly serrate
            23. Branches unarmed, often with corky ridges; fruit a long-stalked,
spherical head of capsules . . . .*Liquidambar*
            23. Branches armed with stout, triangular, broad-based spines, not
corky; fruit a drupe . . . .*Kalopanax*
         22. Leaves deeply or shallowly lobed, less divided, not
star-shaped, 3–5 lobed, the margins regularly or irregularly
serrate, entire, or with smaller lobes; branches smooth or
spiny, but lacking triangular, broad-based spines or
corky ridges
            24. Base of petiole hollow, forming a hood over the lateral bud;
stipules or stipule scars prominent; bark of trunk and branches
peeling off in large thin plates; fruit a suspended ball of
hairy achenes . . . .*Platanus*
            24. Base of petiole solid, not hollow, subtending the lateral
(axillary) bud; stipules or stipule scars inconspicuous or
absent; bark tight, or if peeling, not peeling in large thin
plates; fruit various
               25. Leaves conspicuously broadened into a sheath at the
base; southern . . . . ×*Fatshedera*
               25. Leaves broadened only slightly, if at all, at the base,
lacking sheaths
                  26. Leaves with lobes entire, the petiole greater than 12 cm long;
fruit of separate, papery, leafy follicles . . . .*Firmiana*
                  26. Leaves with lobes irregularly serrate, the petiole
less than 10 cm long; fruit various
                     27. Branchlets white-tomentose; lower surface of leaves
white-tomentose, petiole often flattened;
fruit a capsule . . . .*Populus*
                     27. Branchlets green or brown, not
white-tomentose; lower surface of leaves

green, not white-tomentose, petiole terete; fruit a capsule, berry,
or cluster of drupelets
28. Petioles and twigs glandular; fruit aggregate, a cluster of
drupelets, berrylike . . . .*Rubus*
28. Petioles and twigs glabrous or pubescent, lacking glands; fruit
a berry or capsule
29. Older bark separating into numerous thin layers;
spines absent . . . .*Physocarpus*
29. Older bark close, not separating in layers; spines present
or absent
30. Leaf base cuneate; stipules or stipule scars present; spines
absent; fruit a capsule . . . .*Hibiscus*
30. Leaf base cordate or if cuneate, stipules absent or
adnate to the petiole; spines often present; fruit
a berry . . . .*Ribes*
16. Leaves pinnately lobed or irregularly lobed or tulip-shaped with
broad truncate apex, at least not distinctly palmately lobed
(except sometimes in *Sassafras*)
31. Leaves with 3 main veins from near the base, mostly with 1 or a
few lateral lobes
32. Lobes entire; leaves spicy-aromatic; bark of branchlets smooth, green;
fruit a blue drupe . . . .*Sassafras*
32. Lobes serrate; leaves lacking strong aroma; bark of branchlets
smooth or rough, yellow, brown, or black; fruit a pome,
multiple, or berrylike
33. Sap clear, not milky; leaf lobes pointed; thorns usually present;
fruit a pome . . . .*Sorbus*
33. Sap milky; leaf lobes rounded or pointed; thorns absent;
fruit multiple
34. Leaves densely pubescent, like velvet on lower surface;
petioles 5–10 cm long; twigs densely pubescent,
like velvet . . . .*Broussonetia*
34. Leaves downy, pubescent, or occasionally glabrous on lower
surface; petioles usually 2–4 cm long; twigs glabrous
or pubescent . . . .*Morus*
31. Leaves with a single large main vein (midrib), variously lobed
35. Leaves irregularly lobed or tulip-shaped with broad
truncate apex
36. Leaves tulip-shaped, with truncate apex and 2 broad lateral lobes on
each side of midrib, the lobe margins entire; buds covered by
membranous stipules . . . .*Liriodendron*
36. Leaves acute or acuminate, irregularly lobed, usually with
more than 2 broadly rounded or acute lobes, the lobe
margins coarsely or finely serrate; buds scaly

37. Leaves with 2 lobes at the base of each margin transitioning
to reduced doubly serrate teeth
    38. Branches and branchlets with thorns; leaves acute
to acuminate . . . . ***Crataegus***
    38. Branches and branchlets lacking thorns; leaves
long acuminate . . . . ***Neillia***
37. Leaves irregularly lobed, the lobes irregularly serrate
    39. Plants with long, stiff, sharp thorns; leaves cuneate to truncate
at the base . . . . ***Crataegus***
    39. Plants lacking thorns; leaves cuneate, rounded, truncate,
or cordate at the base
        40. Leaves cordate at the base . . . . ***Stephanandra***
        40. Leaves cuneate to rounded or truncate at the base . . . . ***Malus***
35. Leaves pinnately lobed
    41. Leaves with numerous deep, rounded lobes on each side of the midrib,
sweet-scented; shrubs; fruit a small, nut-like achene enclosed in
a bur . . . . ***Comptonia***
    41. Leaves with few to several rounded or pointed lobes on each
side of the midrib, lacking a sweet scent; small or large trees;
fruit an acorn or pome
        42. Leaves 10–20 cm long; buds clustered at the ends of the twigs;
thorns absent; fruit an acorn . . . . ***Quercus***
        42. Leaves 4–10 cm long; buds scattered along the twig, not clustered at
the ends; thorns often present; fruit a small pome . . . . ***Crataegus***
6. Leaves entire or toothed, not lobed
    43. Leaves entire (alternate step 43, p. 35)
    44. Plants less than 10 cm tall . . . . ***Diapensia***
    44. Plants greater than 10 cm tall
        45. Leaves usually with pair of tendrils at base of petiole; stem with
vascular bundles scattered (seen in cross section); spines or prickles
usually present; stem usually green and climbing; fruit a blue, black,
or red berry . . . . ***Smilax***
        45. Leaves lacking tendrils at base of petiole; stem with
vascular bundles in a ring in cross section; spines present
or absent; stems climbing or erect, green or dark;
fruit various
            46. Stipules glandular, persistent; leaves ciliate at base of blade
and apex of petiole; shrubs growing near
water; southern . . . . ***Sebastiana***
            46. Stipules, if present, not glandular, deciduous or persistent;
leaves smooth at the basal margins and at apex of petiole,
hairs usually lacking; trees, vines, or shrubs
                47. Leaves persistent, often leathery, often revolute
(alternate step 47, p. 30)

48. Stems climbing, creeping, prostrate, or forming dense low mats
  49. Stems climbing; southern
    50. Leaves ovate or elliptic, with 1 main vein (midrib) from
      the base . . . .*Kadsura*
    50. Leaves deltoid, orbicular, or ovate, with 3–5 main veins from
      the base . . . .*Cocculus*
  49. Stems creeping or prostrate, not climbing
    51. Stems covered with brown hairs
      52. Leaves less than 2 cm long, elliptic to oblong, with coarse, brown,
        scalelike hairs on lower surface; fruit a white berry . . . . . .*Gaultheria*
      52. Leaves 2–8 cm long, oval-oblong to suborbicular, scabrous and pilose
        or glabrous; fruit a capsule . . . .*Epigaea*
    51. Stems glabrous or pubescent, not covered with brown hairs
      53. Twigs and petioles covered with brown scales or scurf; fruit
        a capsule . . . .*Rhododendron*
      53. Twigs and petioles glabrous, glaucous, or pubescent, not
        covered with brown scales or scurf; fruit a pome, berry,
        or drupe
        54. Leaves glaucous on the lower surface . . . .*Vaccinium*
        54. Leaves green, with or without black dots on the
          lower surface
          55. Leaves with black dots (glands) on the lower surface,
            obovate-oblong . . . .*Vaccinium*
          55. Leaves lacking glands, obovate-spatulate
            56. Leaves oblanceolate, subsessile; fruit a drupe . .*Arctostaphylos*
            56. Leaves ovate, the petiole 1–4 mm long; fruit
              a pome . . . .*Cotoneaster*
48. Stems erect
  57. Stems or trunks hollow, often shiny, flattened or thinly grooved
    on 1 or both sides; southern
    58. Stems less than 1 cm thick . . . .*Sasa*
    58. Stems greater than 3 cm thick . . . .*Phyllostachys*
  57. Stems or trunks solid, dull, terete, lacking prominent grooves
    59. Leaves clustered at ends of twigs; leaf scar encircling
      the branchlet . . . .*Trochodendron*
    59. Leaves usually scattered, not clustered at ends of twigs
      (sometimes in *Kalumia* and *Skimmia*); leaf scar usually
      extending less than halfway around the branchlet
      60. Branchlets usually brown, speckled white; leaves broadly obovate to
        orbicular, obtuse or emarginate, lustrous on the upper surface,
        subsessile, petiole 1–2 mm long; southern . . . .*Licania*
      60. Branchlets mostly of uniform color, lacking conspicuous
        speckles; leaves variously shaped, acute, acuminate, or
        obtuse, dull or lustrous, petiole short, long, or lacking

61. Stipule scars encircling the twig .... *Magnolia*
61. Stipule scars usually absent or not encircling the twig
    62. Leaves with rusty hairs or scales on lower surface; fruit a capsule
        63. Lower surface of leaves covered with dense woolly,
            rusty-brown hairs .... *Ledum*
        63. Lower surface of leaves with rusty scales, not
            densely woolly .... *Chamaedaphne*
    62. Leaves green, pale, or white on lower surface, lacking
        rusty-brown hairs or scales; fruit various
        64. Stipules very small, black, spine-like, persistent .... *Ilex*
        64. Stipules absent or green and membranous, not small, black,
            or spine-like
            65. Leaves mostly less than 5 cm long; shrubs
                66. Twigs green; leaves cuspidate to spinose; reproductive
                    structure a berry or drupe
                    67. Leaves sessile or subsessile, mostly greater than 2 cm long;
                        fruit a black drupe .... *Sarcococca*
                    67. Leaves with petioles about 5 mm long, the blades mostly
                        less than 2 cm long; fruit a red or yellow berry born on the
                        leaf (cladophyll) .... *Ruscus*
                66. Twigs brown, red-brown, or gray; leaves acute,
                  acuminate, obtuse, or rounded; reproductive structure
                  a cone or capsule
                  68. Leaves ciliate .... *Pieris*
                68. Leaves lacking marginal hairs
                    69. Leaves spinose, lanceolate; reproductive structure
                      a cone
                      70. Leaves greater than 6 mm wide; southern .... *Araucaria*
                      70. Leaves less than 6 mm wide .... *Cunninghamia*
                    69. Leaves rounded, obtuse, or acute, linear, narrowly
                    elliptic, oblong to obovate, not lanceolate;
                    reproductive structure a capsule
                    71. Branchlets, petioles, and veins of lower leaf surface
                      stellate-pubescent; southern .... *Loropetalum*
                    71. Branchlets, petioles, and veins of lower
                    leaf surface glabrous or, if pubescent,
                    not stellate
                    72. Leaves white on lower surface, revolute .... *Andromeda*
                    72. Leaves pale green on lower surface, revolute
                      or margins flattened
                      73. Leaves 6–13 mm long, subsessile ...... *Leiophyllum*
                      73. Leaves mostly 20–50 mm long, with
                        distinct petioles .... *Kalmia*

65. Leaves mostly 5–25 cm long; shrubs or trees
  74. Leaves with petiole greater than 2 cm long, often blue-glaucous on the lower surface; fruit a glaucous, black, or blue-black drupe; southern . . . .*Daphniphyllum*
  74. Leaves with petiole mostly less than 2 cm long; often pale but not blue-glaucous on lower surface; fruit various
    75. Leaves or buds clustered at end of twig; bark red-brown, brown, or black, peeling or not; tall shrubs or trees
      76. Leaves scattered along branch, not clustered at twig ends, buds clustered at ends of twigs; bark gray, brown, or black, tight, not peeling; fruit an acorn . . . .*Quercus*
      76. Leaves clustered near the end of the season's growth; buds usually scattered along branch; bark often red-brown, usually peeling off; fruit a 5-valved capsule
        77. Leaves rugulose, pale or rusty on lower surface; growth of preceding season with numerous scale scars below the leaves . . . .*Rhododendron*
        77. Leaves smooth, not rugulose, light green on lower surface; growth of preceding season with 2 scale scars at base of leaves . . . .*Kalmia*
    75. Leaves and buds scattered (often clustered in *Skimmia*) along branch; bark gray, yellow-brown, brown, or black; small or tall shrubs, small or tall trees
      78. Leaves sessile or subsessile; southern . . . .*Podocarpus*
      78. Leaves with a distinct petiole, at least 2 mm long
        79. Young twigs dark rusty-tomentose; flower buds large, rusty-pubescent, stalked . . . .*Michelia*
        79. Young twigs glabrous to pubescent, not dark rusty-tomentose; flower buds, if present, small, glabrous, or pubescent
          80. Leaves spicy-aromatic, the petiole less than 1 cm long; branchlets light red-brown or rusty; fruit a berry . . . . .*Laurus*
          80. Leaves lacking strong aroma, the petiole long or short; branchlets variously colored; fruit a head of follicles, capsule, drupe, or pome
            81. Midrib of lower leaf surface usually prominent only about half its length, the apex rounded or obtuse, the petiole often less than 4 mm long . . . .*Ternstroemia*
            81. Midrib of lower leaf surface prominent its entire length or nearly so, the apex rounded, acute, or mucronate, petiole usually greater than 5 mm long
              82. Twigs green or yellow-brown; leaves dotted with black glands . . . .*Skimmia*
              82. Twigs brown or gray; leaves lacking glands

83. Leaves with broad white midrib projecting from lower surface, greater than 1mm wide toward the base; fruit a head of follicles; southern . . . . *Illicium*
83. Leaves with narrow brown, green, or pale midrib, less than 1mm broad toward the base; fruit a capsule, drupe, or pome
    84. Leaves obovate, rounded at the apex; fruit a capsule; southern . . . . *Pittosporum*
    84. Leaves oblong to lanceolate or elliptic, acute, cuspidate, or mucronate; fruit a capsule, drupe, or pome
        85. Leaves mucronate; fruit a small red pome . . . . *Stranvaesia*
        85. Leaves acute or cuspidate; fruit a capsule or dark blue drupe
            86. Lateral veins on lower leaf surface prominent or obvious; fruit a drupe; southern . . . . *Persea*
            86. Lateral veins on lower leaf surface inconspicuous or barely visible; fruit a capsule
                87. Leaves acute or acuminate-cuspidate, sharply pointed . . . . *Kalmia*
                87. Leaves cuspidate to acuminate, blunt; southern . . . . . . *Cleyera*
47. Leaves deciduous, mostly thin, occasionally leathery, usually flat, not revolute
    88. Leaves and young twigs covered with silver or silver and brown scales; thorns often present; fruit drupe-like
        89. Leaves lanceolate to ovate, less than 5 times longer than wide . . . . *Elaeagnus*
        89. Leaves linear, greater than 10 times longer than wide . . . . . *Hippophae*
    88. Leaves and young twigs lacking silver or brown scales; thorns lacking; fruit various
        90. Leaves and bark spicy-aromatic; fruit a red or blue drupe
            91. Leaves with 3 principal veins from near the base, often lobed; bark of branchlets green; fruit blue . . . . *Sassafras*
            91. Leaves with 1 principal vein (midrib), never lobed; bark of branchlets brown; fruit red . . . . *Lindera*
        90. Leaves and bark lacking strong scent, at least not spicy-aromatic; fruit various
            92. Leaves with 3–7 large veins from broadly cordate or oblique base; fruit a legume, drupe, or capsule
                93. Stems climbing; vines; fruit a capsule . . . . *Aristolochia*
                93. Stems erect; trees or shrubs; fruit a legume or drupe
                    94. Leaf base cordate; fruit a legume . . . . *Cercis*
                    94. Leaf base truncate, cuneate, or oblique; fruit a drupe
                        95. Leaf base only slightly oblique, nearly symmetrical . . . . *Aphananthe*
                        95. Leaf base asymmetrical . . . . *Celtis*

92. Leaves with single large vein or midrib from base, the base mostly
cuneate, truncate, or rounded, sometimes cordate; fruit various
  96. Vines or scrambling shrubs; fruit a red berrylike pome
    97. Mostly low shrubs with spreading, arching horizontal branches; leaves
mostly less than 25 mm long . . . .*Cotoneaster*
    97. Scrambling shrubs or high-climbing vines; leaves mostly
greater than 25 mm long
      98. Stems and branchlets shiny red-brown; leaves with margins undulate,
the veins prominent, straight, parallel; fruit a drupe . . . . .*Berchemia*
      98. Stems and branchlets dull light or dark brown; leaves with
margins even, the veins inconspicuous, curving; fruit a berry
        99. Leaves oblong or lanceolate with cuneate base; leaves and
stems lacking strong odor; stems often with short
thorn-like branches . . . .*Lycium*
        99. Leaves ovate to elliptic, the base often cordate or lobed;
stems lacking thorn-like branches
          100. Crushed leaves and stems with strong, disagreeable odor;
short scrambling shrub or vine; leaf base often cordate
or lobed . . . .*Solanum*
          100. Leaves and stems lacking strong odor; high climbing vine;
leaves narrowed at the base . . . .*Schisandra*
  96. Erect shrubs or trees; fruit various
    101. Leaves rhombic or deltoid, acuminate, with glands at the base of
the blade . . . .*Sapium*
    101. Leaves shaped variously but not rhombic or deltoid,
obtuse to acuminate, usually lacking glands at the base of
the blade
      102. Twigs and leaves with milky sap; thick, stout thorns usually present;
fruit multiple, globose, large (to 12 cm in diameter), heavy,
yellow-green . . . .*Maclura*
      102. Twigs and leaves with clear sap; thorns lacking (spines or
thorns present in *Berberis* and *Bumelia*); fruit various
        103. Stems with branched or simple spines or thorns, leaves
often fascicled; fruit red, purple, or black, berrylike or
a drupe
          104. Pith chambered . . . .*Prinsepia*
          104. Pith solid
            105. Leaves usually less than 7 cm long, often spiny on the
margins; inner bark and wood yellow; fruit
red; southern . . . .*Berberis*
            105. Leaves usually greater than 7 cm long, lacking marginal
spines; inner bark brown; fruit black . . . .*Bumelia*
        103. Stems lacking spines; leaves usually scattered along the
branch; fruit various

106. Base of petiole covering the axillary bud or stipules sheathing
and enclosing the stem; fruit a capsule or drupe
  107. Leaves with membranous, sheathing stipules; herb to 3 m tall that has
appearance of a shrub . . . .*Polygonum*
  107. Leaves lacking membranous, sheathing stipules; shrubs or trees
    108. Leaves 10–25 cm long; bark tight, thin, not fibrous or leathery; fruit
a capsule . . . .*Styrax*
    108. Leaves 5–7 cm long; bark fibrous and leathery; fruit a drupe . . . .*Dirca*
106. Base of petiole solid, not hollow, subtending the exposed axillary
lateral bud; fruit various
  109. Bark of stems and branches fibrous and leathery; twigs very
flexible, never snapping (even when bent back 180 degrees or
more); fruit a drupe
    110. Leaves greater than 6 cm long, greater than 2 cm wide . . .*Edgeworthia*
    110. Leaves less than 6 cm long, less than 2 cm wide . . . .*Daphne*
  109. Bark of stems and branches tight, smooth or rough, thin, not
leathery; twigs stiff or brittle, at least not so flexible;
fruit various
    111. Lateral veins running parallel to the margins of the leaf and ending
near the apex; white vein threads stretched and visible when leaf is
severed in cross section; fruit a drupe . . . .*Cornus*
    111. Lateral veins fading before reaching the margin of the leaf or
not running to the apex; leaf breaking clean, with no threads
of veins evident; fruit various
      112. Pith chambered (divided by woody plates or diaphragms)
        113. Leaves 20–30 cm long; buds dark, silky; fruit a very
large berry . . . .*Asimina*
        113. Leaves 5–15 cm long; buds brown, glabrous, not silky;
fruit a drupe or berry
          114. Leaves mucronate or apiculate; shrubs; branchlets black,
sinewy, often glaucous . . . .*Oemleria*
          114. Leaves acute to acuminate, not mucronate or
apiculate; small trees; branchlets black to brown, not
sinewy or glaucous
            115. Leaf veins, especially the midrib on upper surface, often
pink or pale; buds with only 2 scales visible; fruit
a berry . . . .*Diospyros*
            115. Leaf veins green to brown, not pink or pale; buds with
several scales visible; fruit a drupe . . . .*Nyssa*
      112. Pith solid or hollow, not chambered
        116. Stipules and stipule scars encircling the twig; leaves 10–30 cm
long, acuminate . . . .*Magnolia*
        116. Stipule scars absent or not encircling the twig; leaves
mostly less than 10 cm long

117. Stipules small, black, spine-like; fruit a red berry . . . .*Ilex*
117. Stipules, if present, small or large, foliar or membranous, green
     or brown, not spine-like or black; fruit a berry, capsule, drupe,
     legume, pome, or nut
  118. Buds clustered at the end of twigs (In the absence of buds in
       Rhododendron this condition ascertained by the falsely
       whorled arrangement of twigs and leaves.); fruit a capsule
       or acorn
    119. Leaves mostly clustered near the end of twigs, thin, the
         margins ciliate; shrubs; fruit an elongated capsule on a
         glandular-pubescent pedicel . . . .*Rhododendron*
    119. Leaves scattered on the twigs, thick, the margins entire or lobed,
         not ciliate; trees; fruit an acorn . . . .*Quercus*
  118. Buds scattered along twigs, not clustered at twig tips; fruit a
       capsule, drupe, legume, berry, or pome
    120. Buds each covered by a single hood-like scale; fruit a capsule . . . .*Salix*
    120. Buds each covered by at least 2 scales; fruit a capsule, drupe,
         legume, berry, or pome
      121. Leaf blades 8–20 cm long
        122. Exposed bud scales 2; pith often chambered (divided by
             diaphragms or plates); bark breaking into regular blocks on
             older trunks; fruit a berry . . . .*Diospyros*
        122. Exposed bud scales several, imbricated; pith continuous;
             bark smooth, scaly, or ridged; fruit a drupe or pome
          123. Leaf blades less than twice as long as wide, usually cordate or
               obtuse at base; fruit a pome . . . .*Pyrus*
          123. Leaf blades greater than twice as long as wide,
               tapering at both ends; fruit a drupe; southern
            124. Buds densely pubescent; tree . . . .*Leitneria*
            124. Buds glabrous; parasitic shrub . . . .*Pyrularia*
      121. Leaf blades 1–10 cm long
        125. Petioles 1–4 cm long; fruit a pome or drupe-like, red
             or yellow
          126. Leaves often mucronate; fruit drupe-like, red . . .*Nemopanthus*
          126. Leaves obtuse or rounded; fruit a pome or drupe-like,
               yellow or red
            127. Leaves ovate to oblong, pubescent on lower surface, apex
                 obtuse; fruit a pome, yellow . . . .*Cydonia*
            127. Leaves obovate, glabrous, apex rounded; fruit
                 drupe-like, red . . . .*Cotinus*
        125. Petioles less than 1 cm long or absent; fruit a capsule,
             drupe, berry, or legume
          128. Lower surface of leaves covered with yellow, resinous
               dots; fruit a capsule or drupe

129. Leaves when crushed sweet-scented, often serrate near apex; fruit a waxy, gray to blue drupe . . . .*Myrica*

129. Leaves odorless or at least not sweet-scented; fruit a capsule or blue to black drupe, lacking waxy coating

    130. Leaves glabrous or lightly pubescent; fruit a capsule . . . .*Lyonia*

    130. Leaves glandular; fruit a fleshy drupe . . . .*Gaylussacia*

128. Lower surface of leaves glabrous or pubescent, lacking yellow, resinous dots; fruit a capsule, drupe, berry, or legume

    131. Twigs pale green or gray-green; fruit a capsule; southern . . . . .*Securinega*

    131. Twigs light brown, light gray, red-brown, dark brown, green to bright green; fruit a capsule, drupe, berry, or legume

        132. Branchlets green or red-brown, minutely white-speckled or hairy, or the buds subglobose and spreading; fruit a berry . . . .*Vaccinium*

        132. Branchlets brown, gray, or black, not minutely white-speckled; buds ovoid to ellipsoid; fruit a capsule, legume, drupe, or berry

            133. Leaves mostly greater than 5 cm long, the petiole mostly greater than 3 mm long; fruit a capsule or drupe

                134. Twigs glaucous; leaves subentire, glaucous on lower surface; fruit a capsule; southern . . . .*Zenobia*

                134. Twigs glabrous or pubescent, lacking bloom; leaves entire, glabrous or pubescent, lacking bloom on lower surface; fruit a capsule or drupe

                    135. Leaves often mucronulate or apiculate, the petiole mostly greater than 15 mm long; fruit a capsule . . . .*Exochorda*

                    135. Leaves acute or abruptly acuminate, the petiole usually less than 15 mm long; fruit a capsule or drupe

                        136. Leaves $2^{1}/_{2}$–3 times longer than wide, the secondary veins inconspicuous, or the tertiary and ultimate veins as prominent as the secondary veins; bark of branchlets gray and shredding; fruit a dry indehiscent drupe . . . .*Cyrilla*

                        136. Leaves less than $2^{1}/_{2}$ times longer than wide, the secondary veins obvious; bark of branchlets black, gray-brown, or tan, shredding or not; fruit a drupe or capsule

                            137. Bark of branchlets black, tight, not shredding; leaves usually with 6–7 pairs of prominent lateral veins; fruit a drupe . . . .*Rhamnus*

                            137. Bark of branchlets gray or brown, shredding; leaves usually with 4–6 pairs of less prominent lateral veins; fruit a globular or urn-shaped capsule in naked racemes . . . .*Lyonia*

            133. Leaves 1–4 cm long, sessile or the petiole less than 3 mm long; fruit a legume, capsule, or berry

138. Branches striate, angled; leaves lanceolate; fruit a legume . . . . .*Genista*
138. Branches uniform, terete; leaves ovate or elliptic; fruit a
  capsule or berry
  139. Leaves narrowly oblong, leathery; fruit a berry; southern . . . .*Litsea*
  139. Leaves elliptic to ovate, thin; fruit a capsule (achenes
    in *Solidago*)
    140. Shrubs, about 1 m tall; leaves brown on lower surface, the veins raised
      in an alligator-skin pattern on both surfaces; southern . . . . . .*Solidago*
    140. Large shrubs, greater than 1 m tall; leaves pale or green on
      lower surface, the veins inconspicuous
      141. Leaves with chaffy, lacerate, or cleft scales on midrib of
        lower surface, distinctly alternate, the petiole
        2–3 mm long . . . .*Menziesia*
      141. Leaves lacking scales, subopposite, sessile . . . .*Lagerstroemia*
43. Leaves toothed
  142. Leaves toothed only at the rounded apex
    143. Leaves greater than 2 cm wide . . . .*Exochorda*
    143. Leaves less than 2 cm wide
      144. Leaves attenuate at the base . , ...*Myrica*
      144. Leaves cuneate or rounded at the base . . . .*Ilex*
  142. Leaves toothed at least halfway toward the base
    145. Leaves with hard, sharp spines terminating the teeth
      146. Leaves scattered along the branchlet, shiny, not fibrous, less than
        10 cm long . . . .*Ilex*
      146. Leaves densely overlapping, dull, tough, fibrous, greater than
        20 cm long . . . .*Yucca*
    145. Leaves toothed but lacking hard, sharp spines; fruit various
      147. Plants with spines or thorns on the stems, branches, or
        twigs (alternate step 147, p. 36)
        148. Pith chambered . . . .*Prinsepia*
        148. Pith solid
          149. Twigs, branchlets, and branches bearing stipular or
            branched spines; fruit a berry or drupe
            150. Twigs and branches armed with branched spines; inner bark
              and wood yellow; leaves with 1 main vein or midrib; shrubs;
              fruit a red berry . . . .*Berberis*
            150. Twigs bearing stipular spines, 1 hooked and
              recurved, the other straight at each node; inner
              bark brown or orange-brown; leaves with 3–5
              main veins from the base; shrubs or small trees;
              fruit a drupe, yellow-brown, dark red, or brown
              151. Leaves rounded to slightly narrowed, often slightly oblique
                at the base, the petiole usually less than 2 mm long;
                fruit fleshy . . . .*Ziziphus*

151. Leaves rounded, symmetrical at the base, the petiole usually greater than 5 mm long; fruit dry . . . .*Paliurus*

149. Twigs and branches bearing thorns; fruit a drupe or pome

152. Leaves mostly less than 5 cm long; fruit a pome

153. Stipules reniform, large; thorns woody; leaves deciduous . . . .*Chaenomeles*

153. Stipules absent, or if present, small, not reniform; thorns leafy; leaves persistent . . . .*Pyracantha*

152. Leaves mostly greater than 5 cm long; fruit a drupe or pome

154. Petioles with glands; fruit a drupe . . . .*Prunus*

154. Petioles lacking glands; fruit a drupe or pome

155. Leaves sessile or with petiole less than 2 mm long . . . .*Mespilus*

155. Leaves with petioles greater than 2 mm long

156. Thorns present on twigs and branches; thorns lacking lateral buds and leaves . . . .*Crataegus*

156. Thorns present on branches only; thorns usually leafy or with lateral buds . . . .*Pyrus*

147. Plants lacking spines or thorns on stems and branches

157. Stems climbing or twining; vines; fruit a capsule or berry

158. Lateral veins of leaves straight and parallel; leaves serrulate, the petiole less than 5 mm long; fruit a berry . . . .*Berchemia*

158. Lateral veins curved, not straight; leaves crenate-serrate, the petiole greater than 5 mm long; fruit a berry or orange capsule

159. Leaves mucronate or apiculate, finely serrate; fruit a berry . . . .*Actinidia*

159. Leaves cuspidate, acuminate, or acute; fruit a berry or capsule

160. Leaves cordate; fruit a berry . . . .*Ampelopsis*

160. Leaves cuneate to attenuate; fruit a berry or capsule

161. Leaves crenate; fruit a yellow or orange capsule . . . . .*Celastrus*

161. Leaves undulate-denticulate; fruit a red berry

162. Leaves greater than half as wide as long . . . .*Schisandra*

162. Leaves less than half as wide as long; southern . . . .*Kadsura*

157. Stems erect; shrubs or trees; fruit various

163. Stems low, 10–20 cm tall, almost herbaceous, from subterranean creeping stems; leaves persistent, glossy, coriaceous, clustered near the end of the season's growth

164. Leaves ovate, with wintergreen flavor; fruit red, berrylike . . . .*Gaultheria*

164. Leaves oblanceolate or with main veins of lighter green, lacking wintergreen flavor; fruit a capsule . . . .*Chimaphila*

163. Stems or trunks greater than 20 cm tall; leaves various

165. Leaves with 3–5 main veins from near the base, the lateral 2–4
    nearly equal to the midrib or at least thicker and more prominent
    than other lateral veins (alternate step 165, p. 38)
    166. Low, much-branched, almost herbaceous shrubs; fruit
        a capsule . . . .*Ceanothus*
    166. Trees or tall shrubs; fruit a capsule, berry, drupe, or multiple
        and fleshy
        167. Sap milky; fruit multiple and fleshy
            168. Leaves glossy and smooth on upper surface . . . .*Morus*
            168. Leaves rough or scabrous on upper surface
                169. Leaf base oblique, the lower leaf surface densely pubescent,
                    like velvet . . . .*Broussonetia*
                169. Leaf base symmetrical, the lower leaf surface
                    pubescent, sometimes even densely so, but not
                    like velvet . . . .*Morus*
        167. Sap clear, not milky; fruit a capsule, berry, or drupe
            170. Leaves with deltoid, truncate, or cordate, symmetrical
                base, the petiole terete or flattened; fruit a capsule, berry,
                or drupe
                171. Leaves coarsely or remotely toothed, the petiole terete, greater
                    than 5 cm long, nearly as long or longer than the blade;
                    fruit a berry . . . .*Idesia*
                171. Leaves finely and irregularly serrate or crenate or,
                    if coarsely toothed, deltoid or doubly toothed, the
                    petiole often flattened, less than 5 cm long, usually
                    much shorter than the blade; fruit a capsule or drupe
                    172. Leaves ovate, truncate, serrate, or crenate, the
                        petioles $^1/_4$ to $^1/_3$ the length of the blade; fruit
                        a drupe . . . .*Hovenia*
                    172. Leaves ovate to deltoid, truncate to cordate, crenate or coarsely
                        or doubly toothed, the petiole at least half the length of the
                        blade; fruit a capsule . . . .*Populus*
            170. Leaves with oblique base (nearly symmetrical in
                Pteroceltis), the petiole terete; fruit a drupe
                173. Leaves about as wide as long, cordate, acute or short acuminate;
                    pith solid, not chambered; bark of trunk smooth or furrowed,
                    lacking corky ridges; fruit a small nut-like drupe attached to a
                    winglike leaf . . . .*Tilia*
                173. Leaves longer than wide, cuneate to truncate, long
                    acuminate or taper-pointed; pith chambered or not;
                    bark of trunk with high, corky ridges; fruit a drupe,
                    winged or not
                    174. Leaves rounded at the base; fruit lacking wings . . . .*Celtis*
                    174. Leaves truncate to broadly cuneate; fruit winged . . .*Pteroceltis*

165. Leaves with only 1 main vein (midrib) from the base that is significantly thicker than other veins
  175. Buds distinctly stalked; leaves nearly as wide as long
    176. Leaves serrate or doubly serrate, sometimes undulate, the base symmetrical, not oblique; buds club-shaped, not woolly, often somewhat sticky; fruit of small nutlets in a persistent cone-like woody structure . . . .*Alnus*
    176. Leaves undulate, with coarse broad teeth or with veins extended into abrupt linear teeth, the base oblique; buds sickle-shaped and woolly; fruit a 2-celled woody capsule
      177. Leaves cuspidate, the margins interrupted by short, linear extensions of the veins or teeth . . . .*Corylopsis*
      177. Leaves acute or obtuse, undulate-margined or coarsely dentate, with wide triangular rounded or acute teeth
        178. Leaves coarsely toothed
          179. Leaves obovate to oblong, rounded to broadly cuneate at the base, coarsely toothed in apical half, the teeth broad, triangular, acute . . . .*Fothergilla*
          179. Leaves ovate to suborbicular, truncate to cordate at the base, coarsely toothed nearly the entire margin, the teeth acute, not broadly triangular . . . .*Parrotiopsis*
        178. Leaves with undulate margins or with broad rounded teeth
          180. Leaves undulate or toothed to well below the middle, as long or slightly longer than wide, with narrowed, oblique base . . . .*Hamamelis*
          180. Leaves undulate or toothed usually only in apical third, often twice as long as wide, with rounded, nearly even base . . . .*Parrotia*
  175. Buds sessile, not stalked (except flower buds of *Chamaedaphne*); leaves longer than wide (except in *Styrax obassia*)
    181. Leaf base oblique
      182. Leaves mostly doubly toothed; the teeth lacking glands; fruit a samara . . . .*Ulmus*
      182. Leaves singly toothed; the teeth gland-tipped; fruit a drupe
        183. Twigs roughened by many crowded, coarse, raised lenticels; leaf base only slightly oblique, nearly symmetrical; fruit smooth . . . .*Aphananthe*
        183. Twigs with some lenticels, but not roughened; leaf base asymmetrical; fruit with raised ribs . . . .*Planera*
    181. Leaf base symmetrical or slightly oblique
      184. Buds naked; leaves subopposite, secondary veins prominently curving to be roughly parallel to the margin; fruit a drupe . . . .*Rhamnus*

184. Buds with scales; leaves distinctly alternate, secondary veins
usually inconspicuous, not running parallel to the margin;
fruit various
  185. Buds covered by a single hood-like scale; fruit a capsule . . . .***Salix***
  185. Buds covered by 2 or more scales; fruit various
    186. Leaves acuminate, the petiole 1–2 cm long and deeply grooved, the base
of the blade folded or twisted so as to form a small pit at the apex of
the petiole; branchlets dark brown to black; fruit berrylike, green
to yellow . . . .***Stachyurus***
    186. Leaves rounded to acuminate, the petiole absent, short, or
long, the groove absent or of varying depths, the petiolie
apex and base of the blade not obviously folded or twisted,
thus lacking a pit; branchlets variously colored; fruit various
      187. Leaves singly toothed, often coarsely dentate or serrate,
never doubly toothed (alternate step 187, p. 40)
        188. Leaves nearly as wide as long, the petiole flattened or terete
          189. Petioles flattened or marginal teeth on the blade incurved;
plant lacking thorns; fruit a capsule . . . .***Populus***
          189. Petioles terete, not flattened; marginal teeth on the
blade straight or arched, not incurved; plant thorny
or not; fruit a pome, drupe, or samara
            190. Leaves mostly greater than 15 cm long, the lower surface
densely tomentose; twigs usually densely rusty-tomentose;
branchlets pubescent; fruit a pome; southern . . .***Eriobotrya***
            190. Leaves less than 15 cm long, glabrous or pubescent,
not densely rusty-tomentose; twigs and branchlets
glabrous or pubescent; fruit a pome, drupe,
or samara
              191. Leaves cordate, the teeth of roughly the same size on each
margin; fruit a drupe . . . .***Davidia***
              191. Leaves truncate to rounded or slightly cuneate,
the teeth long and short on the same margin;
fruit a pome or samara
                192. Buds brown, dull; leaves acute, with several long and
short teeth between each lobelike tooth; plants often
thorny; fruit a red pome . . . .***Crataegus***
                192. Buds black, shiny; leaves cuspidate or abruptly
acuminate, with 2 short teeth between each long tooth;
plants lacking thorns; fruit a samara . . . .***Euptelea***
        188. Leaves much longer than wide, the petiole terete
          193. Leaves toothed only in apical ½ to ⅔
            194. Leaves gray-green, with glands on lower surface, with 3–7
larger (greater than 1 mm long) teeth in apical half of
the blade . . . .***Baccharis***

194. Leaves green, lacking glands on lower surface, remotely toothed, about 5 short teeth (less than 1 mm long) in the apical $^2/_3$ on each margin; southern . . . .*Kadsura*
193. Leaves with many evenly spaced teeth, at least 1 mm long, covering the full length of each margin
  195. Twigs and young branchlets thin, less than 1 mm in diameter, consistently zigzagging, dull, gray or dark brown; leaves dentate, the width of each tooth equal to the length, often about 2 mm
    196. Branchlets often horizontal, short, stiff, spine-like . . . . .*Hemiptelea*
    196. Branchlets usually ascending, longer, more flexible, not resembling spines . . . .*Zelkova*
  195. Twigs and young branchlets greater than 1 mm in diameter, straight or irregularly zigzagging, smooth or shiny red-brown to black; leaves dentate or denticulate, the width of each tooth longer or shorter than the length
    197. Leaves long acuminate, rounded at the base, the teeth about 1 mm long; fruit a samara . . . .*Eucommia*
    197. Leaves acute to acuminate, cuneate to rounded, truncate or cordate at the base, the teeth greater than 1 mm long; fruit a nut or drupe
      198. Leaves cordate; fruit a drupe . . . .*Davidia*
      198. Leaves cuneate to rounded or truncate at the base; fruit a nut
        199. Terminal buds clustered; leaves oblong-lanceolate; bark rough, furrowed; fruit an acorn . . . .*Quercus*
        199. Terminal buds scattered, not clustered; bark smooth or rough and furrowed; leaves oblong-lanceolate to oblong-ovate; fruit a prickly bur
          200. Leaves oblong-ovate; buds long and sharp-pointed, usually at least 5 times as long as wide; terminal bud present; bark smooth, gray; bur with 2 triangular nuts . . . .*Fagus*
          200. Leaves oblong-lanceolate; buds ovate and blunt, usually much less than 3 times as long as wide; terminal bud absent; bark rough, furrowed; bur with 1–3 nuts . . . .*Castanea*
187. Leaves doubly or singly toothed, finely serrate, serrulate, denticulate, crenate, crenulate, or nearly entire, not coarsely serrate or dentate
201. Branches green, ridged; leaves doubly toothed . . . .*Kerria*
201. Branches gray, brown, or black, or if green, then leaves singly toothed
  202. Pith chambered; fruit a capsule
    203. Leaves with stellate pubescence on the lower surface; fruit dry, a 4-winged capsule . . . .*Halesia*

203. Leaves glabrous; fruit a 2-valved capsule ....*Itea*
202. Pith continuous; fruit various
    204. Leaf base broad, rounded, cordate, truncate, or sometimes
        slightly acute (alternate step 204, p. 43)
        205. Petioles with 1 or more glands near the blade end ....*Prunus*
        205. Petioles lacking glands
            206. Leaves glaucous on upper surface, persistent ....*Pieris*
            206. Leaves lacking bloom on upper surface, deciduous or
                sometimes persistent
                207. Leaves with at least the basal margins triply toothed or
                    more (3–8), the initial teeth sometimes appearing as
                    lobes; fruit of achenes or follicles
                    208. Branchlets gray, gray-brown, or brown, their bark shredding,
                        peeling; leaves with larger teeth toward the base appearing as
                        lobes; fruit of follicles ....*Neillia*
                    208. Branchlets often red or red-brown, their bark smooth, not
                        prominently shredding; leaves mostly regularly toothed,
                        lacking large teeth that appear as lobes at the base; fruit of
                        achenes; southern ....,,....*Neviusia*
                207. Leaves singly or doubly toothed only; fruit various
                    209. Leaves evenly and mostly singly toothed, serrate,
                      dentate, or crenate, usually in more than 2 ranks;
                      fruit a capsule, winged nutlet, drupe, or pome
                      (alternate step 209, p. 42)
                    210. Leaves suborbicular; buds nearly concealed by the base
                      of the enlarged petiole; fruit a capsule ....*Styrax*
                    210. Leaves broadly ovate to lanceolate; buds exposed;
                      fruit a capsule, winged nutlet, drupe, or pome
                      211. Leaves broadly ovate or deltoid; petioles usually flattened;
                      fruit a capsule ....*Populus*
                    211. Leaves ovate, ovate-lanceolate, elliptic, oblong, oblan-
                      ceolate, or obovate; petioles terete, never flattened;
                      fruit a capsule, winged nutlet, drupe, or pome
                      212. Leaves apiculate, strongly glaucous on lower surface;
                      fruit a capsule ....*Zenobia*
                    212. Leaves acute, acuminate, obtuse, or rounded,
                      not apiculate, green or pale but not glaucous
                      on lower surface; fruit various
                      213. Leaves with veins raised on both surfaces, dark
                        brown, sometimes rugose on upper surface; fruit a
                        winged nutlet ....*Eucommia*
                    213. Leaves with veins raised on lower surface
                      only, green to brown, flat on upper surface;
                      fruit a pome, drupe, or capsule

214. Leaves oblong, oblanceolate, or obovate; fruit a pome . . . .*Photinia*
214. Leaves ovate, ovate-lanceolate, or elliptic; fruit a pome, drupe,
     or capsule
  215. Leaves 1–4 cm long, the lateral veins straight at least halfway toward the
     margin, sharply serrate; fruit a berrylike pome . . . .*Amelanchier*
  215. Leaves greater than 5 cm long, the veins curving from the
     midrib to the margin, obscurely crenulate or serrulate;
     fruit a pome, drupe, or capsule
    216. Leaves with waxy, shiny coating on upper surface, the petiole usually
     greater than 4 cm long and at least half the length of the blade;
     fruit a pome . . . .*Pyrus*
    216. Leaves dull, lacking waxy coating on upper surface, the
     petiole less than 4 cm long and less than ¼ the length of
     the blade; fruit a pome, drupe, or capsule
      217. Leaves singly toothed in more or less regular intervals, the teeth
     sometimes terminating in minute linear, hair-like points, petiole
     usually less than 1 cm long; bark often peeling in thin, patchy
     plates, leaving a lacy or patchy appearance; fruit
     a capsule . . . .*Stewartia*
      217. Leaves somewhat irregularly toothed or appearing
     doubly toothed, the teeth lacking hair-like points,
     petiole usually greater than 2 cm long; bark uniform,
     thick, tight; fruit a drupe or pome
        218. Leaves less than twice as long as wide and the teeth usually
     greater than 0.5 mm long, or with larger nearly
     lobelike teeth . . . .*Malus*
        218. Leaves mostly 3 times as long as wide, the teeth greater than
     0.5 mm long . . . .*Ehretia*
209. Leaves unevenly and mostly doubly serrate or dentate, mostly in
     2 ranks; fruit a samara, nut, or appendaged nutlet
  219. Branches of mature trees with corky ridges; fruit a samara . . . .*Ulmus*
  219. Branches of mature trees smooth or rough but lacking corky
     ridges; fruit a samara, nut, or appendaged nutlet
    220. Older branches with short spurs bearing crowded leaves or leaf scars
     and terminated by a single bud; bark on younger trunks smooth or
     peeling off in rolls, the lenticels elongated horizontally; twigs and
     inner bark often with wintergreen flavor; fruit a samara, in
     cone-like catkins . . . .*Betula*
    220. Branches lacking short spurs; bark tight, never peeling off
     in rolls, the lenticels, if present, not prominent; twigs and
     inner bark lacking wintergreen flavor; fruit a nut or nutlets
     enclosed by a bur or appendaged by bracts
      221. Leaves ovate or ovate-oblong, not much longer than wide, acute;
     twigs mostly with bristly hairs; shrubs; fruit a nut within a
     husk-like involucre . . . .*Corylus*

221. Leaves oblong-ovate, much longer than wide, short-acuminate, taper-pointed; twigs glabrate; trees; fruit a nutlet with bracts attached

   222. Lateral veins unbranched; buds with flattened sides; bark light gray, smooth, and sinewy-fluted; fruit a nutlet with bract-like appendage, several grouped in a flexuous spike ....*Carpinus*

   222. Lateral veins branched near the leaf margins; buds terete; bark light brown or gray-brown, rough, breaking off in scaly plates; fruit a nutlet enclosed in an inflated bag, several grouped in a cone-like spike ....*Ostrya*

204. Leaf base acute or tapering

   223. Petioles with 1 or more glands at the blade end, or the lower surface of leaf waxy and with a dense row of light brown hairs along each side of midrib; fruit a drupe ....*Prunus*

   223. Petioles without glands at the blade end, the lower surface of leaf rough or shining, not waxy, the midrib lacking rows of hairs; fruit various

     224. Terminal bud long, sharp-pointed, resinous; first bud scale of lateral buds anterior; fruit a capsule ....*Populus*

     224. Terminal bud absent, or if present, mostly blunt, resinous or not; first bud scale of lateral buds interior or lateral; fruit various

       225. Leaves thick, coriaceous, persistent, crenulate-undulate to crenulate-serrulate; fruit a pome, drupe, or capsule; southern

         226. Leaves about 15 cm long, crenulate on upper margins, entire on lower half of the blade ....*Gordonia*

         226. Leaves mostly less than 15 cm long, toothed for most of the margin

           227. Leaves obtuse, rounded emarginate, or subacute, irregularly and inconspicuously toothed, obovate to ovate; fruit a pome ....*Raphiolepis*

           227. Leaves acuminate, undulate-crenate or crenulate-serrulate; fruit a drupe or capsule

             228. Leaves undulate-crenulate, elliptic, mostly less than 3 cm wide; fruit a red or black drupe ....*Ardisia*

             228. Leaves crenulate-serrate, ovate, sometimes elliptic, mostly greater than 2 cm wide; fruit a capsule ....*Camellia*

       225. Leaves thin, often deciduous, serrulate to dentate; fruit various

         229. Leaves with a sour taste, 10–15 cm long, oblong-lanceolate; fruit a 5-valved capsule, numerous in panicled racemes ...*Oxydendrum*

         229. Leaves lacking sour taste, less than 10 cm long, or if longer then not oblong-lanceolate; fruit various

230. Midrib of leaf with dark or black glands on the upper surface, these
     sometimes minute; fruit a berrylike pome . . . .*Aronia*
230. Midrib of leaf lacking dark glands on the upper surface;
     fruit various
 231. Lower leaf surface covered with yellow glands, scurf, or dots
  232. Stems with longitudinal ridges; leaves and twigs scurfy; buds globose,
   resinous; fruit an achene . . . .*Baccharis*
  232. Stems smooth, lacking longitudinal ridges; buds longer than
   wide, usually not strongly resinous; fruit a drupe, winged
   nutlet, or capsule
   233. Leaves, when crushed, sweet-scented, often entire toward base; fruit
    a waxy drupe or a winged nutlet . . . .*Myrica*
   233. Leaves lacking sweet scent; leaves toothed throughout;
    fruit a drupe or capsule
    234. Leaves coriaceous, persistent, lower surface covered with
     yellow-brown scurf; fruit a capsule . . . .*Chamaedaphne*
    234. Leaves thin, not coriaceous, deciduous, lower surface covered
     with yellowish resin globules; fruit a drupe . . . .*Gaylussacia*
 231. Lower surface of leaves green or pale, not covered with yellow
  glands or dots
  235. Twigs stout, mostly 3–5 mm in diameter when mature; branches often
   with stout spurs; fruit a pome . . . .*Pyrus*
  235. Twigs slender, less than 3 mm in diameter when mature;
   branches mostly lacking stout spurs; fruit a capsule, drupe,
   berry, or follicle
   236. Branchlets finely white-speckled or hairy, green or red-brown;
    fruit a berry . . . .*Vaccinium*
   236. Branchlets usually gray, brown, or black, not finely
    white-speckled or hairy; fruit a capsule, follicle,
    or drupe
    237. Twigs stellate-pubescent; fruit a 3-valved capsule . . . .*Clethra*
    237. Twigs glabrous or pubescent but not stellate-pubescent;
     fruit a capsule, follicle, or drupe
     238. Leaves with clusters of red-brown hairs along the midrib on the
      lower surface; fruit a capsule . . . .*Enkianthus*
     238. Leaves glabrous or pubescent, lacking red-brown
      clusters of hairs; fruit a capsule, drupe, or follicle
      239. Branchlets stout, pubescent, mottled gray and brown,
       shedding in small flakes; leaves obovate, the base cuneate;
       fruit a drupe . . . .*Symplocos*
      239. Branchlets slender, glabrous or pubescent,
       uniformly gray, brown, or black, usually not
       shedding; leaves variously shaped; fruit a capsule,
       drupe, or follicle

240. Leaves obovate, the base cuneate to long-tapering; fruit a capsule
   241. Leaves cuneate, dull on upper surface, less than 10 cm long . . . . . *Clethra*
   241. Leaves very long-tapering at the base, bright green on upper surface,
       nearly 15 cm long . . . . *Franklinia*
240. Leaves lanceolate to ovate, not obovate, base cuneate to truncate,
   but not long-tapering; fruit a capsule, drupe, or follicle
   242. Stipules small, sharp, nearly black, persistent; fruit a drupe . . . . *Ilex*
   242. Stipules larger, foliar, membranous or lacking; fruit a drupe,
      capsule, or follicle
      243. Leaves serrate, sometimes coarsely so, singly or doubly
         toothed, often entire toward the base, the midrib prominent
         the entire length of the lower surface
         244. Leaves 8–11 cm long, long acuminate; fruit a
            5-valved capsule . . . . *Leucothoe*
         244. Leaves less than 8 cm long, acute; fruit a drupe
            or follicle
            245. Leaves oblanceolate, the petioles 7–13 cm long; fruit a
               drupe, solitary . . . . *Prunus*
            245. Leaves obovate, oblanceolate, elliptic, or elliptic-ovate, the
               petioles less than 5 cm long; fruit a follicle, in corymbs
               or panicles . . . . *Spiraea*
      243. Leaves serrulate, crenulate, or subentire, singly toothed, the
         midrib losing prominence toward the apex on the
         lower surface
         246. Leaves mostly 10–17 cm long, with clusters of beige hairs in the axils
           on lower surface; fruit a capsule . . . . *Pterostyrax*
         246. Leaves mostly less than 10 cm long, glabrous or pubescent,
           lacking axillary clusters of hairs; fruit a capsule or drupe
           247. Leaves subentire in apical half, entire below the middle,
             the teeth remotely and irregularly spaced
             248. Leaves strongly glaucous on lower surface; twigs usually
               glaucous; fruit a capsule; southern . . . . *Zenobia*
            248. Leaves glabrous to pubescent, lacking a bloom;
               twigs glabrous or pubescent, usually not glaucous;
               fruit a drupe or capsule
               249. Leaves crenulate, the petiole grooved, enlarged at the base to
                   partially conceal the bud; bud scales glabrous, yellow-green
                   to yellow-brown becoming dark brown or black, the
                   margins often red-brown; fruit a drupe . . . . *Symplocos*
               249. Leaves serrulate, the petiole terete, lacking groove, not
                 hiding the bud; bud scales pubescent, uniformly brown;
                 fruit a capsule . . . . *Styrax*
           247. Leaves toothed nearly to the base, the teeth evenly and
             closely spaced

250. Leaves ovate, the veins straight and prominent, the blade less than $2^1/_2$ times longer than wide; fruit a drupe . . . .*Rhamnus*
250. Leaves elliptic, the veins curved and less impressed, the blade greater than $2^1/_2$ times longer than wide; fruit a drupe or capsule
   251. Leaves crenulate, acute to acuminate; fruit a capsule . . . .*Leucothoe*
   251. Leaves subentire or serrulate, obtuse, acute or mucronate; fruit a capsule or drupe
      252. Leaves mostly greater than 5 cm long, ovate, lanceolate (obovate in *Lyonia ligustrina*), acute, or mucronate; bark of branchlets gray or light brown and shredding or shining red-brown with longitudinal ridges; fruit a 5-valved capsule, in open clusters . . . .*Lyonia*
      252. Leaves mostly less than 4 cm long, obovate, elliptic, or elliptic-ovate, obtuse (mucronate in *G. frondosa*); bark of branchlets black, close or tight; fruit a drupe . . . .*Gaylussacia*

KEY V. PLANTS WITH ALTERNATE COMPOUND LEAVES

1. Leaves greater than 1 m long; southern . . . .*Sabal*
1. Leaves less than 1 m long
  2. Leaves bi- or tri-pinnately compound
   3. Leaflets toothed or lobed, not entire
    4. Ultimate leaflets 3; fruit a cluster of follicles . . . .*Paeonia*
    4. Ultimate leaflets 5 or more; fruit a follicle, drupe, legume, or berry
     5. Ultimate leaflets more than 11; branches often bearing stout thorns; fruit a legume
      6. Leaflets crenulate to subentire, less than 3 cm long; thorns often present; fruit a legume . . . .*Gleditsia*
      6. Leaflets coarsely toothed or lobed, greater than 3 cm long; thorns lacking; fruit a capsule . . . .*Koelreuteria*
     5. Ultimate leaflets fewer than 11; branches lacking thorns; fruit a follicle, berry, or drupe
      7. Vines; fruit a berry . . . .*Ampelopsis*
      7. Trees or shrubs; fruit a berry or follicle
       8. Stem prickly; wood brown; fruit a drupe . . . .*Aralia*
       8. Stem smooth or rough, lacking prickles; wood brown or yellow; fruit a berry or follicle
        9. Ultimate leaflets mostly 7–9, with petiolules 3–5 mm long; wood brown; trees; fruit a berry; southern . . . .*Melia*
        9. Ultimate leaflets mostly 5, sessile or subsessile; wood yellow; shrubs; fruit a follicle . . . .*Xanthorhiza*
   3. Leaflets entire; fruit a legume
    10. Leaflets greater than 5 cm long . . . .*Gymnocladus*
    10. Leaflets less than 5 cm long

11. Ultimate leaflets fewer than 15, greater than 1 cm long; fruit a
capsule or berry
 12. Leaflets less than 2 cm long, rounded or notched at the apex; fruit
 a capsule . . . .*Ruta*
 12. Leaflets greater than 2 cm long, acute or acuminate; fruit
 a berry . . . .*Nandina*
11. Ultimate leaflets more than 15, less than 1 cm long or linear and
less than 2 mm wide
 13. Foliage aromatic, gray-green; fruit of achenes . . . . **Artemisia**
 13. Foliage lacking strong odor; fruit a legume
  14. Ultimate leaflets about 20 . . . .*Chamaecrista*
  14. Ultimate leaflets more than 30
   15. Leaflets fewer than 50, with petiolules 1–2 mm long, the leaflet
   pairs 5–10 mm apart; southern . . . .*Sesbania*
   15. Leaflets more than 50, sessile, the leaflet pairs less than
   2 mm apart
    16. Leaves less than 20 cm long; branchlets and
    stems spiny; southern . . . .*Mimosa*
    16. Leaves greater than 20 cm long; branchlets and
    stems unarmed . . . .*Albizia*
2. Leaves once compound
 17. Leaflets 2, each leaflet deeply divided . . . .*Paeonia*
 17. Leaflets more than 2, divided or not
  18. Leaflets 3 (alternate step 18, p. 48)
   19. Stipules present; leaflets sometimes more than 3; plants
   often climbing
    20. Stems and petioles lacking prickles; leaflets lobed or not;
    fruit a head of achenes or a legume
     21. Leaflets greater than 9 cm long, 2–3 lobed;
     fruit a legume . . . .*Pueraria*
     21. Leaflets less than 5 cm long, 0–3 lobed; fruit a legume
     or head of achenes
      22. Leaflets 3–5 toothed near apex, oblong-lanceolate; fruit a head
      of achenes . . . .*Potentilla*
      22. Leaflets entire, elliptic to lance-elliptic; fruit a
      legume; southern . . . .*Lespedeza*
    20. Stems or petioles bearing prickles; leaflets lacking lobes;
    fruit berrylike
     23. Stipules adnate to the petiole about half its length
     or more . . . .*Rosa*
     23. Stipules free, not adnate to the petiole . . . .*Rubus*
   19. Stipules lacking
    24. Petioles winged; branches green, with stout
    branched thorns . . . .*Poncirus*

24. Petioles smooth, lacking wings; branches green or brown,
     lacking thorns
  25. Branches with prominent longitudinal ridges, green; leaves nearly
     sessile; fruit a legume . . . .*Cytisus*
  25. Branches smooth, not ridged, brown; leaves with obvious
     petiole; fruit a drupe or legume
    26. Lateral buds visible in the axils of leaves; lateral leaflets
         symmetrical or not; stems with or without aerial rootlets;
         fruit a drupe or legume
      27. Lateral leaflets asymmetrical; stems often with aerial rootlets;
             fruit a pale or white drupe . . . .*Toxicodendron*
      27. Lateral leaflets symmetrical; stems lacking aerial rootlets; fruit
             a legume . . . .*Laburnum*
    26. Lateral buds imbedded in the bark, not visible; lateral
         leaflets nearly symmetrical; stems lacking aerial rootlets;
         fruit a drupe or samara
      28. Petioles 1–3 cm long; leaflets crenate; fruit a red drupe . . . . . .*Rhus*
      28. Petioles 5–10 cm long; leaflets entire or crenulate; fruit
             a samara . . . .*Ptelea*
18. Leaflets more than 3
  29. Leaves palmately compound
    30. Stems smooth, lacking spines or prickles; high-climbing
         woody vines; fruit a berry
      31. Vine lacking tendrils; leaflets entire, retuse, with petiolules
             3–10 mm long . . . .*Akebia*
      31. Vine with branched tendrils; leaflets serrate,
             acuminate, subsessile . . . .*Parthenocissus*
    30. Stems with spines or prickles; plants erect; fruit berrylike or a
         cluster of drupelets
      32. Stems biennial; leaflets pale or dull on lower surface; fruit aggregate,
             a cluster of drupelets . . . .*Rubus*
      32. Stems perennial; leaflets glossy on lower surface;
             fruit berrylike . . . .*Eleutherococcus*
  29. Leaves pinnately compound
    33. Buds hidden (base of petiole hollow, forming a hood-like
         covering over the lateral buds or these embedded under the
         base of the petiole) (alternate step 33, p. 49)
      34. Leaflets small, 1–5 cm long
        35. Leaflets about 1 cm long, silky; low shrub with shreddy bark; fruit
                 a head of achenes . . . .*Potentilla*
        35. Leaflets greater than 1 cm long, glabrous or pubescent,
                 not silky; bark tight, not shredding; shrubs, or small or
                 large trees; fruit a legume
          36. Thorns, spines, and prickles lacking

37. Leaves less than 2 cm long, emarginate ....*Colutea*
37. Leaves greater than 2 cm long, acute ....*Sophora*
36. Thorns, spines, or prickles usually present
    38. Leaflets entire; twigs usually with stipular spines or prickles; pith
        pink-white ....*Robinia*
    38. Leaflets somewhat serrate or crenate; twigs usually with simple or
        branched thorns; pith salmon-colored ....*Gleditsia*
34. Leaflets large, 5–10 cm long
  39. Leaflets tipped with hard spines ....*Mahonia*
  39. Leaflets acute or obtuse, not tipped with hard spines
    40. Leaflets alternate, ovate; sap clear, not milky; fruit
        a legume ....*Cladrastis*
    40. Leaflets opposite or subopposite, ovate to lanceolate or
        oblong; sap milky or clear; fruit a drupe or nut
      41. Leaves mostly even-pinnate, the rachis often winged; leaflets
          mostly subopposite, the apex rounded, obtuse, or acute; sap clear;
          fruit a nut ....*Pterocarya*
      41. Leaves odd-pinnate, the rachis winged or not;
          leaflets opposite, the apex usually acuminate; fruit
          a drupe
        42. Leaflets doubly serrate ....*Platycarya*
        42. Leaflets entire or coarsely and singly serrate
          43. Leaflets narrowly oblong to lanceolate; fruit red,
            glandular-pubescent ....*Rhus*
          43. Leaflets ovate, obovate, suborbicular, or elliptic;
            fruit yellow to white, glabrous or sparsely
            pubescent; poisonous ....*Toxicodendron*
33. Buds evident, not embedded under or surrounded by base of
  the petiole
  44. Leaflets tipped with hard spines ....*Mahonia*
  44. Leaflets merely acute or obtuse, not tipped with spines
    45. Spines or thorns present on the stems and often the midrib
      of leaves
      46. Stipules lacking; leaflets dotted with pellucid glands; wood yellow;
        fruit a capsule ....*Zanthoxylum*
      46. Stipules present; leaflets lacking pellucid glands; wood
        brown; fruit berrylike (an aggregate of drupelets), achene,
        or hip
        47. Stipules adnate to the petiole half its length or more; leaflets
          evenly serrate ....*Rosa*
        47. Stipules free from the petiole; leaflets usually unevenly coarsely
          toothed or often doubly serrate ....*Rubus*
    45. Spines, thorns, and prickles absent
      48. Leaflets toothed or lobed

49. Leaflets often more than 20, with rank odor when crushed, with glands on
   the lower surface of the small basal lobes; fruit a samara . . . . . *Ailanthus*
49. Leaflets usually fewer than 20, odiferous but not rankly so,
   glabrous, pubescent, or glandular-hairy but lacking glandular
   basal lobes; fruit a follicle, nut, capsule, winged nutlet, or
   berrylike pome
   50. Petiole base clasping the twig; wood yellow; fruit
      a follicle . . . .*Xanthorhiza*
   50. Petiole base not or only partially enclosing the twig; wood
      brown to nearly black; fruit a capsule, nut, winged nutlet, or
      berrylike pome
      51. Leaflets lobed, sometimes all the way to the midrib, coarsely toothed;
         fruit a winged capsule . . . .*Koelreuteria*
      51. Leaflets finely or coarsely toothed but not lobed; fruit a
         nut, winged nutlet, or berrylike pome
         52. Stipules present, although sometimes ephemeral but
            then buds red; leaves glabrous or pubescent, not
            glandular-hairy, lacking resinous odor; branches and
            twigs light and slender; fruit a pome
            53. Leaves with 4–6 leaflets at the base . . . . ×*Sorbaronia*
            53. Leaves with more than 15 leaflets
               54. Leaves singly toothed; terminal bud present; trees
                  or tall shrubs . . . .*Sorbus*
               54. Leaves doubly toothed; terminal bud absent; low
                  suffrutescent shrubs . . . .*Sorbaria*
         52. Stipules lacking; buds yellow, brown, or black; leaves
            often glandular-hairy, often with strong resinous odor;
            branches and twigs heavy and stout; fruit a nut or winged nutlet
            55. Leaflets doubly serrate; fruit a winged nutlet in the axil of
               stiff bracts . . . .*Platycarya*
            55. Leaflets singly serrate; fruit a nut
               56. Pith chambered; husk of nut indehiscent . . . .*Juglans*
               56. Pith solid; husk of nut dehiscing into 4 valves . . . .*Carya*
48. Leaflets entire or with small, glandular lobes at the base
   57. Stems twining, vines; leaves with rank odor, the leaflets with glandular
      lobes at the base; fruit a legume . . . .*Wisteria*
   57. Stems erect, shrubs or trees; leaves usually lacking strong odor,
      the leaflets entire, lacking glandular lobes; fruit various
      58. Leaflets subtended by stipels; fruit a legume
         59. Leaflets ovate, apiculate or bristle-tipped . . . .*Indigofera*
         59. Leaflets oblong, sometimes bristle-tipped . . . .*Amorpha*
      58. Leaflets lacking subtending stipels; fruit a legume, nut,
         capsule, drupe, or head of achenes
         60. Leaflets 1–3 cm long; short or tall shrubs

61. Leaflets glabrous or nearly so, spaced on a long rachis, mostly 2 cm long;
　　tall shrub; fruit a legume . . . .***Caragana***
61. Leaflets silky, crowded on a short rachis, about 1 cm long; low dense shrub;
　　fruit a head of achenes . . . .***Potentilla***
60. Leaflets greater than 4 cm long, glabrous; trees or tall shrubs
　　62. Pith chambered; fruit a large brown nut . . . .***Juglans***
　　62. Pith solid; fruit a drupe, capsule, or legume
　　　　63. Leaves with winged rachis; fruit a small white to gray drupe . . . . .***Rhus***
　　　　63. Leaves with rachis lacking wings; fruit a capsule, drupe,
　　　　　　or legume
　　　　　　64. Leaflets mostly less than 4 cm long; fruit a legume . . . .***Maackia***
　　　　　　64. Leaflets mostly greater than 4 cm long; fruit a capsule
　　　　　　　　or drupe
　　　　　　　　65. Leaflets with stalks 5–10 mm long, cuneate at the base; twigs
　　　　　　　　　　red-brown to dark brown; fruit a capsule . . . .***Toona***
　　　　　　　　65. Leaflets sessile or subsessile, oblique or truncate at the base; twigs
　　　　　　　　　　light brown to light gray; fruit a drupe; southern . . . .***Pistacia***

KEY VI. PLANTS WITH OPPOSITE OR WHORLED LEAF SCARS

1. Stems climbing or twining; vines (alternate step 1, p. 52)
　　2. Twigs and young branchlets green; leaf scars
　　decurrent; southern . . . .***Jasminum***
　　2. Twigs and branchlets usually shades of gray, brown, or black;
　　leaf scars usually not prominently decurrent
　　　　3. Stems with 6 or more prominent longitudinal ridges, often nearly
　　　　herbaceous; fruit a cluster of hairy achenes . . . .***Clematis***
　　　　3. Stems smooth, lacking prominent longitudinal ridges; fruit a
　　　　capsule or berry
　　　　　　4. Bundle scars in a closed or nearly closed ring; fruit a large
　　　　　　fusiform capsule
　　　　　　　　5. Stems often with aerial rootlets at the nodes;
　　　　　　　　　　tendrils absent . . . .***Campsis***
　　　　　　　　5. Stems lacking aerial rootlets; leaf tendrils
　　　　　　　　　　sometimes persisting . . . .***Bignonia***
　　　　　　4. Bundle scars in a crescent-shaped line, usually 3
　　　　　　　　6. Stems climbing by aerial rootlets; fruit a
　　　　　　　　　　capsule; southern . . . .***Decumaria***
　　　　　　　　6. Stems twining, lacking aerial rootlets; fruit a berry
　　　　　　　　　　or capsule
　　　　　　　　　　7. Bud scales slightly pubescent to glabrous on outer surface; nodes
　　　　　　　　　　　　only slightly swollen . . . .***Schizophragma***
　　　　　　　　　　7. Bud scales densely red-pubescent on both surfaces;
　　　　　　　　　　　　nodes swollen at least $1\frac{1}{2}$ times the branchlet diameter

    8. Opposing leaf scars distinct, lacking a connecting line or ridge; fruit a capsule; southern . . . .*Periploca*

    8. Opposing leaf scars joined by a line (scars) left by the stipules; fruit a berry . . . .*Lonicera*

1. Stems erect; trees or shrubs

    9. Bark of trunk red, fibrous, shedding; branchlets and twigs red or red-brown; trees; female reproductive structure a cone . . . . . *Metasequoia*

    9. Bark of trunk gray, brown, or black, tight or scaly; branchlets and twigs gray, brown, red-brown, or black; trees or shrubs; female reproductive structure a fruit or cone

        10. Buds and twigs covered with silver to brown scales; shrubs . . . .*Shepherdia*

        10. Buds and twigs glabrous or pubescent, lacking scales; trees or shrubs

          11. Branchlets bearing short or spur shoots 2–15 mm long, with crowded whorls of leaf scars or persistent leaf sheaths; female reproductive structure a cone or follicles

            12. Branchlets black, the spurs appressed; reproductive structure of 4–6 follicles . . . .*Cercidiphyllum*

            12. Branchlets light brown to gray, the spurs divergent; reproductive structure a cone

              13. Spurs mostly 5–15 mm long; twigs stout; cones 4–6 cm long . . . .*Pseudolarix*

              13. Spurs mostly less than 5 mm long; twigs slender; cones mostly less than 4 cm long . . . .*Larix*

          11. Branchlets lacking spur shoots with whorls of leaf scars or persistent leaf sheaths; female reproductive structure a fruit

            14. Shrubs to 50 cm tall; stipules persistent, with stiff long-pointed awns; twigs and branchlets with 2 longitudinal bands of hairs; southern . . . .*Serissa*

            14. Trees or shrubs mostly greater than 50 cm tall; stipules deciduous or lacking; twigs and branchlets glabrous or pubescent, hairs not arranged in 2 longitudinal bands (except in *Weigela* and *Diervilla*)

              15. Bundle scars 1, or appearing as 1 (sometimes very numerous and almost confluent, thus forming a transverse, lunate, or U-shaped line) (alternate step 15, p. 55)

                16. Leaf scars mostly subopposite, often barely visibly so

                  17. Twigs and branchlets gray-green, light gray to light brown or red-brown, often angled . . . .*Fontanesia*

                  17. Twigs and branchlets dark gray to black, terete, rarely angled . . . .*Rhamnus*

16. Leaf scars mostly opposite
    18. Twigs densely rusty, stellate-pubescent; buds naked, densely rusty to
        yellow pubescent . . . .***Buddleia***
    18. Twigs glabrous or pubescent; buds scaly or naked
        19. Plants strongly aromatic; branchlets light gray to light brown
            or green
            20. Twigs green, quadrangular . . . .***Lavandula***
            20. Twigs light gray to light brown, terete . . . .***Vitex***
        19. Plants mostly lacking a noticeable odor; branchlets
            variously colored
            21. Leaf scars strongly decurrent or raised; fruit a capsule
                22. Pith hollow or chambered; twigs olive-green or green-brown; buds
                    fusiform; bud scales thin . . . .***Forsythia***
                22. Pith solid or only incompletely chambered; twigs gray,
                    brown, red-brown, or green; buds globose to ovoid; bud
                    scales thin or fleshy
                    23. Buds superposed; branchlets and
                        twigs rugulose . . . .***Abeliophyllum***
                    23. Buds separate; branchlets and twigs smooth
                        24. Bud scales fleshy; terminal bud usually absent; twigs gray to
                          brown, lacking wings . . . .***Syringa***
                      24. Bud scales thin, not fleshy; terminal bud present;
                        twigs pale green, bright green, or red-brown
                          25. Twigs bright green or red-brown; often with corky wings;
                            fruit orange-red . . . .***Euonymus***
                        25. Twigs pale green, tight, hard, not corky;
                            fruit black . . . .***Jasminum***
            21. Leaf scars scarcely, or not at all, raised or decurrent;
              fruit various
                26. Bud scales loose and open; buds never sunken in the bark; bark
                  exfoliating; low shrubs; fruit a capsule . . . .***Hypericum***
                26. Bud scales mostly close and firm or else absent; buds
                  occasionally sunken in the bark; bark mostly close,
                  not separating; fruit a berry, drupe, capsule, samara,
                  or nutlets
                  27. Opposing leaf scars connected by a distinct raised line
                    formed by the stipular scars
                    28. Twigs pubescent; branchlets dark gray, greater than 4 mm
                      in diameter, the bark tight, the pith white; fruit a
                      capsule; southern . . . .***Pinckneya***
                    28. Twigs glabrous; branchlets light brown or
                      pink-brown to gray, usually less than 4 mm in
                      diameter, the bark shredding or not, the pith light
                      brown or white; fruit a berry or head of nutlets

29. Buds submerged in the bark; branchlets usually light brown or pink-brown with close bark, the pith white; leaf scars often whorled, not raised on bases of petioles; bundle scars U-shaped; fruit a head of nutlets . . . .***Cephalanthus***

29. Buds axillary; branchlets red-brown, often with shredding bark, pith light brown or white; leaf scars strictly opposite, raised on persistent bases of petioles; bundle scars nearly circular; fruit a berry . . . . ***Symphoricarpos***

27. Opposing leaf scars separate, not connected by a distinct stipular line (an indistinct line present in some species of *Fraxinus*)

30. Buds often superposed, naked, or the smaller ones covered with a pair of nearly valvate scales; twigs usually scurfy . . . . ***Callicarpa***

30. Buds covered with several scales; buds standing alone, not superposed, twigs glabrous or pubescent

31. Twigs bright green or red-brown, often glaucous, usually 4-angled; trees or shrubs . . . .***Euonymus***

31. Twigs brown, gray, black, or occasionally olive, not glaucous, terete

32. Twigs sometimes short, stiff, horizontal, abruptly terminating or terminating in a spine; buds appressed, black or gray, acute; fruit a drupe or berry

33. Twigs dark gray to black, some terminating in a spine; buds black; fruit a berry . . . .***Rhamnus***

33. Twigs mostly light gray or light brown, often abruptly terminating; buds gray; fruit a drupe . . . .***Forestiera***

32. Twigs variable, not stiff, horizontal, or terminating abruptly or in a spine; buds divaricate, variously colored, obtuse or acute; fruit a samara, capsule, or drupe

34. Branchlets usually with shredding bark . . . . ***Heptacodium***

34. Branchlets with bark close or tight

35. Twigs slender, less than 2 mm in diameter . . . . ***Ligustrum***

35. Twigs stout, greater than 2 mm in diameter

36. Twigs light, with large pith, light gray to light brown; shrubs to 3 m tall, often dying back in winter . . . . . .***Clerodendrum***

36. Twigs dense or heavier, with small pith, light brown to dark brown; trees or shrubs greater than 5 m tall

37. Buds scurfy, brown or black; bundle scars often almost separate and very numerous, forming a long U-shaped line; trees; fruit a samara . . . .***Fraxinus***

37. Buds glabrous or scarcely pubescent, not scurfy, brown or light brown; bundle scars forming a straight or curved line; shrubs or small trees; fruit a capsule or drupe

38. Bud scales fleshy, green or red-brown; twigs glabrous; fruit
a capsule . . . .*Syringa*

38. Bud scales thin, not fleshy, brown; twigs usually hirsute; fruit
a drupe . . . .*Chionanthus*

15. Bundle scars several, separate

  39. Bundle scars in a closed, or nearly closed, ellipse (*Staphylea*
sometimes has as few as 4 bundle scars, which nevertheless form
an ellipse when connected by a line.)

    40. Stipule scars present; ellipse of bundle scars transverse; leaf scars broadly
crescent-shaped; older twigs finely white-striped . . . .*Staphylea*

    40. Stipule scars absent; ellipse longitudinal; leaf scars orbicular;
older twigs solid brown, gray, or black (*Broussonetia*
sometimes with opposite leaf scars but has milky sap)

      41. Pith chambered or hollow; ellipse of bundle scars not quite closed;
leaf scars opposite; fruit a short capsule . . . .*Paulownia*

      41. Pith solid; ellipse closed; leaf scars opposite or whorled; fruit a
long capsule . . . .*Catalpa*

  39. Ellipse of bundle scars open, the 3 or more bundle scars quite
distinct, in a lunate or U- or V-shaped line, or the leaf scar
C-shaped and nearly closed around the bud

    42. Leaf scars large, C-shaped or U-shaped, nearly completely encircling
the bud; branchlets and twigs stout, with scattered
large lenticels . . . .*Phellodendron*

    42. Leaf scars smaller, heart- or shield-shaped or lunate, not
C-shaped, not encircling the buds; twigs slender or
sometimes stout

      43. Buds sunken in the bark or imbedded under the leaf scars
(usually bursting through in late winter); shrubs

        44. Twigs spicy-aromatic . . . .*Calycanthus*

        44. Twigs lacking odor

          45. Buds imbedded in the bark, superposed; twigs branched
and striated . . . .*Iva*

          45. Buds imbedded under the leaf scars; twigs
unbranched, of uniform color

            46. Twigs glabrous or pubescent,
not stellate-pubescent . . . .*Philadelphus*

            46. Twigs mostly stellate-pubescent . . . .*Deutzia*

      43. Buds axillary; shrubs or trees

        47. Stems with hollow pith between the nodes; twigs gray
or gray-white; fruit a berry

          48. Twigs gray or light gray . . . .*Lonicera*

          48. Twigs green, occasionally red-brown; southern . . . . .*Leycesteria*

        47. Stems with solid pith; twigs black, brown, red,
blue-green, or gray; fruit various

49. Twigs usually with 2 longitudinal bands of hairs . . . .*Weigela*
49. Twigs glabrous or uniformly pubescent
    50. Bud scales of axillary buds 0–3 pairs or solitary (sometimes 1–2
        pairs of extra bracteoles beneath)
        51. Bud scale 1 . . . .*Salix*
        51. Bud scales 0 or in pairs of 2, 4, or 6
            52. Twigs and young branchlets stout, usually 3–5 mm in diameter; buds
                red-brown, densely hirsute . . . .*Tetradium*
            52. Twigs and young branchlets slender, usually less than 3 mm
                in diameter; buds glabrous or pubescent
                53. Buds lacking scales, densely tomentose, the leaves
                    serving as bud scales; shrubs; fruit drupe-like
                    54. Buds small and slender, usually less than
                        3 mm long . . . .*Rhamnus*
                    54. Buds large and stout, usually greater than
                        3 mm long . . . .*Viburnum*
                53. Buds scaly, pubescent, glabrous, or silky, not tomentose;
                  shrubs or trees; fruit various
                    55. Buds scurfy, linear-lanceolate, often curved;
                      fruit drupe-like . . . .*Viburnum*
                    55. Buds glabrous or pubescent, not scurfy, ovoid to
                      lanceolate; fruit a samara, drupe, or drupe-like
                      56. Junction of the upper leaf scars forming a raised
                        projection; bud scales in 3 pairs; twigs olive,
                        red-brown, or blue, often glaucous or polished;
                        fruit a samara . . . .*Acer*
                    56. Junction of the leaf scars in plane with the twig,
                      not projecting, often notched; bud scales
                      in 1–3 pairs
                      57. First pair of bud scales shorter than the
                        bud; shrubs
                        58. Branchlets light yellow-brown, the bark exfoliating on
                          older branchlets; fruit bristly . . . .*Kolkwitzia*
                        58. Branchlets brown or gray to dark brown, the bark
                          remaining intact; fruit glabrous . . . .*Viburnum*
                    57. First pair of bud scales as long as the bud
                      (until swelling begins)
                      59. Bud scales 2 (paired); a pair of petiole bases
                        persisting beneath the terminal buds; fruit
                        a drupe . . . .*Cornus*
                    59. Bud scales 4, in pairs of 2; persistent petiole
                      bases lacking; fruit drupe-like or a samara
                      60. Second pair of bud scales hairy; twigs pubescent or
                        glabrous; older bark white-striped; shrubs or trees;
                        fruit a samara . . . .*Acer*

60. Second pair of bud scales glabrous or glutinous; twigs glabrous;
older bark solid gray or black, not white-striped; shrubs;
fruit drupe-like . . . . *Viburnum*
50. Bud scales of axillary buds 4 to many pairs
   61. Bundle scars usually 5, sometimes many more; leaf scars broad;
twigs stout
      62. Terminal bud present; pith small; trees; fruit a capsule with large
nut-like seeds . . . . *Aesculus*
      62. Terminal bud absent; pith very large; shrubs;
fruit berrylike . . . .*Sambucus*
   61. Bundle scars usually 3, rarely 5; leaf scars narrow; twigs stout
or slender
      63. Twigs often stellate-pubescent . . . .*Deutzia*
      63. Twigs glabrous or pubescent, not stellate-pubescent
         64. Opposing leaf scars connected by a decurrent, usually
hairy ridge; shrubs; fruit a capsule
            65. Connecting ridge densely pubescent or ciliate; branchlets
uniformly densely pubescent, lacking decurrent longitudinal
bands of hairs . . . . , ... *Abelia*
            65. Connecting ridge subglabrous to pubescent or ciliate;
branchlets with short- to long-pubescent decurrent
longitudinal bands
               66. Branchlets with short pubescent or barely
visible bands . . . .*Diervilla*
               66. Branchlets, especially young branchlets, with prominent
decurrent bands of thick pubescence . . . . *Weigela*
         64. Opposing leaf scars separate, lacking connecting
decurrent ridge
            67. Leaf scars ciliate, buds bronze-brown; terminal bud absent; shrubs;
fruit a cluster of black nutlets . . . . . , ...*Rhodotypos*
            67. Leaf scars entire, not ciliate, or if so, then terminal bud
present; shrubs or trees; fruit a berry, samara, capsule,
or cluster of achenes
               68. Bud scales of previous year's buds persisting at the
base of twigs; shrubs
                  69. Leaf scars opposite; buds often superposed;
fruit a berry . . . .*Lonicera*
                  69. Leaf scars whorled or opposite; buds separate, not
superposed; fruit a capsule
                     70. Older branchlets with bark mostly intact,
fruit glabrous . . . .*Hydrangea*
                     70. Older branchlets often with flaking or peeling
bark; fruit bristly or glabrous
                       71. Twigs and branchlets glabrous or pubescent;
fruit bristly . . . .*Kolkwitzia*

71. Twigs and branchlets with a few scattered long hairs;
    fruit glabrous . . . .***Heptacodium***
68. Bud scales of previous year's buds deciduous; trees or shrubs
  72. Shrubs; branchlets, usually terminating in stiff sharp spines, light brown
      or light gray; fruit a berry; southern . . . .***Punica***
  72. Trees or shrubs; branchlets lacking spines or, if terminating in a
      sharp spine, dark brown or black; fruit a samara, berry,
      or achenes
    73. Shrubs less than 3 m tall; buds pubescent; twigs usually sparsely
        pubescent; flowering in winter; fruit of many achenes . .***Chimonanthus***
    73. Trees or shrubs usually greater than 3 m tall; buds and twigs
        glabrous or pubescent; flowering in spring or summer; fruit
        a berry or samara
      74. Upper margin of leaf scar strongly concave; inner bark brown, sweet,
          at least not bitter; fruit a samara; trees . . . .***Acer***
      74. Upper margin of leaf scar nearly straight; inner bark yellow, bitter;
          fruit a berry; shrubs or small trees . . . .***Rhamnus***

KEY VII. PLANTS WITH ALTERNATE LEAF SCARS

1. Trees with fibrous bark; female reproductive structure
   a globose cone . . . .***Taxodium***
1. Trees, shrubs, or vines; bark mostly not fibrous; female reproductive
   structure a fruit or an ovoid drupe-like cone
  2. Leaf scars, except on young shoots, densely clustered in rings on
     short spur-like branches; bundle scars 1 or 2; bark often
     resinous; gymnosperms
    3. Spur shoots with leaf sheaths deciduous; twigs stout; female reproductive
       structure drupe-like . . . .***Ginkgo***
    3. Spur shoots with leaf sheaths persistent; twigs slender or stout;
       female reproductive structure a cone
      4. Spurs mostly 5–15 mm long; twigs stout; cones
         4–6 cm long . . . .***Pseudolarix***
      4. Spurs mostly less than 5 mm long; twigs slender; cones mostly less
         than 4 cm long . . . .***Larix***
  2. Leaf scars mostly scattered along the twigs; bundle scars 1 to
     many; angiosperms
    5. Bundles in the stem scattered; stem brier-like, green, often prickly;
       petiole base persisting; vascular bundles scattered in stem cross section;
       climbing shrubs; fruit a berry; monocotyledons . . . .***Smilax***
    5. Bundles in the stem in a ring; stems usually not brier-like,
       green, or prickly; petiole base usually deciduous; vascular
       bundles in a ring in stem cross section; shrubs, trees, or
       climbing or twining vines; fruit various; dicotyledons

6. Stems climbing or twining; vines (alternate step 6, p. 60)
   7. Tendrils present; fruit a berry
      8. Woody partitions through the brown pith usually present at the nodes;
         outer bark usually forming loose strips . . . .*Vitis*
      8. Woody partitions absent at the nodes; pith continuous; outer
         bark solid
         9. Pith white . . . . *Ampelopsis*
         9. Pith brown . . . .*Parthenocissus*
   7. Tendrils absent; fruit a drupe, berry, capsule, or legume
      10. Stipule scars with short, dense fringe of white hairs at the margin, large,
          mostly broader than the leaf scar; branchlets with short white and long
          yellow hairs; pith chambered . . . .*Pueraria*
      10. Stipule scars, if present, pubescent or glabrous at the margin,
          small; branchlets glabrous or uniformly pubescent; pith solid
          or chambered
          11. Buds in slightly supra-axillary depressions or hidden; leaf
              scars usually orbicular; fruit a drupe
              12. Twigs light brown or light gray to black;
                  pith partitioned . . . .*Actinidia*
              12. Twigs green; pith solid
                  13. Twigs pubescent; fruit red; southern . . . .*Cocculus*
                  13. Twigs glabrous or nearly so; fruit blue or black
                      14. Stems slender, green, often suffrutescent; drupe less than 1 cm
                          in diameter, blue . . . .*Menispermum*
                      14. Stems stout, woody; drupe at least 2 cm long,
                          black; southern . . . .*Calycocarpum*
          11. Buds axillary, not in depressions; leaf scars variously shaped;
              fruit various
              15. Stems prickly; leaf scars very narrow . . . .*Rosa*
              15. Stems smooth or rough, lacking prickles (thorns often
                  present in *Lycium*)
                  16. Bundle scars 3–7, distinct
                      17. Twigs gray, twining; buds glabrous . . . .*Akebia*
                      17. Twigs green or brown, twining or not;
                          buds tomentose
                          18. Twigs green, twining; buds clustered;
                              fruit a capsule . . . .*Aristolochia*
                          18. Twigs brown, climbing by rootlets; buds solitary; fruit a
                              white drupe; poisonous . . . .*Toxicodendron*
                  16. Bundle scar 1, or several confluent scars appearing
                      as 1
                      19. Pith hollow; stem often almost herbaceous, green or nearly
                          white, with a strong odor; twigs angled; leaf scars raised; fruit
                          a red berry . . . .*Solanum*

19. Pith solid or chambered; stem woody, brown, lacking strong
odor; twigs terete; leaf scars more or less flush with plane of
twig; fruit a legume, berry, or capsule
  20. Buds silky; leaf scars often 2-horned on the lower side, projecting, not
decurrent; fruit a legume . . . .*Wisteria*
  20. Buds glabrous or nearly so; leaf scars raised or not, decurrent
or not, lacking horns or projections; fruit a berry or capsule
    21. Stems ridged, often thorny; buds blunt, often clustered; fruit a
red berry . . . .*Lycium*
    21. Stems terete, not ridged, not thorny; buds more pointed,
solitary; fruit a black berry or orange capsule
      22. Pith dark brown . . . .*Schisandra*
      22. Pith white or light brown
        23. Buds divaricate; leaf scars in plane of twig, not raised; pith
white; fruit an orange capsule with red pulp around
the seeds . . . .*Celastrus*
        23. Buds appressed; leaf scars raised; pith light brown; fruit a
black berry . . . .*Berchemia*
6. Stems mostly erect; not climbing or twining, rarely prostrate; trees
or shrubs
  24. Branchlets slender, red-brown to purple brown, with twig scars that appear
as leaf scars, often subtended by a persistent scalelike leaf . . . .*Tamarix*
  24. Branchlets stout or slender, variously colored, with scars that are
leaf scars (thus lacking a subtending leaf or leaf scar)
    25. Plants prostrate or very low with arching, spreading main branches or
stems; leaf scars less than 1 mm in diameter, the bundle scars small,
not easily visible; fruit a persistent red to blue pome with
2–5 nutlets . . . .*Cotoneaster*
    25. Plants usually erect, not prostrate or spreading, the main
branches or stems not arching; leaf scars usually larger than
1 mm in diameter; fruit various
      26. Plants procumbent or low shrubs to 1 m tall; leaf scars prominently
raised, with persistent, hardened stipules; branchlets green, with
prominent longitudinal ridges; fruit a legume . . . .*Genista*
      26. Trees or shrubs, usually not procumbent; leaf scars raised
or not, the stipules deciduous or not; branchlets variously
colored, ridged or not; fruit various
        27. Buds deltoid or cordate, very wide at the base, the first pair of
scales thickened or swollen at the base; twigs with longitudinal
ridges or slightly winged from the decurrent leaf scars; bark
of the branchlets shedding and turning gray; fruit a
capsule; southern . . . .*Lagerstroemia*
        27. Buds variously shaped, the first pair of scales thin, not
swollen at the base; twigs with or without ridges or

wings from the decurrent leaf scars; bark of the branchlets
shedding or not; fruit various
28. Bundle scar 1 (sometimes spread out into a transverse line)
(alternate step 28, p. 64)
  29. Buds usually superposed, yellow-brown or rusty, often stalked;
trees or shrubs greater than 4 m tall
    30. Bundle scar raised as a broadly curved ridge; buds rusty or red-brown;
branchlets light brown to gray . . . . *Pterostyrax*
    30. Bundle scar a raised rounded or bilobed knob; buds yellow-brown or
yellow; branchlets gray to black . . . . *Styrax*
  29. Buds solitary, separate, brown, sessile
    31. Twigs covered with red-brown scales or stellate-pubescent or
buds naked
      32. Buds naked, lacking scales; young twigs or buds minutely
stellate-pubescent; bud scales lacking; fruit a capsule . . . . *Clethra*
      32. Buds with scales; young twigs covered with red-brown
scales; fruit a drupe
        33. Branchlets mostly terminating bluntly, if sharply pointed then
not thick; fruit red, silver, or yellow, sometimes persisting
in winter . . . . *Elaeagnus*
        33. Branchlets often terminating in long, sharp, thick thorns; fruit
orange-yellow, often persisting through winter . . . . . . , *Hippophae*
    31. Young twigs pubescent or glabrous, not covered with scales,
not stellate-pubescent; buds usually with scales
      34. Stipules or stipule scars present; shrubs
        35. Stipules deciduous
          36. Buds superposed; twigs often striate with gray and green or
brown stripes; fruit a legume . . . . *Amorpha*
          36. Buds solitary, not superposed; twigs yellow or brown,
lacking striations; fruit of 3 nutlets or multiple in a
large ball
            37. Low, weak, unarmed shrubs, dying nearly to the ground each
winter; buds above the leaf scar; fruit of 3 small nutlets, the
bases usually persisting . . . . *Ceanothus*
            37. Trees, the branches bearing thorns; buds lateral to
the thorns; fruit multiple, forming a large ball
(Compare also with *Celtis* in which bundle scars are
fused into 1.) . . . . *Maclura*
        35. Stipules persistent (sometimes very small)
          38. Stipules large, sheathing the stem; bark brown, shredding; fruit
of many hairy achenes . . . . *Potentilla*
          38. Stipules minute, divergent, not sheathing the stem;
bark smooth, rough, or shredding; fruit a legume
or berry

39. Twigs bright green, ridged, angled, or winged, wand-like, usually dying
at the tip; fruit a legume . . . .*Cytisus*
39. Twigs gray-brown or gray, mostly smooth or rough, lacking
ridges, angles, or wings, dying at the tip or not; fruit a berry
or legume
  40. Leaf scars raised on persistent petiole base; fruit a legume . . .*Caragana*
  40. Leaf scars flush with the stem, not raised; fruit a red or
black berry . . . .*Ilex*
34. Stipule scars or stipules absent; shrubs or trees
  41. Bark of branchlets and branches green; aromatic; trees; fruit a
berry or drupe
    42. Branches with an axillary thorn at each node, the buds lateral to the
thorns; bark of branchlets finely punctate with oil glands, fragrant
citrus aroma; fruit a berry . . . .*Poncirus*
    42. Branches unarmed; branchlets lacking punctate oil glands but with
spicy aroma; fruit a blue drupe . . . .*Sassafras*
  41. Bark of branchlets gray or brown to black or red, lacking strong
odor; trees or shrubs; fruit a drupe, capsule, samara, follicle,
or berry
    43. Twigs finely white-speckled or granulose, green or red-brown;
shrubs; fruit a berry or drupe
      44. Twigs and branchlets brittle or stiff, not flexible or leathery; fruit a
red, blue, or black berry . . . .*Vaccinium*
      44. Twigs and branchlets leathery, very flexible; fruit a red or
yellow drupe or dry and drupe-like
        45. Twigs glabrous or lightly pubescent, red-brown to brown; leaf scars
lacking fringe of hairs . . . .*Daphne*
        45. Twigs densely white hairy, gray-green leaf scars often with
fringe of hairs . . . .*Edgeworthia*
    43. Twigs uniformly colored, neither white-speckled nor
granulose; shrubs or trees; fruit a berry, follicle, samara,
or capsule
      46. Trees
        47. Pith continuous
          48. Visible bud scales 4–6; fruit a 5-celled capsule borne in
spreading racemes . . . .*Oxydendrum*
          48. Visible bud scales 2; fruit a large berry . . . .*Diospyros*
        47. Pith chambered
          49. Buds superposed; fruit a dry, 4-winged capsule . . . .*Halesia*
          49. Buds solitary; fruit a samara or large berry
            50. Buds with 2 visible scales; fruit a large berry . . . . .*Diospyros*
            50. Buds with at least 8 visible scales; fruit
a samara . . . .*Eucommia*
      46. Shrubs

51. Buds, except on strong, young, long shoots, clustered at the end of twigs; terminal bud often very large; fruit a capsule . . . .***Rhododendron***
51. Buds scattered along twig, not clustered; terminal bud small; fruit a drupe, berry, follicle, or capsule
  52. Winter twigs often with red or red-brown catkin-like racemes of flower buds . . . .***Leucothoe***
  52. Winter twigs lacking catkin-like racemes; fruits or flower buds, if present, in corymbs, fascicles, racemes, or solitary
    53. Twigs with 2 kinds of buds: larger flower buds with several visible scales and shorter buds with 2 visible scales, all red-yellow to yellow-brown; twigs red to red-brown, mostly pubescent; fruit a berry . . . .***Gaylussacia***
    53. Twigs with only 1 kind of bud; twigs red-brown to black, pubescent or glabrous; fruit a follicle, berry, or capsule
      54. Visible bud scales mostly 2 or 3; fruit a capsule
        55. Shrubs to 9 m tall; twigs mostly gray, sometimes light brown . . . .***Enkianthus***
        55. Shrubs usually less than 3 m tall; twigs black, gray, red-brown, or light brown
          56. Shrubs 2–3 m tall; branchlets red-brown to light gray; flowers clustered or in axillary racemes . . . .***Lyonia***
          56. Shrubs less than 2 m tall; branchlets light brown or gray to black; flowers solitary, clustered, or in corymbs
            57. Branchlets gray to nearly black; flowers in few-flowered terminal corymbs . . . .***Menziesia***
            57. Branchlets light brown; flowers solitary or in clusters; southern . . . .***Securinega***
      54. Visible bud scales several; fruit a capsule, follicle, or berry
        58. Leaf scars mostly flush with the twig; fruit a capsule or berry
          59. Twigs glabrous or pubescent, sometimes glaucous; shrubs; fruit a berry, solitary orin clusters or racemes above the leaf scar . . . .***Vaccinium***
          59. Twigs glabrous, often glaucous; fruit a capsule, in racemes of corymbs on stems lacking leaf scars; southern . . . .***Zenobia***
        58. Leaf scars, at least the upper edge, projecting slightly from the surface of the twig; fruit a berry or follicle
          60. Leaf scars very rough and uneven; bark usually bronze-brown, tending to exfoliate; stems wand-like or much branched and recurved with slender twigs; fruit a follicle . . . .***Spiraea***
          60. Leaf scars nearly smooth and even; bark on older branches gray, smooth, often with a white exfoliating crust; stems much branched, neither wand-like nor recurved; fruit a dark red berry . . . .***Nemopanthus***

28. Bundle scars more than 1
  61. Young twigs pale green, glabrous, glaucous, with many glandular dots, aromatic; leaf scars linear, about 1 mm high by 6 mm wide; subshrub less than 1 m tall . . . .*Ruta*
  61. Young twigs variously colored, glabrous or pubescent, glaucous or not, lacking glandular dots, with or without odor; leaf scars variously shaped, usually not linear; shrubs or trees
    62. Bundle scars more than 3, in any arrangement except a single lunate line (alternate step 62, p. 66)
      63. Branches and stems with thorns or prickles
        64. Branches with thick, long, sharp thorns; buds flat-topped and depressed; branches light olive-gray; fruit multiple, large and fleshy . . . .*Maclura*
        64. Branches with prickles, usually broad-based, short, not long and thorn-like; buds pointed or rounded, neither flat-topped nor depressed; branches variously colored; fruit drupe-like
          65. Leaf scars extending less than halfway around the stem; bundle scars 5–10 . . . .*Eleutherococcus*
          65. Leaf scars nearly completely encircling the twig; bundle scars about 5–20
            66. Twigs and branchlets with well-spaced, short, triangular spines; trees or shrubs to 10 m tall
              67. Bundle scars about 15 . . . .*Kalopanax*
              67. Bundle scars 5–10 . . . .*Aralia*
            66. Twigs and branchlets densely covered with many lanceolate spines; shrubs or trees
              68. Bundle scars about 20; leaf scars nearly completely encircling the twig . . . .*Aralia*
              68. Bundle scars about 15; leaf scars extending $^{2}/_{3}$ of the way around the twig . . . .*Oplopanax*
      63. Branches and stems lacking thorns or prickles
        69. Leaves usually deciduous above the base, the enlarged leaf base persisting, forming a raised leaf scar; shrubs 1–3 m tall; fruit a legume; southern . . . .*Sesbania*
        69. Leaves deciduous at the base, forming a simple flushed or raised leaf scar; trees or shrubs; fruit various
          70. Stipule scars and stipules absent
            71. Leaf scars nearly circular; twigs slender; axillary buds not visible; flower buds terminal, clustered, similar to young catkins; low or decumbent shrubs; fruit a drupe . . . .*Rhus*
            71. Leaf scars inversely triangular or oblong; twigs stout; axillary buds usually visible; flower buds scattered or not conspicuously terminal; shrubs or trees; fruit a drupe or nut

72. Bark of twigs mottled; lateral buds small, solitary; poisonous shrubs; fruit a white or pale drupe . . . .***Toxicodendron***
72. Bark of twigs a uniform color, not mottled; lateral buds large, superposed; trees; fruit a nut . . . .***Carya***
70. Stipule scars or stipules present
  73. Bundle scars in a ring; sap milky; buds green; fruit a fig, a fleshy many-seeded aggregate; southern . . . .***Ficus***
  73. Bundle scars arranged variously, but not in a ring; sap clear or milky; buds brown, gray, yellow, or black; fruit various
    74. Bud scales 1, or if 2, united into a cap; fruit cone-shaped
      75. Bud scales 1, with a scar on the back; leaf scars mostly lunate; fruit a fleshy "cone" . . . .***Magnolia***
      75. Bud scales 2, valvate; leaf scars mostly orbicular; fruit a dry "cone," at least its axis persisting . . . .***Liriodendron***
    74. Bud scales several, imbricated; fruit various
      76. Buds densely rusty-tomentose, globose to flattened or nearly saucer-shaped; bundle scars numerous, scattered through the suborbicular leaf scar; fruit of several large (to 5 cm long) papery follicles . . . .***Firmiana***
      76. Buds pubescent or glabrous, usually ovoid, not globose; bundle scars arranged in lines, circles, or shapes, usually not scattered; fruit various
        77. Buds indistinct, white-tomentose with branched hairs; fruit a capsule . . . .***Hibiscus***
        77. Buds prominent, glabrous or pubescent, not white-tomentose; fruit a nut, drupe, berrylike, or multiple
          78. Bundle scars in a thick circle; twigs light gray to gray-green or brown-green; shrub to 3 m tall; fruit a legume . . . . .***Indigofera***
          78. Bundle scars arranged variously, not in a thick circle; trees or shrubs; fruit a nut, acorn, drupe, berrylike, or multiple
            79. Terminal bud present; fruit a nut or acorn
              80. Buds linear-lanceolate, sharp-pointed, scattered along the twig; trees; fruit a 3-angled nut, 2 together in a spiny, 4-valved involucre . . . .***Fagus***
              80. Buds ovoid, obtuse, or rounded, clustered near the end of the twig; twigs often angled or fluted; trees or large shrubs; fruit an acorn . . . .***Quercus***
            79. Terminal bud absent; fruit a nut, drupe, or multiple
              81. Visible bud scales 4 or more; catkins often present; trees or shrubs
                82. Buds deltoid, appressed or slightly spreading; trees; fruit multiple, berrylike . . . .***Morus***

82. Buds ovoid, spreading; tall shrubs; fruit a nut with
    an involucre . . . .*Corylus*
  81. Visible bud scales 2 or 3; catkins absent in winter; trees
    83. Twigs scabrous, often mottled; sap milky; fruit
      multiple, berrylike . . . .*Broussonetia*
    83. Twigs glabrous or puberulent, not mottled; sap clear; fruit
      drupe-like or a nut
      84. Buds red-brown, lopsided due to the short first scale on 1 side; twigs
        red-brown or olive, usually zigzagging; pith terete; fruit drupe-like,
        attached to a leafy bract . . . .*Tilia*
      84. Buds light olive-brown, symmetrical; twigs light olive-brown,
        nearly straight; pith 5-sided or angled; fruit a nut with
        prickly involucre . . . .*Castanea*
62. Bundle scars 3 or more in a single lunate line, or, in *Juglans,*
Pterocarya, and sometimes *Idesia* and *Toona,* often 3 U-shaped
groups that form a lunate line (If leaves are not deciduous at the
base, then bundle scars must be counted in a section cut through
the base of the petiole.)
  85. Pith minute, less than $^1/_{10}$ of the twig diameter; stipule scars so
    small as to be nearly invisible most of the year; leaf scars 2-ranked;
    branchlets zigzagging; fruit a samara, winged nut, or drupe
    86. Pith chambered
      87. Twigs light gray to light brown; fruit winged . . . .*Pteroceltis*
      87. Twigs light brown or gray to black; fruit lacking a wing . . . . . .*Celtis*
    86. Pith solid
      88. Twigs roughened by many crowded, coarse,
        raised lenticels . . . .*Aphananthe*
      88. Twigs mostly smooth, the lenticels few, not
        significantly raised
        89. Branchlets often becoming short, stiff,
          and spine-like . . . .*Hemiptelea*
        89. Branchlets longer, flexible, not appearing as thorns
          or spines
          90. Buds globose or subovoid; twigs dark red-brown to
            light green . . . .*Planera*
          90. Buds ovoid; twigs brown to black
            91. Fruit an asymmetrical drupe . . . .*Zelkova*
            91. Fruit a 1-seeded samara . . . .*Ulmus*
  85. Pith larger than $^1/_{10}$ the diameter of the twig; stipule scars present
    or not; leaf scars alternate or spiral but rarely 2-ranked;
    branchlets linear, not zigzagging; fruit various
    92. Stipule scars or stipules present (alternate step 92, p. 71)
      93. Branches with spines
        94. Branchlets bristly hairy; fruit a legume; southern . . . .*Mimosa*

94. Branchlets glabrous or pubescent, not bristly
    95. Spines solitary, of similar lengths; shrub to 3 m; fruit a pome
        or drupe
        96. Pith solid . . . .*Chaenomeles*
        96. Pith chambered . . . .*Prinsepia*
    95. Spines in pairs, 1 either recurved or longer than the other of
        the pair; small trees; fruit a drupe with or without wings
        97. Twigs light brown, green, or light gray; fruit
            lacking wings . . . .*Zizyphus*
        97. Twigs dark gray or dark brown to black; fruit winged . . . . .*Paliurus*
93. Branches lacking spines
    98. Stipule scars extending entirely around the stem; leaf scar nearly
        encircling the bud; bud with 1 scale; bark distinctively mottled,
        flaking, pale to brown; fruit a stalked globular head of
        hairy achenes . . . .*Platanus*
    98. Stipule scars not extended or extending only partway around
        the stem; leaf scar subtending but not encircling the bud; buds
        naked or with 1 or more scales; bark various, but usually not
        mottled; fruit various
    99. Buds stalked
        100. Buds black, round, imbedded in branchlet, pubescent, the scales
            difficult to distinguish; twigs dark brown, the lenticels conspicuous,
            pale; fruit a legume . . . .*Sophora*
        100. Buds variously colored, mostly not black, not embedded,
            glabrous or pubescent, the scales mostly evident; twigs
            variously colored, the lenticels inconspicuous or lacking
            101. Buds scurfy or glutinous; pith 3-sided; fruit cone-like . . . . *Alnus*
            101. Buds densely tomentose or glabrous; pith terete: fruit a
               woody capsule
               102. Buds glabrous or subglabrous . . . .*Corylopsis*
               102. Buds tomentose
                  103. Buds with 2 scales, white-tomentose
                    104. Buds and twigs stellate-pubescent . . . .*Fothergilla*
                    104. Buds and twigs with unbranched hairs only . . . .*Parrotia*
                  103. Buds naked, yellow-tomentose
                    105. Buds and young twigs with unbranched and
                      stellate hairs . . . .*Parrotiopsis*
                  105. Buds and young twigs with unbranched
                      hairs only . . . .*Hamamelis*
    99. Buds sessile
        106. Buds naked . . . .*Rhamnus*
        106. Buds with 1 or more scales
            107. Buds asymmetrical, the most basal bud scale short and
               protruding on 1 side . . . .*Tilia*

107. Buds symmetrical, the most basal scale not noticeably different
     in position or size than the others
  108. Bud scale 1, hood-like, the suture down the inner face of the bud . . .*Salix*
  108. Bud scales more than 1
    109. Stipule scars small, usually triangular; leaf scars orbicular, the
       bundle scars usually about 12, often in 3 circles of 4; buds
       small, globose . . . .*Idesia*
    109. Stipule scars usually linear, not triangular; leaf scars variously
       shaped, the bundle scars variously arranged; buds small or
       large, usually ovoid
      110. Stipule scars about 2 mm wide, at least 5 times wider than long, the
         3 stipular bundle scars often prominent; branchlets dark red-brown
         to black with many prominent light brown to orange-brown
         rod-shaped lenticels; fruit red, club-shaped, 7–8 mm in diameter,
         on fleshy stalks . . . .*Hovenia*
      110. Stipule scars less than 5 times wider than long, the bundle
         scars mostly not conspicuous; branchlets various shades
         of brown, gray, and green, the lenticels usually
         inconspicuous, circular; fruit varied in shape and color
        111. Exposed bud scales 2 or 3 (rarely 4)
           112. Exposed bud scales 3, the 2 lateral ones large, their margins
              nearly meeting, the third scale carinate, the swollen or keeled
              portion protruding between the 2 lateral scales; buds broadly
              deltoid, appressed; leaf scars sometimes with a fourth or fifth
              bundle scar above the others; fruit a capsule with 3 large
              white seeds . . . .*Sapium*
           112. Exposed bud scales 2–3, of varying shapes and
              arrangements; bundle scars usually 3; buds globose
              to deltoid, appressed or divergent; fruit various
             113. Branchlets pubescent, often with shreddy bark; leaf scars
                raised; stipules hardened, often persistent; shrub to
                5 m tall . . . .*Colutea*
             113. Branchlets glabrous or pubescent, mostly lacking
                shreddy bark; leaf scars raised or flush with the plane
                of the twig; stipules deciduous; shrubs or trees
                114. Branchlets dark brown to black; flower buds usually
                  present in crowded racemes; fruit berrylike . . . .*Stachyurus*
                114. Branchlets variously colored; flower buds not
                  evident; fruit a legume, follicle, drupe-like,
                  or cone-like
                  115. Twigs usually with evident longitudinal ridges,
                    with or without inconspicuous lenticels;
                    shrubs; fruit a legume
                    116. Buds superposed . . . . *Amorpha*

116. Buds solitary, separate ....*Lespedeza*
115. Twigs smooth, lacking longitudinal ridges, often with
    prominent lenticels; trees or shrubs; fruit cone-like,
    drupe-like, follicle, or legume
   117. Buds red; shrubs to 2 m tall; fruit a follicle ....*Stephanandra*
   117. Buds dark brown to black; trees; fruit a legume, drupe-like,
      or cone-like
     118. Buds less than 1 mm long, often wider than long; twigs light
        red-brown; fruit a legume ....*Albizia*
     118. Buds mostly greater than 1 mm long, usually longer than
        wide; twigs yellow or brown to black
       119. Buds globose or ovoid, lopsided due to a short scale on 1 side;
         twigs zigzagging; lacking wintergreen odor or flavor; fruit
         drupe-like, dry, attached to a leafy bract ....*Tilia*
       119. Buds elongated, acute; twigs straight, not zigzagging
         (spurs with a terminal bud); often with wintergreen flavor;
         fruit cone-like ....*Betula*
111. Exposed bud scales 4 to many
  120. Shrubs, usually with sharp prickles; stems biennial; fruit an aggregate
    of drupelets ....*Rubus*
  120. Shrubs or trees with smooth or rough stems and trunks,
    lacking prickles; stems perennial; fruit simple
   121. Bark of twigs and branchlets green
     122. Buds silky; small trees; fruit a legume ....*Laburnum*
     122. Buds glabrous to slightly pubescent; finely branched shrubs; fruit
       an achene ....*Kerria*
   121. Bark of twigs and branchlets gray, yellow-brown, brown,
    red-brown, or black, not bright green
     123. Bud tips appressed, brown; pith mostly chambered; trees; fruit
       a drupe ....*Celtis*
     123. Bud tips divaricate or at least not appressed; pith
       continuous, not chambered; trees or shrubs; fruit various
      124. First scale of axillary bud anterior; trees ....*Populus*
      124. First scale of axillary bud lateral; trees or shrubs
       125. Leaf scars strongly decurrent; bark light brown, shreddy;
         fruit a follicle ....*Physocarpus*
       125. Leaf scars slightly or not at all decurrent; bark gray,
         black, brown, or yellow-brown, tight or shedding;
         fruit various
        126. Bud scales evident in 2 vertical rows; bundle scars
          typically sunken in a smooth, corky layer which
          covers the leaf scars; catkins and spurs absent
         127. Buds globose or short conical; twigs dark red ...*Planera*
         127. Buds ovoid; twigs brown, not dark red ....*Ulmus*

126. Bud scales spiral, not in 2 rows; bundle scars evident at leaf scar
    surface, not sunken in a corky layer; bark smooth, thin, papery,
    shedding or curly, often with horizontal lines, or sinewy; catkins
    and spurs present or not
  128. Growth of previous season densely covered at base with leaf scars;
      numerous short spurs on older wood with many leaf scars and a
      terminal bud; bark of twigs often with wintergreen flavor;
      lenticels elongated . . . .*Betula*
  128. Growth of previous season with leaf scars scattered, not
      clustered at base, lacking short spurs; bark of twigs lacking
      wintergreen flavor; lenticels small or lacking (except
      in *Prunus*)
    129. Low weak shrubs; terminal bud present; fruit a 3-lobed capsule, the
        base usually persistent . . . .*Ceanothus*
    129. Trees or shrubs; terminal bud absent; fruit various
      130. Buds globose; twigs glandular, pubescent, or glabrous;
          aromatic or resinous
        131. Twigs glandular, pubescent; fruit symmetrical . . . .*Comptonia*
        131. Twigs lacking glands, pubescent or glabrous;
            fruit asymmetrical . . . .*Myrica*
      130. Buds ovoid; twigs glabrous or pubescent, lacking glands,
          resin, and aroma
        132. Stipule scars positioned nearly between the upper edge of the leaf
            scar and the twig; fruit a drupe . . . .*Prunus*
        132. Stipule scars lateral to the leaf scars; fruit a pome,
            samara, winged nut, drupe, follicles, or achenes
          133. Large or small trees; buds 4-angled, striate or not,
              pale brown to red-brown; bark gray or pale brown,
              smooth or scaly; fruit winged nuts or samaras
              in catkins
            134. Bark of trunk scaly, dark gray (red-brown when young);
                buds usually 3–7 mm long, the scales often striate;
                staminate catkins usually present in winter . . . .*Ostrya*
            134. Bark of trunk sinewy-fluted, light gray; buds usually
                2–4 mm long, the scales smooth, continuous, not striate;
                staminate catkins usually present only in spring, not
                in winter . . . .*Carpinus*
          133. Shrubs or small trees; buds terete, uniform, pale
              brown to black; bark dark brown or black, smooth
              or rough; fruit a drupe, pome, achenes, or follicles
            135. Young twigs densely pubescent; buds often clustered or
                hidden; fruit a pome . . . .*Mespilus*
            135. Young twigs glabrous to pubescent; buds scattered,
                visible; fruit a drupe, pome, achenes, or follicles

136. Bud scales dark red-brown to black, often with white hairs at the margins; stipule scars minute, linear, leading into inconspicuous decurrent lines; twigs brown to red-brown or purple; fruit of 2–4 achenes . . . . . ***Neviusia***
136. Bud scales red, gray, light brown, dark brown, or black, mostly glabrous; stipule scars evident; twigs gray, green, olive, or red-brown; fruit a pome, drupe, or follicles
   137. Buds superposed, dark red-brown; branchlets gray; fruit usually of 2 follicles . . . . ***Neillia***
   137. Buds separate, solitary, light brown or gray to black; branchlets various shades of green, gray, and brown to black; fruit a drupe or pome
      138. Bud scales dark brown or black; twigs gray or green; fruit a drupe . . . . ***Rhamnus***
      138. Bud scales red, gray, or light brown; twigs olive or red-brown; fruit a pome . . . . ***Cydonia***
92. Stipule scars and stipules absent, or, rarely, modified into spines
   139. Pith chambered or with woody partitions in solid pith
      140. Pith solid with transverse woody partitions
         141. Bundle scars 5–7; lateral buds globose, densely dark-tomentose; terminal bud naked; fruit large, berrylike . . . . ***Asimina***
         141. Bundle scars 3; lateral buds ovoid, nearly or quite glabrous; terminal bud scaly; fruit a small drupe . . . . , , , , ***Nyssa***
      140. Pith chambered
         142. Branchlets black, sinewy, glaucous, or lustrous; leaf scars usually greater than 5 times wider than high (if thinner, then extending only partway around the twig); shrubs; fruit a drupe . . . . ***Oemleria***
         142. Branchlets variously colored, not sinewy, glabrous, or pubescent, usually not glaucous or lustrous; leaf scars usually less than 5 times wider than high (if wider, then nearly encircling the twig) or as high as or higher than wide; trees or shrubs; fruit a follicle, nut, drupe, or capsule
           143. Wood bright yellow; leaf scars almost encircling the twigs; buds solitary, not superposed; bud scales thin; low shrubs; fruit a follicle . . . . ***Xanthorhiza***
           143. Wood white to dark brown, not yellow; leaf scars extending only partway around twigs; buds superposed; bud scales thick or thin; trees or shrubs; fruit a drupe, nut, or capsule
              144. Buds large, at least 5 mm long, with thick scales; trees; fruit a nut
                 145. Twigs gray, the lenticels oval or elongate, large; pith pale or gray to light brown; fruit a winged nutlet . . . . . . ***Pterocarya***
                 145. Twigs dark gray to black, the lenticels orbicular, small; pith light brown to dark chocolate brown; fruit a nut . . . ***Juglans***

Keys to Genera: Inclusive

144. Buds small, much less than 5 mm long, with thin scales; shrubs; fruit a drupe or a 2-valved capsule

   146. Branchlets uniformly black or dark gray; fruit a drupe, in dense axillary clusters . . . .*Symplocos*

   146. Branchlets light brown to dark brown or red-brown; fruit a capsule, in narrow terminal racemes . . . .*Itea*

139. Pith continuous and homogenous

   147. Pith large, greater than $^2/_3$ the branchlet or twig diameter; fruit a drupe, a cluster of follicles, or woody capsules with seeds winged in upper half

      148. Pith white, very large, nearly 90 percent of the twig or branchlet diameter; buds hidden, at least partially covered by the leaf scar; shrub; fruit a cluster of follicles . . . .*Paeonia*

      148. Pith brown or white, about $^2/_3$ the twig or branchlet diameter; buds visible above the leaf scar; trees; fruit a drupe or capsule

         149. Branchlets light brown or light gray, occasionally darker gray, the lenticels small, mostly inconspicuous; leaf scars lunate, wider than high; fruit a drupe; southern . . . .*Pistacia*

         149. Branchlets red-brown to dark brown or dark gray, the lenticels prominent; leaf scars triangular, as high or higher than wide; fruit a capsule with winged seeds . . . .*Toona*

   147. Pith mostly less than half the twig or branchlet diameter; fruit various

      150. Young twigs slender, yellow-brown to light brown, stellate-pubescent; branchlets gray-brown, stellate-pubescent (often appearing as brown or black dots from particles caught in the hairs), the older ones mostly with shreddy bark . . . .*Buddleia*

      150. Young twigs variously colored, glabrous or pubescent, mostly not stellate; branchlets variously colored, glabrous or pubescent, not stellate, mostly not with shreddy bark

         151. Bundle scars more than 3 (alternate step 151, p. 74)

            152. Stems or branchlets prickly, often very stout; fruit drupe-like

               153. Leaf scars extending less than halfway around the stem; bundle scars 5–10 . . . .*Eleutherococcus*

               153. Leaf scars extending greater than halfway around the twigs; bundle scars 5–20

                  154. Twigs and branchlets with well-spaced, short, triangular spines; trees or shrubs to 10 m tall

                     155. Bundle scars about 15 . . . .*Kalopanax*

                     155. Bundle scars 5–10 . . . .*Aralia*

                  154. Twigs and branchlets densely covered with many lanceolate spines; shrubs or trees

                     156. Bundle scars about 20; leaf scars nearly completely encircling the twig . . . .*Aralia*

156. Bundle scars about 15; leaf scars extending $^2/_3$ of the way around the twig . . . . ***Oplopanax***

152. Stem and branchlets smooth, rough or ridged, not prickly, slender or stout; fruit various

157. Young twigs pale green, glabrous, glaucous, with many gladular dots, aromatic; leaf scars linear, about 1 mm high by 6 mm wide; subshrub less than 1 m tall . . . . ***Ruta***

157. Young twigs variously colored, glabrous or pubescent, glaucous or not, lacking glandular dots, with or without odor; leaf scars variously shaped, usually not linear; shrubs or trees

158. Pith with transverse woody partitions; lateral buds globose, densely dark-tomentose, the terminal bud naked; fruit large, berrylike . . . . ***Asimina***

158. Pith solid but lacking transverse woody partitions; lateral buds variously shaped and colored, the terminal bud, if present, naked or scaly; fruit various

159. Leaf scars deeply V-shaped, extending nearly around the bud; buds often tomentose or silky; twigs slender or stout; fruit a drupe, follicle, legume, or berry

160. Pith white, $^1/_2$–$^3/_4$ the diameter of the branchlet or twig; fruit a drupe, orange becoming black . . . . ***Ehretia***

160. Pith brown or white, small or large; fruit a legume, berry, follicle, or drupe, red, yellow, brown, or green

161. Bundle scars projecting out of the nearly white leaf scar; buds superposed; fruit a legume . . . . ***Cladrastis***

161. Bundle scars flush with other leaf scar tissue, not projecting; buds solitary, not superposed; fruit a drupe, follicle, or berry

162. Wood bright yellow; terminal bud present; fruit a follicle . . . . ***Xanthorhiza***

162. Wood gray to light brown; terminal bud absent; fruit a drupe or berry

163. Twigs pliable, slender; bark tough and fibrous, not resinous; fruit a drupe . . . . ***Dirca***

163. Twigs rigid, stout or slender; fruit a red drupe or berry

164. Bark with resinous juice; fruit a red drupe . . . . . . ***Rhus***

164. Bark lacking resinous juice; fruit a yellow berry; southern . . . . ***Melia***

159. Leaf scars semicircular, deltoid, or lunate, extending not more than halfway around the depressed bud; buds pubescent or glabrous; twigs mostly very stout; fruit various

165. Branchlets with prominent yellow-brown lenticels; bud scales 2, obtuse; fruit a bladdery capsule with 3 black seeds . . . ***Koelreuteria***

165. Branchlets with lenticels pale, inconspicuous, or lacking; bud scales more than 2; fruit various

166. Bundle scars run together, projecting from the leaf scar as a curved ridge; buds rusty stellate-pubescent; fruit a ribbed or winged, often stellate-pubescent capsule . . . .***Pterostyrax***

166. Bundle scars arranged variously, usually not raised significantly; buds glabrous or pubescent, usually lacking stellate hairs; fruit various

167. Bud scales dark purple-brown, often lustrous, the inner margins ciliate; twigs often glaucous; fruit a samara . . . .***Euptelea***

167. Bud scales dark red to brown, lacking ciliate fringe; twigs pubescent or glabrous, usually not glaucous; fruit a samara, legume, drupe, pome, or capsule

168. Buds lanceolate, long-acuminate, usually at least 4 times longer than wide, the outer 2 scales covering the full length of the bud; fruit a capsule

169. Buds red-brown, glabrous, pubescent only toward the apex; branchlets usually less than 3 mm in diameter, the pith very light brown . . . .***Stewartia***

169. Buds white to gray-pubescent; branchlets 3–5 mm in diameter, the pith brown to dark brown . . . .***Franklinia***

168. Buds globose to ovoid, acute to obtuse, several scales visible; fruit a samara, legume, drupe, or pome

170. Buds dark red; bark lacking resinous juice or strong odor; base of petiole often persistent; terminal bud present; fruit a red pome . . . .***Sorbus***

170. Buds brown; bark often with resinous juices or foul odor; base of petiole deciduous; terminal bud present or absent; fruit a samara, drupe, or legume

171. Buds superposed, imbedded in the bark; pith salmon-colored; terminal bud absent; twigs pale gray or brown; fruit a legume . . . .***Gymnocladus***

171. Buds solitary, exposed; pith brown; terminal bud present or absent; twigs usually yellow-brown to brown; fruit a samara or drupe

172. Bark lacking resinous juices, not toxic, but with foul odor; terminal bud absent; fruit a samara . . . .***Ailanthus***

172. Bark with resinous, poisonous juices, with little or no odor; terminal bud present; shrub; fruit a drupe . . .***Toxicodendron***

151. Bundle scars 3

173. Leaf scar a narrow line extending about halfway around the stem, not decurrent; plant often prickly; fruit red to red-brown, berrylike . . . .***Rosa***

173. Leaf scar broad or narrow, but not forming a line, often decurrent; plants prickly or not; fruit various

174. Twigs prickly or buds red-tomentose
   175. Bud scales blunt, terminated by a scar . . . .***Berberis***
   175. Bud scales acute or blunt, not terminated by a scar; inner bark
       dull, not yellow
      176. Stems often recurved or pendulous, with scattered prickles; fruit a
         berrylike cluster of drupelets . . . .***Rubus***
      176. Stems and twigs straight, not conspicuously recurved or
         pendulous, the prickles 1 to 3 at or below each leaf scar or
         scattered; fruit a berry or follicle
         177. Prickles 1 to 3 below each leaf scar, the latter decurrent; twigs pale;
            buds lanceolate, with thin scales; fruit a berry . . . .***Ribes***
         177. Prickles 2 at each leaf scar or scattered; buds depressed,
            red-tomentose or nearly black; fruit a follicle . . . .***Zanthoxylum***
174. Twigs smooth or rough, not prickly, sometimes with thorns;
   buds yellow to brown or black, gabrous or pubescent,
   not red-tomentose
   178. Leaf scars semicircular or broadly lunate, large, greater than
      2 mm in diameter (alternate step 178, p. 76)
      179. Bark with resinous juice; fruit a drupe
         180. Buds acute; fruit on a plumose stalk . . . .***Cotinus***
         180. Buds rounded; fruit sessile or on a glabrous or
            pubescent stalk . . . . , , , .***Rhus***
      179. Bark lacking resinous juice; fruit a capsule, berry, or drupe,
        but not borne on a plumose pedicel
         181. First bud scale of lateral buds anterior; bud
            scales puberulent . . . .***Populus***
         181. First bud scale of lateral buds lateral, at least not anterior;
            bud scales pubescent or glabrous
            182. Bud scales persisting at base of twigs; wood very light and soft;
               fruit a drupe; southern . . . .***Leitneria***
            182. Bud scales deciduous, not persisting at base of twigs;
               wood light to dark, hard; fruit a drupe, berry, or head
               of capsules
               183. Buds pubescent, 3 scales visible, the lateral 2 larger; fruit a
                  berry; southern . . . .***Melia***
               183. Buds glabrous, or at most, the scales ciliate, more
                 than 3 scales visible; fruit a drupe or head of
                 many capsules
                 184. Bud scales ciliate, polished; buds, when broken, resinous and
                   aromatic; leaf scars raised; branches often with corky winged
                   ridges of bark; pith large, angular; fruit a head of
                   many capsules . . . .***Liquidambar***
               184. Bud scales glabrous, lacking cilia; buds lacking
                 resin or strong odor; leaf scars mostly in plane of

twig, not prominently raised; branches lacking corky ridges; pith
pith terete, not angular; fruit a drupe

185. Branchlets slender, less than 5 mm in diameter, with bitter almond
odor when rubbed; leaf scars usually well-spaced along
the branchlet . . . .*Prunus*

185. Branchlets stout, usually greater than 5 mm in diameter even near the
tip, lacking strong odor when rubbed; leaf scars crowded on short thick
twigs or spurs . . . .*Davidia*

178. Leaf scars narrowly lunate, or small, less than 2 mm in diameter

186. Internodes very unequal, branches much exceeding the central axis;
fruit a drupe . . . .*Cornus*

186. Internodes nearly equal, branches shorter than the central axis;
fruit various

187. Buds superposed

188. Leaf scars deeply **V**-shaped, partly surrounding the bud,
or buds bursting through the leaf scars; fruit a legume
or samara

189. Buds in the upper angle of **V**-shaped leaf scars, pubescent,
mostly superposed; twigs lacking thorns, spines, and prickles;
fruit a samara . . . .*Ptelea*

189. Buds bursting through the leaf scars, pubescent or
glabrous, superposed or solitary; twigs usually bearing
thorns, spines or prickles; fruit a legume

190. Buds glabrous; branchlets often bearing stout branched thorns;
twigs mostly terete . . . .*Gleditsia*

190. Buds pubescent; branchlets usually bearing
unbranched nodal spines or prickles; twigs often
strongly angled . . . .*Robinia*

188. Leaf scars variously shaped, but not deeply **V**-shaped, free
from the buds; fruit a legume, drupe, or achenes

191. Twigs furrowed, green or gray-brown; fruit a head of
hairy achenes . . . .*Baccharis*

191. Twigs even, not furrowed, finely white-speckled; fruit a
drupe or legume

192. Twigs spicy-aromatic; most branchlets bearing globose
collateral flower buds; shrubs; fruit a red drupe . . . . . .*Lindera*

192. Twigs lacking aroma; flower buds axillary, not easily
distinguished from vegetative buds; small trees;
fruit a legume . . . .*Cercis*

187. Buds solitary, not superposed

193. Bark bright green; finely branched, low shrubs . . . .*Kerria*

193. Bark brown, gray, red-brown, yellow-brown, or black;
trees or shrubs

194. Buds much elongated, often pointed; thorns absent

195. Bark, at least the older, shredding; leaf scars often strongly decurrent; bud scales very thin, often glandular; shrubs; fruit a berry . . . .*Ribes*

195. Bark close, not shreddy; leaf scars not or only slightly decurrent; bud scales thick or thin, usually lacking glands; fruit a berrylike pome

    196. Second bud scale usually less than half the length of the bud; bud scales lacking glands on the teeth, thin, closely appressed, mostly brown with a black tip; shrubs or small trees . . . .*Amelanchier*

    196. Second bud scale about half the length of the bud or more; bud scales glandular-toothed, rather thick, somewhat spreading or divaricate, red to red-brown, lacking a black tip; shrubs

        197. Branchlets slender, less than 3 mm in diameter . . . .*Aronia*

        197. Branchlets stout, usually greater than 5 mm in diameter . . . .×*Sorbaronia*

194. Buds ovoid, globose, or depressed, blunt; branches sometimes with thorns (buds narrowly ovoid and lateral twigs thorn-like in *Pyrus coronaria* and *Mespilus*)

    198. Twigs with raised lenticels or scattered black dots, usually pubescent; fruit a red, berrylike pome . . . .*Photinia*

    198. Twigs lacking lenticels and black dots, or lenticels not raised, glabrous to pubescent; fruit various

        199. Buds with 2–3 scales, the scales glabrous and lustrous or sometimes with a mucronate tuft of white hairs; branchlets black; fruit a legume . . . .*Maackia*

        199. Buds usually with more than 4 scales, the scales pubescent or glabrous; branchlets red-brown, brown, gray, or black; fruit various

            200. Buds less than 1 mm long, densely rusty-tomentose; twigs densely rusty-tomentose; branchlets dark gray to black; leaf scars clustered on knobs, spurs, or very short branchlets; native southern plants of sandy or wet areas . . . .*Bumelia*

            200. Buds usually greater than 1 mm long, pubescent or glabrous; twigs glabrous or pubescent; branchlets various shades of color; leaf scars usually not clustered on knobs or spurs (sometimes in *Pyrus*)

                201. Branchlets clustered near the end of each season's growth, often with minute golden resin-granules; scales of the terminal bud acute, the others rounded; shrubs; fruit a small waxy drupe or glabrous drupe-like nut . . . .*Myrica*

                201. Branchlets more evenly distributed, lacking resin-granules; scales of terminal and lateral buds similar, usually rounded; shrubs or trees; fruit a pome, berry, capsule, drupe, or aggregate of berries

202. Leaves deciduous above the base of petiole . . . .*Rubus*
202. Leaves deciduous at the base of petiole
  203. Shrubs less than 3 m tall
    204. Branches many, short; rare southern native of swamps . . . .*Litsea*
    204. Branches few, longer; common, of various habitats . . . .*Ribes*
  203. Shrubs or trees greater than 3 m tall
    205. Branchlets less than 3 mm in diameter; buds black or purple-black,
      glabrous; fruit a capsule . . . .*Exochorda*
    205. Branchlets greater than 5 mm in diameter; buds brown to
      black, glabrous to pubescent; fruit a pome or drupe
      206. Twigs densely tomentose; buds often in clusters,
        often hidden . . . .*Mespilus*
      206. Twigs glabrous to pubescent, not densely tomentose;
        buds scattered, visible
        207. Scales of terminal bud narrowly ovate, thick; buds narrowly
          ovoid; twigs dark red or bronze-gray; lateral twigs mostly sharp
          and thorn-like; small trees . . . .*Malus*
        207. Scales of terminal bud broadly ovate, thinner,
          appressed, dentate or entire; buds usually broadly
          ovoid; twigs gray or olive to dark brown; lateral twigs
          lacking or bearing thorns, sharp or dull; trees
          or shrubs
          208. Axillary buds flattened and closely appressed, broadly ovoid,
            mostly pubescent; twigs mostly dark, rarely olive; trees with
            broad, spreading crowns; fruit a pome with papery
            carpel walls . . . .*Malus*
          208. Axillary buds plump and divaricate, sharp-pointed
            or rounded, pubescent or glabrous; twigs gray to
            dark brown; shrubs or trees with narrow or spreading
            crowns; fruit a drupe or a pome with stony or
            papery carpel walls
            209. Buds acute, mostly glabrous, conical, sharp-pointed;
              small trees . . . .*Pyrus*
            209. Buds obtuse, rarely acute, pubescent or glabrous,
              globose to ovoid; small trees or shrubs
               210. Buds about 1 mm long, globose; fruit
                a drupe . . . .*Symplocos*
              210. Buds less than 1 mm long, ovoid; fruit a pome
                211. Branchlets smooth, shining; buds mostly depressed;
                  branches mostly with thorns; terminal bud present or
                  absent; small trees or shrubs . . . .*Crataegus*
                211. Branchlets dull; buds raised or only somewhat
                  depressed; branches unarmed; terminal bud
                  present; shrubs . . . .*Sorbaria*

ABRIDGED KEY: NATIVE AND COMMONLY NATURALIZED WOODY PLANTS ONLY

KEY VIII. PLANTS WITH NEEDLELIKE, SCALELIKE, OR AWL-SHAPED LEAVES

1. Low, prostrate, or creeping shrubs or mats or parasitic and growing on trees, 10–100 cm tall
   2. Leaves alternate; stems forming low mats (or creeping and horizontal in *Taxus*)
      3. Leaves covered with downy hairs . . . .***Hudsonia***
      3. Leaves glabrous or nearly so
         4. Leaves revolute, falsely whorled, 4–7 mm long; leaf scars decurrent; fruit a drupe
            5. Leaves denticulate . . . .***Corema***
            5. Leaves entire . . . .***Empetrum***
         4. Leaves flattened, not revolute, not appearing whorled, 2–20 mm long; leaf scars decurrent or not; fruit a capsule or drupe-like
            6. Leaves imbricated, subulate, 2–8 mm long; leaf scars not decurrent; fruit a capsule
               7. Leaves spreading, mostly 4–8 mm long, about 1 mm wide . . . .***Pyxidanthera***
               7. Leaves crowded, appressed, mostly 2–4 mm long, less than 1 mm wide . . . .***Harrimanella***
            6. Leaves spaced apart on the branchlet, not imbricated, linear, 5–20 mm long; leaf scars decurrent; fruit a capsule or drupe-like

8. Leaves acuminate; reproductive structure drupe-like,
fleshy, red . . . . *Taxus*
8. Leaves obtuse; fruit a dry capsule . . . . *Phyllodoce*
2. Leaves opposite or whorled, scalelike or minute; stems erect
or creeping
9. Plants growing on trees, parasitic; leaves barely visible . . *Arceuthobium*
9. Plants erect or creeping, growing from the ground;
leaves small, usually obvious
10. Stems creeping or horizontal; leaves subulate or broader
or scalelike . . . . *Juniperus*
10. Stems erect
11. Leaves scalelike or subulate . . . . *Juniperus*
11. Leaves sagittate, oblong, or linear
12. Leaves opposite, sagittate . . . . *Calluna*
12. Leaves whorled, oblong or linear . . . . *Erica*
1. Trees or erect or ascending shrubs, greater than 100 cm tall
13. Leaves mostly scalelike (linear or juvenile leaves often also
present), no greater than twice as long as wide; twigs and
branchlets usually hidden by the brown or green leaves
14. Leaves alternate
15. Leaves linear, horizontal or divergent; branchlets exposed, rough,
gray; southern . . . . *Ceratiola*
15. Leaves subulate, awl-shaped, or scalelike, parallel or ascending;
branchlets exposed or not . . . . *Tamarix*
14. Leaves opposite or whorled
16. Branchlets terete or 4-angled, the groups of branchlets (sprays) in
several different planes, spreading at various angles from trunk or
main stems . . . . *Juniperus*
16. Branchlets flattened, the groups of branchlets (sprays) in
several different planes or mostly horizontal planes
17. Seam where margins of lateral leaves meet on upper surface of
twig usually visible . . . . *Chamaecyparis*
17. Seam where margins of lateral leaves meet on upper surface of
twig usually hidden . . . . *Thuja*
13. Leaves needlelike, at least 4 times as long as wide; twigs visible,
not hidden or clothed by the leaves
18. Leaves with apex extended into a long spine, often 2–3 cm long; female
reproductive structure a legume . . . . *Ulex*
18. Leaves rounded or blunt to pungent at the apex, lacking a
spine; female reproductive structure a woody cone
19. Leaves in clusters or fascicles
(solitary only on young shoots)
20. Leaves in fascicles of 2–5 . . . . *Pinus*
20. Leaves in clusters or whorls of 10 or more . . . . *Larix*

19. Leaves all solitary
    21. Leaves curved, awl-shaped, keeled, with thick ridge on the keel
        22. Leaves whorled or, if opposite, usually greater than
            10 mm long . . . .*Juniperus*
        22. Leaves opposite, less than 10 mm long . . . .*Chamaecyparis*
    21. Leaves straight, flat, or angled
        23. Buds lance-ovoid, at least 2½ times as long as wide, sharp-pointed, the
            terminal buds usually greater than 5 mm long . . . .*Pseudotsuga*
        23. Buds globose or ovoid, less than twice as long as wide, usually
            less than 5 mm long
            24. Leaves 4-sided or angular in cross section, easily rolled between the
                fingers; branchlets very rough because of persistent woody
                projections at the base of each leaf . . . .*Picea*
            24. Leaves flattened, not easily rolled between the fingers; twigs
                smooth or lacking prominent woody projections
                25. Leaves less than 1 mm wide; ultimate twigs less than
                    0.5 mm wide . . . .*Taxodium*
                25. Leaves greater than 1 mm wide; twigs greater than
                    1 mm wide
                    26. Leaves opposite or whorled; female reproductive structure
                        fleshy, berrylike or drupe-like . . . .*Juniperus*
                    26. Leaves alternate; female reproductive structure a woody
                      cone or drupe-like
                      27. Leaves of conspicuously different lengths and
                        orientations on the same twig
                        28. Leaves pulling away from branchlet cleanly, leaving
                          indented orbicular leaf scars; branches and
                          shoots stiff . . . .*Abies*
                      28. Leaves pulling away from branchlet with bark attached;
                        leading shoots flexible, often pendulous . . . .*Tsuga*
                      27. Leaves of similar lengths and orientations on the
                      same twig
                        29. Leaves mostly less than 25 mm long
                        30. Leaves pulling away from branchlet cleanly, leaving
                          indented orbicular leaf scars; cones erect, disintegrating
                          on the tree . . . .*Abies*
                      30. Leaves pulling away from branchlet with bark
                        attached; cones pendulous and falling intact
                        or fleshy and drupe-like
                        31. Leaves on woody projections to 2 mm long on all sides
                        of the twig; trees . . . .*Picea*
                      31. Leaves decurrent, in mostly horizontal planes,
                        appearing to be opposite, giving twig a flattened
                        appearance; shrubs . . . .*Taxus*

29. Leaves mostly greater than 25 mm long
    32. Leaves pulling away from branchlet cleanly, leaving indented orbicular leaf scars; cones erect, disintegrating on the tree . . . .*Abies*
    32. Leaves pulling away from branchlet with bark attached; cones pendulous and falling intact or fleshy and drupe-like
        33. Leaves borne on woody projections with white longitudinal bands on all surfaces or upper surface only; trees . . . .*Picea*
        33. Leaves decurrent, with white or pale longitudinal bands on lower surface only; shrubs . . . .*Taxus*

KEY IX. PLANTS WITH OPPOSITE OR WHORLED SIMPLE LEAVES

1. Leaves subopposite . . . .*Rhamnus*
1. Leaves distinctly opposite or whorled
    2. Leaves lobed
        3. Leaves mostly pinnately lobed
            4. Margin of lobes entire; sap clear; shrubs; fruit a capsule . . . . .*Syringa*
            4. Margin of lobes serrate; sap milky or clear; trees or tall shrubs; fruit a capsule or head of achenes
                5. Trees; sap milky; fruit a head of achenes . . . .*Broussonetia*
                5. Shrubs; sap clear or milky; fruit a capsule . . . .*Hydrangea*
        3. Leaves palmately lobed
            6. Leaf blades less than 20 cm long; fruit a drupe or samara
                7. Petioles with stipules and glands, or if lacking glands, the lower surface of leaf densely pubescent; fruit a drupe . . . .*Viburnum*
                7. Petioles lacking stipules and glands, or if stipules present, the lower surface of leaf glabrous to pubescent, not densely so; fruit a samara . . . .*Acer*
            6. Leaf blades greater than 20 cm long; fruit a capsule
                8. Leaves with long tapering tip, glabrous or softly pubescent, usually in whorls of 3; pith continuous; fruit a long cylindrical capsule, 20–50 cm long . . . .*Catalpa*
                8. Leaves with obtuse or rounded tip, densely pubescent, like velvet, always opposite; pith chambered or hollow; fruit an ovoid capsule, 3–4 cm long . . . .*Paulownia*
    2. Leaves entire or toothed, lacking lobes
        9. Stems climbing, creeping, prostrate, or forming low mats, or plants parasitic and growing on trees (alternate step 9, p. 000)
            10. Stems climbing (vines); leaves deciduous or persistent
                11. Aerial rootlets present on stems; fruit a capsule; southern . . . .*Decumaria*
                11. Aerial rootlets absent; fruit a berry or capsule
                    12. Leaves lanceolate, long acuminate; fruit a capsule . .*Gelsemium*
                  12. Leaves ovate, acute to cuspidate; fruit a berry . . . .*Lonicera*

10. Stems prostrate, creeping, or forming low mat or plants
   parasitic and growing on trees; leaves persistent
  13. Plants growing on trees, parasitic; fruit a berry
    14. Leaves about 1 mm long, scalelike; stems
      1–3 cm long . . . .*Arceuthobium*
    14. Leaves 20–40 mm long; stems to 40 cm long . . . .*Phoradendron*
  13. Plants prostrate, creeping, not parasitic or growing on trees;
    fruit a berry, follicle, or capsule
    15. Leaves crenate or serrate; plants creeping or erect;
      fruit a capsule
      16. Leaves rounded-ovate, crenate; branches
        brown-pubescent . . . .*Linnaea*
      16. Leaves oblanceolate, sharply serrate;
        branches glabrous
        17. Leaves opposite; stems with longitudinal ridges;
          plants stoloniferous . . . .*Euonymus*
        17. Leaves in false whorls; stems smooth, erect, from
          creeping rootstocks . . . .**Chimaphila**
    15. Leaves entire or slightly undulate; plants creeping or
      prostrate; fruit a berry, capsule, or follicle
      18. Leaves with translucent or pellucid dots (visible when
        held up to light) . . . .*Hypericum*
      18. Leaves opaque, lacking translucent dots
        19. Leaves and young stems strongly pubescent;
          fruit a berry . . . .*Lonicera*
        19. Leaves and stems glabrous or nearly so; fruit a berry,
          follicle, or capsule
          20. Leaf base cordate or rounded; fruit a red berry . . .*Mitchella*
          20. Leaf base narrow or cuneate; fruit a follicle
            or capsule
            21. Leaves greater than 2 cm long, margins flat, not revolute;
              stems slightly woody, creeping; fruit a follicle . . . . . .*Vinca*
            21. Leaves less than 1 cm long, revolute; stems woody, forming
              dense low mats; fruit a capsule . . . .*Loiseleuria*
9. Stems erect; trees or shrubs
  22. Leaves entire (alternate step 22, p. 85)
    23. Leaves with blade mostly greater than 20 cm long, ovate,
      with cordate, subcordate, or truncate base; large trees
      24. Leaves with long-tapering tip, glabrous or softly pubescent, usually
        in whorls; pith continuous; fruit a long cylindrical capsule,
        20–50 cm long . . . .*Catalpa*
      24. Leaves with obtuse or rounded tip, densely pubescent like velvet,
        opposite; pith chambered or hollow; fruit an ovoid capsule,
        3–4 cm long . . . .*Paulownia*

Keys to Genera: Abridged

23. Leaves with blade mostly less than 15 cm long; if longer, shape not
    ovate; shrubs or small trees
    25. Leaves with yellow, glandular dots, often with 1–3 remote, sharp teeth,
        ill-scented; twigs striate . . . .*Iva*
    25. Leaves green, pale, or brown, lacking yellow, glandular dots,
        entire, aromatic or lacking odor; twigs smooth or rough, mostly
        lacking ridges or lines
        26. Leaves and twigs spicy-aromatic . . . .*Calycanthus*
        26. Leaves and twigs odorless or at least not spicy-aromatic
            27. Leaves and twigs covered with silvery or rusty scales . . . .*Shepherdia*
            27. Leaves and twigs lacking silvery or rusty scales
                28. Leaves ovate with cordate, subcordate, or truncate base and
                    long-tapering tip . . . .*Syringa*
                28. Leaves of various shapes, the base acute, or if cordate,
                    the apex not long-tapering
                    29. Leaves mostly less than 1 cm long . . . .*Leiophyllum*
                    29. Leaves greater than 2 cm long
                        30. Leaves with pellucid dots (visible when held up to light);
                            fruit a capsule . . . .*Hypericum*
                        30. Leaves opaque, lacking pellucid dots; fruit various
                            31. Leaves gray-green or pale green, cuneate-attenuate or long
                                attenuate at the base, acute to acuminate or often
                                mucronate, minutely undulate or occasionally with 2–6
                                remotely spaced linear teeth; southern . . . .*Borrichia*
                            31. Leaves green to brown, variously shaped at the
                                base, not long attenuate, mucronate or usually
                                not, always entire
                                32. Leaves sessile or subsessile (petiole less than
                                    1 mm long); branchlets terete, quadrangular,
                                    flattened, or winged
                                    33. Branchlets terete; fruit a drupe;
                                        southern . . . .*Buckleya*
                                    33. Branchlets quadrangular, flattened, or winged;
                                        fruit a capsule . . . . *Kalmia*
                                32. Leaves with a distinct petiole; branchlets
                                    terete, neither quadrangular, flattened,
                                    nor winged
                                    34. Leaves tending to be crowded toward the end of the
                                        season's growth, often in whorls of 3, leathery,
                                        persistent; fruit a capsule . . . .*Kalmia*
                                    34. Leaves scattered in pairs along the season's
                                        growth, thin, deciduous or persistent;
                                        fruit various

35. Lateral buds imbedded in the bark, supra-axillary (located some
distance above the axil); leaves often in whorls of 3; fruit a head
of nutlets . . . .***Cephalanthus***
35. Lateral buds axillary, not imbedded in the bark; leaves opposite;
fruit various
   36. Bark of stems and branches loose, peeling off in long,
     shreddy strips
      37. Twigs crowded on the slender branchlets; bundle scar 1; bud
        scales deciduous . . . .***Symphoricarpos***
      37. Twigs widely spaced on branchlet of medium thickness; bundle scars
        3; bud scales often persisting at base of twigs . . . .***Lonicera***
   36. Bark of stems and older branches smooth, not peeling off in
     long shreds
      38. Leaves with lateral veins running somewhat parallel with the
        margin and meeting near the apex, the veins when torn dangling
        silky strands . . . .***Cornus***
      38. Leaves with lateral veins ending near the margin and not
        running to the apex, the veins tearing clean (sometimes
        leaving dangling strands in *Viburnum*)
        39. Leaves persistent, leathery, the petioles often 2 mm thick
          40. Leaves mostly greater than 14 cm long, pubescent . . .***Viburnum***
          40. Leaves mostly less than 12 cm long, glabrous
            41. Leaves obtuse to cuspidate, ovate, greater than
              4 cm wide . . . .***Viburnum***
            41. Leaves acute to acuminate, elliptic, oblanceolate, or
              lanceolate, less than 3 cm wide; southern . . . .***Osmanthus***
        39. Leaves deciduous, thin, the petioles less than 1 mm thick
          42. Lower leaf surface, especially along the midrib, scurfy, rusty
            brown, woolly, or glandular (margin of leaf often
            finely serrate) . . . .***Viburnum***
          42. Lower leaf surface glabrous, lightly pubescent,
            lacking glands, not scurfy or woolly
            43. Leaf blades greater than 7 cm long . . . .***Chionanthus***
            43. Leaf blades 1–6 cm long
              44. Twigs red, red-brown, or purple-brown, usually
                glaucous; southern . . . .***Nestronia***
              44. Twigs brown to gray, lacking bloom . . . .***Ligustrum***
22. Leaves serrate to serrulate, dentate to crenulate or nearly undulate
   45. Leaves with the bases of the petioles meeting or joined by a distinct
     transverse line
      46. Buds imbedded under the petiole base, not axillary . . . . .***Philadelphus***
      46. Buds axillary, not imbedded
        47. Leaves with yellow, glandular dots; twigs strongly striated . . . . . .***Iva***

47. Leaves uniformly green, lacking yellow, glandular dots; twigs
uniform, not strongly striated
    48. Twigs and branchlets with decurrent ciliate ridges or densely pubescent
longitudinal bands between the nodes; leaves with main lateral veins
fading before reaching the margin; fruit a capsule . . . .*Diervilla*
    48. Twigs and branchlets glabrous or pubescent, lacking
pubescent longitudinal bands or ridges between the nodes;
leaves with main lateral veins often extending to the margin;
fruit a capsule, drupe, or samara
        49. Buds naked or covered by 2 valvate scales . . . .*Viburnum*
        49. Buds covered by several imbricated scales
            50. Leaves coarsely doubly or triply (or more) toothed, the lateral veins
more than 20 on each side of the midrib . . . . *Acer*
            50. Leaves simply toothed or finely and closely doubly
toothed, the lateral veins fewer than 20 on each side of
the midrib
                51. Bud scales from buds of previous year persistent; leaves
with petiole usually longer than the blade;
fruit a capsule . . . .*Hydrangea*
                51. Bud scales from buds of previous year deciduous; leaves with
petiole shorter than the blade; fruit a drupe . . . .*Viburnum*
45. Leaves with petiole bases of opposite leaves distinct, not meeting
or joined by a transverse line
    52. Leaves woolly on lower surface; branchlets gray-brown, dark
brown, or black; fruit a berry or drupe
        53. Twigs and branchlets gray-brown, lacking spines; inner bark brown;
petioles 3–15 mm long . . . .*Callicarpa*
        53. Twigs and branchlets dark brown or black, often terminating in sharp
black spines; inner bark yellow and bitter; petioles
7–25 mm long . . . .*Rhamnus*
    52. Leaves glabrous or pubescent on lower surface, not woolly;
branchlets gray-brown, green, or red-purple; fruit various
        54. Leaves thick, persistent . . . .*Paxistima*
        54. Leaves thin, deciduous
            55. Pith hollow or chambered, not continuous between
the nodes . . . .*Forsythia*
            55. Pith continuous or spongy between the nodes (occasionally
chambered in *Euonymus*)
                56. Leaves woolly on lower surface; branchlets light brown
to gray-brown . . . .*Callicarpa*
                56. Leaves glabrous or pubescent, not woolly; branchlets
gray, green, red, or purple-red
                    57. Twigs mostly light gray, nearly horizontal, usually terminating
in sharp points; fruit a drupe . . . .*Forestiera*

57. Twigs light brown to gray or dark brown, ascending, flexible, not
terminating in sharp points; fruit a samara or capsule
  58. Leaves irregularly, doubly serrate; fruit a samara . . . . ***Acer***
  58. Leaves singly serrate or serrulate; fruit a capsule
    59. Bud scales several . . . . ***Euonymus***
    59. Bud scales 1 . . . . ***Salix***

KEY X. PLANTS WITH OPPOSITE COMPOUND LEAVES

1. Leaves with 2–3 leaflets
  2. Leaflets 2; vine with branched tendrils; fruit a capsule . . . . ***Bignonia***
  2. Leaflets 3; trees, shrubs, or vines; fruit a capsule, samara, or achene
    3. Stems climbing or twining; fruit a hairy achene . . . . ***Clematis***
    3. Stems erect or sometimes pendulous, not climbing or twining;
    trees or shrubs; fruit a samara or a capsule
      4. Twigs bright green or purple, glaucous; leaflets 3–5, lobed or coarsely
      dentate, the teeth large and few; fruit a samara . . . . ***Acer***
      4. Twigs brown or gray, lacking bloom; leaflets 3, serrate, the teeth
      crowded; fruit a glabrous capsule . . . . ***Staphylea***
1. Leaves with more than 3 leaflets
  5. Leaves palmately compound . . . . ***Aesculus***
  5. Leaves pinnately compound
    6. Stems climbing, with or without aerial rootlets; fruit a capsule
    or head of achenes
      7. Leaflets with 1 main vein from the base; fruit a capsule . . . . . ***Campsis***
      7. Leaflets with 3–5 main veins from the base or near the base; fruit a
      head of achenes . . . . ***Clematis***
    6. Stems erect, not climbing; erect trees or shrubs; leaflets usually
    acute or short acuminate and finely serrate (lobed or coarsely
    toothed in *Acer*); fruit a drupe, samara, or capsule
      8. Twigs and branchlets with lenticels large, raised, and corky;
      older branchlets with very large pith; shrubs; fruit a
      berrylike drupe . . . . ***Sambucus***
      8. Twigs and branchlets with lenticels usually small and
      inconspicuous or lacking, not corky; pith small; trees or
      shrubs; fruit a samara or capsule
        9. Leaflets 3–5, lobed or coarsely serrate; twigs often glaucous, green or
        purple; fruit a pair of samaras . . . . ***Acer***
        9. Leaflets 5–13, entire or finely serrate; twigs glabrous
        or pubescent, not glaucous, gray to brown; fruit a samara
        or capsule
          10. Leaflets 7 or more; buds nearly globose, blunt; trees;
          fruit a samara . . . . ***Fraxinus***
          10. Leaflets 5–7; buds narrowly ovoid, acuminate; shrubs or small
          trees; fruit an inflated capsule . . . . ***Staphylea***

Keys to Genera: Abridged

KEY XI. PLANTS WITH ALTERNATE SIMPLE LEAVES

(In some plants the leaves are clustered on short spurs.)

1. Leaves crowed and overlapping at the base, linear to linear
   lanceolate, at least 10 times longer than wide (woody grasses)
     2. Leaves with inconspicuous midvein, lacking strong white midrib,
       1–5 cm wide, the margins smooth . . . .*Phragmites*
     2. Leaves with conspicuous white midrib projecting from the lower
       surface, less than 2 cm wide, margins sharp and serrulate
       3. Leaves less than 1 cm wide, the midrib on lower surface rough, scabrous,
         or cutting to the touch; southern . . . .*Cortaderia*
       3. Leaves 1–2 cm wide, the midrib smooth to the touch
         4. Leaves mostly less than 25 cm long, about 10 times longer
           than wide . . . . *Arundinaria*
         4. Leaves mostly greater than 50 cm long, at least 50 times longer
           than wide . . . . *Miscanthus*
1. Leaves mostly separated, not densely overlapping at the base,
   shaped variously, mostly less than 10 times longer than wide
     5. Leaves lobed (alternate step 5, p. 90)
       6. Stems climbing or twining; vines
         7. Leaves with 1 or more lateral lobes on each side of midrib; leaves and
           stems with rank odor when crushed; fruit a red berry . . . . . .*Solanum*
         7. Leaves palmately lobed; leaves and stems lacking rank odor
           when crushed; fruit a red, black, or blue berry or drupe
           8. Aerial rootlets present on twigs and branchlets; leaf lobes
             mostly entire; veins on upper surface of leaves white or pale;
             fruit a berry . . . .*Hedera*
           8. Aerial rootlets absent; leaf lobes mostly serrate, veins
             green, not white or pale; fruit a berry or drupe
             9. Tendrils present; fruit a berry
               10. Tendrils ending in expanded tips or
                 disklike appendages . . . .*Parthenocissus*
               10. Tendrils narrowed or blunt at tips, not ending in
                 disklike appendages, coiled
                 11. Bark shredding . . . .*Vitis*
                 11. Bark close, not peeling . . . .*Ampelopsis*
             9. Tendrils absent; fruit a drupe
               12. Petiole 1–4 cm long; leaves with middle lobe longer than the
                 lateral ones, pubescent; fruit red; southern . . . .*Cocculus*
               12. Petiole 5–15 cm long; leaves with lobes nearly
                 equal in length, glabrous or nearly so; fruit blue
                 or black
                 13. Leaves peltate, the lobes rounded or acute;
                   fruit blue . . . .*Menispermum*

13. Leaves cordate, the lobes acuminate; fruit
    black; southern . . . .*Calycocarpum*
6. Stems erect or prostrate; trees or shrubs
  14. Leaves distinctly palmately lobed
    15. Leaves orbicular or wider than long, the petiole and both surfaces
        densely covered with spines . . . .*Oplopanax*
    15. Leaves ovate to suborbicular, longer than wide, lacking
        spines or with spines small and scattered
      16. Leaves star-shaped, 5–7 lobed, with evenly serrate margins; branches
          often with corky ridges . . . .*Liquidambar*
      16. Leaves deeply or shallowly lobed, less divided, not
          star-shaped, 3–5 lobed, the margins regularly or
          irregularly serrate, entire, or with smaller lobes;
          branches smooth or rough, lacking corky ridges
        17. Base of petiole hollow, forming a hood over the lateral
            bud; stipules or stipule scars prominent; bark of trunk and
            branches peeling off in large plates; fruit a suspended ball of
            hairy achenes . . . .*Platanus*
        17. Base of petiole solid, not hollow, subtending the
            lateral (axillary) bud; stipules or stipule scars
            inconspicuous or absent; bark tight, or if peeling,
            not peeling in large plates; fruit a capsule, berry, or
            cluster of drupelets
          18. Branchlets white-tomentose; leaves with lower surface white-
              tomentose, petiole often flattened; fruit a capsule . . . . .*Populus*
          18. Branchlets green or brown, not white-tomentose;
              leaves with lower surface green, not white-tomentose,
              petiole terete; fruit a capsule, berry, or cluster
              of drupelets
            19. Petioles and twigs glandular; fruit aggregate, a cluster of
                drupelets, berrylike . . . .*Rubus*
            19. Petioles and twigs glabrous or pubescent, lacking
                glands; fruit a capsule or berry
              20. Older bark separating into numerous thin layers; spines
                  absent; fruit a capsule . . . .*Physocarpus*
              20. Bark close, not separating in layers, spines often present;
                  fruit a berry . . . .*Ribes*
  14. Leaves pinnately lobed or irregularly lobed or tulip-shaped with
      broad truncate apex, at least not distinctly palmately lobed
      (except sometimes in *Sassafras*)
    21. Leaves with 3 main veins from near the base, mostly with 1 or
        a few lateral lobes
      22. Lobes entire; leaves spicy-aromatic; bark of branchlets smooth,
          green; fruit a blue drupe . . . .*Sassafras*

22. Lobes serrate; leaves lacking strong aroma; bark of branchlets smooth or rough, yellow, brown, or black, not green; fruit a pome or multiple

   23. Sap clear, not milky; leaf lobes pointed; thorns usually present; fruit a pome . . . .***Sorbus***

   23. Sap milky; leaf lobes rounded or pointed; thorns absent; fruit multiple

      24. Leaves densely pubescent, like velvet on lower surface, petiole 5–10 cm long; twigs densely pubescent, like velvet . . . .***Broussonetia***

      24. Leaves downy, pubescent, or occasionally glabrous on lower surface, petiole usually 2–4 cm long; twigs glabrous or pubescent . . . .***Morus***

21. Leaves with a single large main vein (midrib), variously lobed

   25. Leaves irregularly lobed or tulip-shaped with broad truncate apex

      26. Leaves tulip-shaped, with truncate apex and 2 broad lateral lobes on each side of midrib, the lobe margins entire; buds covered by membranous stipules . . . .***Liriodendron***

      26. Leaves acute or acuminate, irregularly lobed, usually with more than 2 broadly rounded or acute lobes, the lobe margins coarsely or finely serrate; buds with scales

         27. Plants with long, stiff, sharp thorns; leaves cuneate to truncate at the base . . . .***Crataegus***

         27. Plants lacking thorns; leaves cuneate, rounded, truncate, or cordate at the base . . . .***Malus***

   25. Leaves pinnately lobed

      28. Leaves with numerous deep, rounded lobes on each side of the midrib, sweet-scented; shrubs; fruit a small, nut-like achene enclosed in a bur . . . .***Comptonia***

      28. Leaves with few to several rounded or pointed lobes on each side of the midrib, lacking a sweet scent; small or large trees; fruit an acorn or pome

         29. Leaves 10–20 cm long; buds clustered at the ends of the twigs; thorns absent; fruit an acorn . . . .***Quercus***

         29. Leaves 4–10 cm long; buds scattered along the twig, not clustered at the ends; thorns often present; fruit a small pome . . . .***Crataegus***

5. Leaves entire or toothed, not lobed

   30. Leaves entire (alternate step 30, p. 97)

      31. Plants less than 10 cm tall . . . .***Diapensia***

      31. Plants greater than 10 cm tall

         32. Leaves usually with a pair of tendrils at base of petiole; stem vascular bundles scattered in cross section; spines or prickles usually

present; stem usually green and climbing; fruit a blue, black,
or red berry . . . . *Smilax*
32. Leaves lacking tendrils at base of petiole; stem vascular bundles in
a ring in cross section; spines present or absent; stems climbing or
erect, green or dark; fruit various
33. Stipules glandular, persistent; leaves ciliate at base of blade and apex of
petiole; southern shrubs near water . . . . *Sebastiana*
33. Stipules, if present, not glandular, deciduous or persistent; leaves
smooth at the basal margins and at apex of petiole, hairs usually
lacking; trees, vines, or shrubs
34. Leaves persistent, often leathery, often revolute (alternate
step 34, p. 92)
35. Stems climbing, creeping, prostrate, or forming dense
low mats
36. Stems climbing; southern . . . .*Cocculus*
36. Stems creeping or prostrate, not climbing
37. Stems covered with brown hairs
38. Leaves less than 2 cm long, elliptic to oblong, with
coarse, brown, scalelike hairs on lower surface; fruit
a white berry . . . .*Gaultheria*
38. Leaves 2–8 cm long, oval-oblong to suborbicular,
scabrous and pilose or glabrous; fruit
a capsule . . . .*Epigaea*
37. Stems glabrous or pubescent, not covered with
brown hairs
39. Twigs and petioles covered with brown scales or scurf;
fruit a capsule . . . .*Rhododendron*
39. Twigs and petioles glabrous, glaucous, or pubescent,
not covered with brown scales or scurf; fruit a
pome, berry, or drupe
40. Leaves glaucous on the lower surface . . . .*Vaccinium*
40. Leaves green, with or without black dots on the
lower surface
41. Leaves with black dots (glands) on the lower surface,
obovate-oblong . . . .*Vaccinium*
41. Leaves lacking glands, obovate-spatulate
42. Leaves oblanceolate, subsessile;
fruit a drupe . . . .*Arctostaphylos*
42. Leaves ovate, the petiole 1–4 mm long;
fruit a pome . . . .*Cotoneaster*
35. Stems erect
43. Stipule scars encircling the twig . . . .*Magnolia*
43. Stipule scars usually absent or not encircling the twig
44. Leaves with rusty hairs or scales on lower surface

45. Lower surface of leaves covered with dense woolly,
rusty-brown hairs . . . .*Ledum*
45. Lower surface of leaves with rusty scales, not
densely woolly . . . .*Chamaedaphne*
44. Leaves green, pale, or white on lower surface, lacking rusty
hairs or scales
46. Stipules very small, black, spine-like, persistent . . . .*Ilex*
46. Stipules absent or green and membranous, not small, black,
or spine-like
47. Leaves mostly less than 5 cm long; shrubs
48. Leaves ciliate . . . .*Pieris*
48. Leaves lacking marginal hairs
49. Leaves white on the lower surface, revolute . . . . . . *Andromeda*
49. Leaves pale green on the lower surface, revolute or
margins flattened
50. Leaves subsessile, 6–13 mm long . . . .*Leiophyllum*
50. Leaves with distinct petioles, mostly
20–50 mm long . . . .*Kalmia*
47. Leaves mostly 5–25 cm long; shrubs or trees
51. Leaves and buds scattered along twig; bark gray, yellow-brown,
brown, or black; shrubs . . . .*Kalmia*
51. Leaves or buds clustered at end of twig; bark red-brown,
brown, or black, peeling or not; tall shrubs or trees
52. Leaves scattered along branch, not clustered at twig ends,
buds clustered on the end of twigs; bark brown or black,
tight, not peeling; fruit an acorn . . . .*Quercus*
52. Leaves clustered near end of twig; buds usually
scattered along branch; bark often red-brown,
usually peeling; fruit a 5-valved capsule
53. Leaves rugulose, pale or rusty on lower surface; growth
of preceding season with numerous scale scars below the
leaves; capsule oblong . . . .*Rhododendron*
53. Leaves smooth, not rugulose, light green on lower surface;
growth of preceding season with 2 scale scars at base of
leaves; capsule subglobose . . . .*Kalmia*
34. Leaves deciduous, mostly thin, occasionally leathery, usually flat,
not revolute
54. Leaves and young twigs covered with silver or silver and brown scales;
thorns often present; fruit drupe-like . . . .*Elaeagnus*
54. Leaves and young twigs lacking silver or brown scales; thorns
lacking; fruit various
55. Leaves and bark spicy-aromatic; fruit a red or blue drupe
56. Leaves with 3 principal veins from near the base, often
lobed; bark of branchlets green; fruit blue . . . .*Sassafras*

56. Leaves with 1 principal vein (midrib), never lobed; bark of branchlets
      brown; fruit red . . . .*Lindera*
55. Leaves and bark lacking strong scent, at least not spicy-aromatic;
    fruit various
    57. Leaves with 3–7 large veins from the broadly cordate or oblique
          base; fruit a legume, drupe, or capsule
          58. Stems climbing; vines; fruit a capsule . . . .*Aristolochia*
          58. Stems erect; trees or shrubs; fruit a legume or drupe
                59. Leaf base cordate; fruit a legume . . . .*Cercis*
                59. Leaf base truncate, cuneate, or oblique; fruit a drupe . . . .*Celtis*
    57. Leaves with single large vein or midrib from base, the base
          mostly cuneate, truncate, or rounded, sometimes cordate;
          fruit various
          60. Vines or scrambling shrubs; fruit a berry or drupe
                61. Stems and branchlets shiny red-brown; leaves with margins
                      undulate, the veins prominent, straight, parallel;
                      fruit a drupe . . . .*Berchemia*
                61. Stems and branchlets dull light or dark brown; leaves
                      with margins even, the veins inconspicuous, curving;
                      fruit a berry
                      62. Leaves oblong or lanceolate with cuneate base; leaves and
                            stems lacking strong odor; stems often with short
                            thorn-like branches . . . .*Lycium*
                      62. Leaves ovate to elliptic, the base often cordate or lobed;
                            stems lacking thorn-like branches
                            63. Crushed leaves and stems with strong, disagreeable odor; short
                                  scrambling shrub or vine; leaf base often
                                  cordate or lobed . . . .*Solanum*
                            63. Leaves and stems lacking strong odor; high climbing vine; leaves
                                  narrowed at the base . . . .*Schisandra*
          60. Erect shrubs or trees; fruit various
                64. Leaves rhombic or deltoid, abruptly but long acuminate, with glands
                      at the base of the blade; southern . . . .*Sapium*
                64. Leaves shaped variously but not rhombic or deltoid,
                      obtuse to acuminate, usually lacking glands at the base of
                      the blade
                      65. Twigs and leaves with milky sap; thick, stout thorns usually
                            present; fruit multiple, globose, large (to 12 cm in diameter), heavy,
                            yellow-green . . . .*Maclura*
                      65. Twigs and leaves with clear sap; thorns lacking (spines
                            or thorns present in *Berberis* and *Bumelia*); fruit various
                            66. Stems with branched or simple spines or thorns, leaves
                                  often fascicled; fruit red, purple, or black, berrylike
                                  or a drupe

67. Leaves usually less than 7 cm long, often spiny on the margins; inner bark and wood yellow; fruit red ....*Berberis*
67. Leaves usually greater than 7 cm long, lacking marginal spines; inner bark brown; fruit black; southern ....*Bumelia*
66. Stems lacking spines; leaves usually scattered along the branch; fruit various
  68. Base of petiole covering the axillary bud or stipules sheathing and enclosing the stem; fruit a capsule or drupe
    69. Leaves with membranous, sheathing stipules; herb to 3 m tall that has appearance of a shrub ....*Polygonum*
    69. Leaves lacking membranous, sheathing stipules; shrubs or trees
      70. Leaves 10–25 cm long; bark tight, thin, not fibrous or leathery; fruit a capsule ....*Styrax*
      70. Leaves 5–7 cm long; bark fibrous and leathery; fruit a drupe ....*Dirca*
  68. Base of petiole solid, not hollow, subtending the exposed axillary lateral bud; fruit various
    71. Bark of stems and branches fibrous and leathery; twigs very flexible, never snapping (even when bent back 180 degrees or more); fruit a drupe ....*Daphne*
    71. Bark of stems and branches tight, smooth or rough, thin, not leathery; twigs stiff or brittle, at least not so flexible; fruit various
      72. Lateral veins running parallel to the margins of the leaf and ending near the apex; white vein threads stretched and visible when leaf is severed in cross section; fruit a drupe ....*Cornus*
      72. Lateral veins fading before reaching the margin of the leaf or not running to the apex; leaf breaking clean, with no threads of veins evident; fruit various
        73. Pith chambered (divided by woody plates or diaphragms)
          74. Leaves 20–30 cm long; buds dark, silky; fruit a very large berry ....*Asimina*
          74. Leaves 5–15 cm long; buds brown, glabrous, not silky; fruit a drupe or berry
            75. Leaf veins, especially the midrib on upper surface, often pink or pale; buds with only 2 scales visible; fruit a berry ....*Diospyros*
            75. Leaf veins green to brown, not pink or pale; buds with more than 2 scales visible; fruit a drupe ....*Nyssa*
        73. Pith solid or hollow, not chambered
          76. Stipules and stipule scars encircling the twig; leaves 10–30 cm long, acuminate ....*Magnolia*

76. Stipule scars absent or not encircling the twig; leaves mostly less than 10 cm long
  77. Stipules small, black, spine-like; fruit a red berry . . . .*Ilex*
  77. Stipules, if present, small or large, foliar or membranous, green or brown, not spine-like or black; fruit a berry, capsule, drupe, legume, pome, or acorn (achene in *Solidago*)
    78. Buds clustered at the end of twigs (In the absence of buds in *Rhododendron* this condition ascertained by the falsely whorled arrangement of twigs and leaves.); fruit a capsule or acorn
      79. Leaves mostly clustered near the end of twigs, thin, the margins ciliate; shrubs; fruit an elongated capsule on a glandular-pubescent pedicel . . . .*Rhododendron*
      79. Leaves scattered on the twigs, thick, the margins entire or lobed, not ciliate; trees; fruit an acorn . . . .*Quercus*
    78. Buds scattered along twigs, not clustered at twig tips; fruit a capsule, drupe, legume, berry, or pome
      80. Buds each covered by a single hood-like scale; fruit a capsule . . . .*Salix*
      80. Buds each covered by at least 2 scales; fruit a capsule, drupe, legume, berry, or pome
        81. Leaf blades mostly 8–20 cm long
          82. Exposed bud scales 2; pith often chambered (divided by diaphragms or plates); bark breaking into regular blocks on older trunks; fruit a berry . . . .*Diospyros*
          82. Exposed bud-scales several, imbricated; pith continuous; bark smooth, scaly, or ridged; fruit a drupe or pome
            83. Leaf blades less than twice as long as wide, usually cordate or obtuse at base; fruit a pome . . . .*Pyrus*
            83. Leaf blades greater than twice as long as wide, tapering at both ends; fruit a drupe . . . .*Pyrularia*
        81. Leaf blades 1–10 cm long
          84. Petioles 1–4 cm long; fruit a pome or drupe-like, red or yellow
            85. Leaves often mucronate; fruit drupe-like, red . . . .*Nemopanthus*
            85. Leaves obtuse or rounded; fruit a pome or drupe-like, yellow or red
              86. Leaves ovate to oblong, pubescent on lower surface, apex obtuse; fruit a pome, yellow . . . .*Cydonia*
              86. Leaves obovate, glabrous, apex rounded; fruit drupe-like, red . . . .*Cotinus*
          84. Petioles less than 1 cm long or absent; fruit a capsule, drupe, berry, or legume

87. Lower surface of leaves covered with yellow, resinous dots; fruit a
    capsule or drupe
  88. Leaves when crushed sweet-scented, often serrate near apex; fruit a waxy,
      gray to blue drupe . . . . *Myrica*
  88. Leaves odorless or at least not sweet-scented; fruit a capsule or
      blue to black drupe, lacking waxy coating
    89. Leaves glabrous or lightly pubescent; fruit a capsule . . . . *Lyonia*
    89. Leaves glandular; fruit a fleshy drupe . . . . *Gaylussacia*
87. Lower surface of leaves glabrous or pubescent, lacking yellow,
    resinous dots; fruit a capsule, drupe, berry, or legume
  90. Branchlets green or red-brown, minutely white-speckled or hairy, or
      the buds subglobose and spreading; fruit a berry . . . . *Vaccinium*
  90. Branchlets brown, gray, or black, not minutely white-speckled;
      buds ovoid to ellipsoid; fruit a capsule, legume, drupe, or berry
    91. Leaves greater than 5 cm long, the petiole mostly greater than
        3 mm long; fruit a capsule or drupe
      92. Twigs glaucous; leaves subentire, glaucous on lower surface; fruit a
          capsule; southern . . . . *Zenobia*
      92. Twigs glabrous or pubescent, lacking bloom; leaves entire,
          glabrous or pubescent, lacking bloom on lower surface;
          fruit a capsule or drupe
        93. Leaves $2^{1}/_{2}$–3 times longer than wide, the secondary veins
            inconspicuous, or the tertiary and ultimate veins as prominent
            as the secondary veins; bark of branchlets gray and shredding;
            fruit a dry indehiscent drupe . . . . *Cyrilla*
        93. Leaves less than $2^{1}/_{2}$ times longer than wide, the secondary
            veins obvious; bark of branchlets black, gray-brown, or
            tan, shredding or not; fruit a drupe or capsule
          94. Bark of branchlets black, tight, not shredding; leaves usually with
              6–7 pairs of prominent lateral veins; fruit a drupe . . . . *Rhamnus*
          94. Bark of branchlets gray or brown, shredding; leaves usually with
              4–6 pairs of less prominent lateral veins; fruit a globular or urn-
              shaped capsule in naked racemes . . . . *Lyonia*
    91. Leaves 1–4 cm long, sessile or the petiole less than 3 mm long;
        fruit a legume, capsule, or berry
      95. Branches striate, angled; leaves lanceolate; fruit a legume . . . *Genista*
      95. Branches uniform, terete; leaves ovate or elliptic; fruit a
          capsule or berry
        96. Leaves narrowly oblong, leathery; fruit a berry; southern . . . *Litsea*
        96. Leaves elliptic to ovate, thin; fruit a capsule
          97. Shrubs, about 1 m tall; leaves brown on lower surface,
              the veins raised in an alligator-skin pattern on both
              surfaces; southern . . . . *Solidago*
          97. Large shrubs, greater than 1 m tall; leaves pale or green on lower
              surface, the veins inconspicuous . . . . *Menziesia*

30. Leaves toothed
   98. Leaves toothed only at the rounded apex
      99. Leaves attenuate . . . .*Myrica*
      99. Leaves cuneate or rounded at the base . . . . *Ilex*
   98. Leaves toothed at least halfway toward the base
      100. Leaves with hard, sharp spines terminating the teeth
         101. Leaves scattered along the branchlet, shiny, not fibrous, less
            than 10 cm long . . . .*Ilex*
         101. Leaves densely overlapping, dull, tough, fibrous, greater than
            20 cm long . . . .*Yucca*
      100. Leaves toothed but lacking hard, sharp spines
         102. Plants with spines or thorns on the stems and branches
            103. Leaves mostly less than 5 cm long; twigs, branchlets, and branches
               bearing stipular or branched spines; inner bark and wood yellow;
               fruit a red berry . . . .*Berberis*
            103. Leaves mostly greater than 5 cm long; twigs and
               branches bearing thorns; inner bark and wood brown
               or pale; fruit a drupe or pome
               104. Petioles with glands near the blade end;
                  fruit a drupe . . . .*Prunus*
               104. Petioles lacking glands near the blade end; fruit
                  a pome
                  105. Thorns present on twigs and branches; thorns lacking lateral
                     buds and leaves . . . .*Crataegus*
                  105. Thorns present on branches only; thorns usually leafy or
                     with lateral buds . . . .*Pyrus*
         102. Plants lacking spines and thorns on stems and branches
           106. Stems climbing or twining; vines; fruit a capsule
              or berry
             107. Lateral veins of leaves straight and parallel; leaves serrulate, the
               petiole less than 5 mm long; fruit a berry . . . .*Berchemia*
             107. Lateral veins curved, not straight; leaves crenate-
               serrate, the petiole greater than 5 mm long; fruit a
               berry or capsule
               108. Leaves cordate; fruit a berry . . . .*Ampelopsis*
               108. Leaves cuneate to attenuate; fruit a berry or capsule
                  109. Leaves crenate; fruit a yellow or
                    orange capsule . . . .*Celastrus*
                  109. Leaves undulate-denticulate; fruit a
                    red berry . . . .*Schisandra*
           106. Stems erect; shrubs or trees; fruit various
             110. Stems low, 10–20 cm tall, almost herbaceous, from
               subterranean creeping stems; leaves persistent,
                glossy, coriaceous, clustered near the end of the
               season's growth

111. Leaves ovate, with wintergreen flavor; fruit red, berrylike ....***Gaultheria***
111. Leaves oblanceolate or with main veins of lighter green, lacking
    wintergreen flavor; fruit a capsule ....***Chimaphila***
110. Stems or trunks greater than 20 cm tall; leaves various
  112. Leaves with 3–5 main veins from near the base, the lateral 2–4
      nearly equal to the midrib or at least thicker and more
      prominent than other lateral veins
    113. Low, much-branched, almost herbaceous shrubs;
        fruit a capsule ....***Ceanothus***
    113. Trees or tall shrubs; fruit a capsule, drupe, or multiple
        and fleshy
      114. Sap milky; fruit multiple and fleshy
        115. Leaves glossy and smooth on upper surface ....***Morus***
        115. Leaves rough or scabrous on upper surface
          116. Leaf base oblique, the lower leaf surface densely
              pubescent, like velvet ....***Broussonetia***
          116. Leaf base symmetrical, the lower leaf surface pubescent, even
              densely so, but not like velvet ....***Morus***
      114. Sap clear, not milky; fruit a capsule or drupe
        117. Leaves with deltoid, truncate, or cordate, symmetrical base, the
            petiole terete or flattened; fruit a capsule ....***Populus***
        117. Leaves with oblique base, the petiole terete; fruit a drupe
          118. Leaves about as wide as long, cordate, acute or short-cuminate;
              pith solid, not chambered; bark of trunk smooth or furrowed,
              lacking corky ridges; fruit a small nut-like drupe attached to a
              winglike leaf ....***Tilia***
          118. Leaves longer than wide, cuneate to truncate, long-acuminate
              or taper-pointed; pith chambered or not; bark of trunk with
              high, corky ridges; fruit a drupe, winged or not ....***Celtis***
  112. Leaves with only 1 main vein (midrib) from the base that is
      significantly thicker than other veins
    119. Buds distinctly stalked; leaves nearly as wide as long
      120. Leaves serrate or doubly serrate, sometimes undulate, the base
          symmetrical, not oblique; buds club-shaped, not woolly, often
          somewhat sticky; fruit of small nutlets in persistent cone-like
          woody structure ....***Alnus***
      120. Leaves undulate, with coarse broad teeth or with veins
          extended into abrupt linear teeth, the base oblique; buds
          sickle-shaped and woolly; fruit a 2-celled woody capsule
        121. Leaves coarsely toothed in apical half ....***Fothergilla***
        121. Leaves undulate or toothed to well below
            the middle ....***Hamamelis***
    119. Buds sessile, not stalked (except flower buds of
        Chamaedaphne); leaves longer than wide

122. Leaf base oblique
   123. Leaves mostly doubly toothed; the teeth lacking glands;
      fruit a samara . . . .*Ulmus*
   123. Leaves singly toothed; the teeth gland-tipped; fruit a drupe . . . . .*Planera*
122. Leaf base symmetrical or slightly oblique
   124. Leaves subopposite, secondary veins prominently curving to
      be roughly parallel to the margin; buds naked;
      fruit a drupe . . . .*Rhamnus*
   124. Leaves distinctly alternate, secondary veins usually
      inconspicuous, not running parallel to the margin; buds with
      scales; fruit various
      125. Buds covered by a single hood-like scale; fruit a capsule . . . .*Salix*
      125. Buds covered by 2 or more scales; fruit various
         126. Leaves singly toothed, often coarsely dentate or serrate,
            never doubly toothed
            127. Leaves nearly as wide as long; petioles flattened or terete
               128. Petioles flattened or marginal teeth on the blade incurved;
                  plant lacking thorns; fruit a capsule . . . .*Populus*
               128. Petioles terete, not flattened; marginal teeth on the blade
                  straight or arched, not incurved; plant usually thorny;
                  fruit a pome . . . .*Crataegus*
            127. Leaves much longer than wide; petioles terete
               129. Leaves toothed only in apical $^1/_2$ to $^2/_3$ . . . .*Baccharis*
               129. Leaves with many evenly spaced teeth, at least 1 mm
                 long, covering the full length of each margin
                 130. Terminal buds clustered; bark rough, furrowed;
                  fruit an acorn . . . .*Quercus*
                 130. Terminal buds scattered, not clustered; bark
                  smooth or rough and furrowed; fruit a prickly bur
                   131. Leaves oblong-ovate; buds long and sharp-pointed, usually
                     at least 5 times as long as wide; terminal bud present; bark
                     smooth, gray; bur with 2 triangular nuts . . . .*Fagus*
                 131. Leaves oblong-lanceolate; buds ovate and blunt, usually
                     much less than 3 times as long as wide; terminal bud
                     absent; bark rough, furrowed; bur with
                     1–3 nuts . . . .*Castanea*
         126. Leaves doubly or singly toothed, finely serrate, serrulate,
            denticulate, crenate, crenulate, or nearly entire, not
            coarsely serrate or dentate
            132. Pith chambered; fruit a capsule
               133. Leaves with stellate pubescence on the lower surface; fruit dry, a
                 4-winged capsule . . . .*Halesia*
               133. Leaves glabrous; fruit a 2-valved capsule . . . .*Itea*
            132. Pith continuous; fruit various

134. Leaf base broad, rounded, cordate, truncate, or sometimes
    slightly acute
  135. Petioles with 1 or more glands near the blade end . . . .*Prunus*
  135. Petioles lacking glands
    136. Leaves glaucous on upper surface, persistent . . . .*Pieris*
    136. Leaves lacking bloom on upper surface, deciduous or
      sometimes evergreen
      137. Leaves evenly and mostly singly toothed, serrate, dentate,
        or crenate, usually in more than 2 ranks; fruit a capsule
        or pome
        138. Leaves suborbicular; buds nearly concealed by the base of the
          enlarged petiole; fruit a capsule . . . .*Styrax*
        138. Leaves broadly ovate to lanceolate; buds exposed; fruit
          a capsule or pome
          139. Leaves broadly ovate or deltoid; petioles usually flattened;
            fruit a capsule . . . .*Populus*
          139. Leaves ovate, ovate-lanceolate, elliptic, oblong,
            oblanceolate, or obovate; petioles terete, not
            flattened; fruit a capsule or pome
            140. Leaves apiculate, strongly glaucous on lower
              surface; fruit a capsule; southern . . . .*Zenobia*
            140. Leaves acute, acuminate, obtuse, or rounded, not
              apiculate, green or pale but not glaucous on lower
              surface; fruit a pome
              141. Leaves 1–4 cm long, the lateral veins straight at least
                halfway toward the margin, sharply serrate; fruit a
                berrylike pome . . . .*Amelanchier*
              141. Leaves greater than 5 cm long, the veins curving
                from the midrib to the margin, obscurely
                crenulate or serrulate; fruit a pome
                142. Leaves with waxy, shiny coating on upper surface, the
                  petiole usually greater than 4 cm long and at least half
                  the length of the blade . . . .*Pyrus*
                142. Leaves dull, lacking waxy coating on upper surface, the
                  petiole less than 4 cm long and less than $^{1}/_{4}$ the length
                  of the blade . . . .*Malus*
      137. Leaves unevenly and mostly doubly serrate or dentate,
        mostly in 2 ranks; fruit a samara, nut, or
        appendaged nutlet
        143. Branches of mature trees with corky ridges;
          fruit a samara . . . .*Ulmus*
        143. Branches of mature trees smooth or rough but
          lacking corky ridges; fruit a samara, nut, or
          appendaged nutlet

144. Older branches with short spurs bearing crowded leaves or leaf scars and terminated by a single bud; bark on younger trunks smooth or peeling off in rolls; lenticels elongated horizontally; twigs and inner bark often with wintergreen flavor; fruit a samara, in cone-like catkins . . . .***Betula***

144. Branches lacking short spurs; bark tight, never peeling off in rolls; lenticels, if present, not prominent; twigs and inner bark lacking wintergreen flavor; fruit a nut or nutlets enclosed by a bur or appendaged by bracts

   145. Leaves ovate or ovate-oblong, not much longer than wide, acute; twigs mostly with bristly hairs; shrubs; fruit a nut within a husk-like involucre . . . .***Corylus***

   145. Leaves oblong-ovate, much longer than wide, short-acuminate, taper-pointed; twigs glabrate; trees; fruit a nutlet with bracts attached

      146. Lateral veins unbranched; buds with flattened sides; bark light gray, smooth, and sinewy-fluted; fruit a nutlet with bract-like appendage, several grouped in a flexuous spike . . . .***Carpinus***

      146. Lateral veins branched near the leaf margins; buds terete; bark light brown or gray-brown, rough, breaking off in scaly plates; fruit a nutlet enclosed in an inflated bag, several grouped in a cone-like spike . . . .***Ostrya***

134. Leaf base acute or tapering

  147. Petioles with 1 or more glands at the blade end, or the lower surface of leaf waxy and with a dense row of light brown hairs along each side of mid-rib; fruit a drupe . . . .***Prunus***

  147. Petioles without glands at the blade end, lower surface of leaf rough or shiny, not waxy, lacking rows of hairs; fruit various

   148. Terminal bud long, sharp-pointed, resinous; first bud scale of lateral buds anterior; fruit a capsule . . . .***Populus***

   148. Terminal bud absent, or if present, mostly blunt, resinous or not; first bud scale of lateral buds interior or lateral; fruit various

    149. Leaves thick, coriaceous, persistent, crenulate-undulate to crenulate-serrulate; fruit various; southern . . . .***Gordonia***

    149. Leaves thin, often deciduous, serrulate to dentate; fruit various

     150. Leaves with a sour taste, mostly greater than 10 cm long, oblong-lanceolate; fruit a 5-valved capsule, numerous in panicled racemes . . . .***Oxydendrum***

     150. Leaves lacking sour taste, mostly less than 10 cm long; if longer, then not oblong-lanceolate; fruit various

      151. Midrib of leaf with dark or black glands on the upper surface, these sometimes minute; fruit a berrylike pome . . . .***Aronia***

151. Midrib of leaf lacking dark glands on the upper surface;
    fruit various
    152. Lower leaf surface covered with yellow glands, scurf, or dots
        153. Stems with longitudinal ridges; leaves and twigs scurfy; buds globose,
            resinous; fruit an achene . . . .*Baccharis*
        153. Stems smooth, lacking longitudinal ridges; buds longer than
            wide, usually not strongly resinous; fruit a drupe, winged
            nutlet, or capsule
            154. Leaves, when crushed, sweet-scented, often entire toward base; fruit
                a waxy drupe or a winged nutlet . . . .*Myrica*
            154. Leaves lacking sweet scent; leaves toothed throughout;
                fruit a drupe or capsule
                155. Leaves coriaceous, persistent, lower surface covered with
                    yellow-brown scurf; fruit a capsule . . . .*Chamaedaphne*
                155. Leaves thin, not coriaceous, deciduous, lower surface covered with
                    yellowish resin globules; fruit a drupe . . . .*Gaylussacia*
    152. Lower surface of leaves green or pale, not covered with yellow
        glands or dots
        156. Twigs stout, mostly 3–5 mm in diameter when mature; branches often
            with stout spurs; fruit a pome . . . .*Pyrus*
        156. Twigs slender, less than 3 mm in diameter when mature;
            branches mostly lacking stout spurs; fruit a capsule, drupe,
            berry, or follicle
            157. Branchlets finely white-speckled or hairy, green or red-brown;
                fruit a berry . . . .*Vaccinium*
            157. Branchlets usually gray, brown, or black, not finely
                white-speckled or hairy; fruit a capsule, follicle,
                or drupe
                158. Leaves obovate, the base cuneate; fruit a
                    3-valved capsule . . . .*Clethra*
                158. Leaves lanceolate to ovate, not obovate, base cuneate to
                    truncate; fruit a capsule, drupe, or follicle
                    159. Stipules small, sharp, nearly black, persistent;
                        fruit a drupe . . . .*Ilex*
                    159. Stipules larger, foliar, membranous, or lacking; fruit
                        a drupe, capsule, or follicle
                        160. Leaves serrate, sometimes coarsely so, singly or
                            doubly toothed, often entire toward the base, the
                            midrib prominent the entire length of the
                            lower surface
                            161. Leaves 8–11 cm long, long acuminate; fruit a
                                5-valved capsule . . . .*Leucothoe*
                            161. Leaves less than 8 cm long, acute; fruit a drupe
                                or follicle

162. Leaves oblanceolate, the petioles 7–13 cm long; fruit a
drupe, solitary . . . .*Prunus*

162. Leaves obovate, oblanceolate, elliptic, or elliptic-ovate, the petioles less
than 5 cm long; fruit a follicle, in corymbs or panicles . . . .*Spiraea*

160. Leaves serrulate, crenulate, or subentire, singly toothed, the
midrib losing prominence toward the apex on the lower surface

  163. Leaves subentire in apical half, entire below the middle, the
teeth remotely and irregularly spaced

    164. Leaves strongly glaucous on lower surface; twigs usually glaucous; fruit
a capsule; southern . . . .*Zenobia*

    164. Leaves glabrous to pubescent, lacking a bloom; twigs glabrous or
pubescent, usually not glaucous; fruit a drupe or capsule . . . . . .*Styrax*

  163. Leaves toothed nearly to the base, the teeth evenly and
closely spaced

    165. Leaves ovate, the veins straight and prominent, the blade less than $2^1/_2$
times longer than wide; fruit a drupe . . . .*Rhamnus*

    165. Leaves elliptic, the veins curved and less impressed, the blade
greater than $2^1/_2$ times longer than wide; fruit a drupe
or capsule

      166. Leaves crenulate, acute to acuminate; fruit a capsule . . . .*Leucothoe*

      166. Leaves subentire or serrulate, obtuse, acute, or mucronate;
fruit a capsule or drupe

        167. Leaves mostly greater than 5 cm long, ovate, lanceolate (obovate in
Lyonia ligustrina), acute or mucronate; bark of branchlets gray or
light brown and shredding or shining red-brown with longitudinal
ridges; fruit a 5-valved capsule, in open clusters . . . .*Lyonia*

        167. Leaves mostly less than 4 cm long, obovate, elliptic, or elliptic-
ovate, obtuse (mucronate in *Gaylussacia frondosa*); bark of
branchlets black, close or tight; fruit a drupe . . . .*Gaylussacia*

KEY XII. PLANTS WITH ALTERNATE COMPOUND LEAVES

1. Leaves greater than 1 m long; southern . . . .*Sabal*
1. Leaves less than 1 m long

  2. Leaves bi- or tri-pinnately compound

    3. Leaflets toothed or lobed, not entire

      4. Ultimate leaflets more than 11; branches often bearing stout
thorns; fruit a legume or capsule

        5. Leaflets crenulate to subentire, less than 3 cm long; thorns often
present; fruit a legume . . . .*Gleditsia*

        5. Leaflets coarsely toothed or lobed, greater than 3 cm long; thorns
lacking; fruit a capsule . . . .*Koelreuteria*

      4. Ultimate leaflets fewer than 11; branches lacking thorns;
fruit a follicle, berry, or drupe-like

6. Vines; fruit a berry ....*Ampelopsis*

6. Trees or shrubs; fruit a follicle or drupe-like

    7. Stem prickly; wood brown; fruit drupe-like ....*Aralia*

    7. Stem smooth or rough, lacking prickles; wood brown or yellow; fruit a follicle ....*Xanthorhiza*

3. Leaflets entire

    8. Leaflets greater than 5 cm long ....*Gymnocladus*

    8. Leaflets less than 5 cm long

        9. Foliage aromatic, gray-green; fruit achenes ....*Artemisia*

        9. Foliage lacking strong odor; fruit a legume

            10. Ultimate leaflets about 20 ....*Chamaecrista*

            10. Ultimate leaflets more than 30

                11. Leaflets fewer than 50, with petiolules 1–2 mm long, the leaflet pairs 5–10 mm apart; southern ....*Sesbania*

                11. Leaflets more than 50, sessile, the leaflet pairs less than 2 mm apart ....*Albizia*

2. Leaves once compound

  12. Leaflets 3

    13. Stipules present (leaflets sometimes more than 3); plants often climbing

      14. Stems and petioles lacking prickles; leaflets lobed or not; fruit a head of achenes or legume

        15. Leaflets greater than 9 cm long, 2–3 lobed; fruit a legume ....*Pueraria*

        15. Leaflets less than 5 cm long, 0–3 lobed; fruit a legume or head of achenes

          16. Leaflets 3–5 toothed near apex, oblong-lanceolate; fruit a head of achenes ....*Potentilla*

          16. Leaflets lacking lobes, elliptic to lance-elliptic; fruit a legume; southern ....*Lespedeza*

      14. Stem or petioles bearing prickles; leaflets lacking lobes; fruit berrylike

        17. Stipules adnate to the petiole about half its length or more ...*Rosa*

        17. Stipules free, not adnate to the petiole ....*Rubus*

  13. Stipules lacking; plants usually erect

    18. Branches with prominent longitudinal ridges, green; leaves nearly sessile; fruit a legume ....*Cytisus*

    18. Branches smooth, not ridged, brown; leaves with obvious petiole; fruit a drupe or samara

      19. Lateral buds visible in the axils of leaves; lateral leaflets symmetrical or not; stems with or without aerial rootlets; fruit a drupe ....*Toxicodendron*

19. Lateral buds imbedded in the bark, not visible; lateral leaflets nearly symmetrical; stems lacking aerial rootlets; fruit a drupe or samara

    20. Petioles 1–3 cm long; leaflets crenate; fruit a red drupe . . . .***Rhus***

    20. Petioles 5–10 cm long; leaflets entire or crenulate; fruit a samara . . .***Ptelea***

12. Leaflets more than 3

  21. Leaves palmately compound

    22. Stems with spines or prickles; plants erect; fruit a cluster of drupelets . . . .***Rubus***

    22. Stems smooth, lacking spines or prickles; high-climbing woody vines; fruit a berry

      23. Vine lacking tendrils; leaflets entire, retuse, with petiolules 3–10 mm long . . . .***Akebia***

      23. Vine with branched tendrils; leaflets serrate, acuminate, subsessile . . . .***Parthenocissus***

  21. Leaves pinnately compound

    24. Buds hidden (base of petiole hollow, forming a hood-like covering over the lateral buds or these imbedded under the base of the petiole)

      25. Leaflets small, 1–5 cm long

        26. Leaflets about 1 cm long, silky; low shrub with shreddy bark; fruit a head of achenes . . . .***Potentilla***

        26. Leaflets greater than 1 cm long, glabrous or pubescent, not silky; bark tight, not shredding; shrubs, or small or large trees; fruit a legume

          27. Leaflets entire; twigs usually with stipular spines or prickles; pith pink-white . . . .***Robinia***

          27. Leaflets somewhat serrate or crenate; twigs usually with simple or branched thorns; pith salmon-colored . . . . .***Gleditsia***

      25. Leaflets large, 5–10 cm long

        28. Leaflets alternate, ovate; sap clear, not milky; fruit a legume . . . .***Cladrastis***

        28. Leaflets opposite or subopposite, ovate to lanceolate or oblong; sap milky or clear; fruit a drupe

          29. Leaflets narrowly oblong to lanceolate; fruit red, glandular-pubescent . . . .***Rhus***

          29. Leaflets ovate, obovate, suborbicular, or elliptic; fruit yellow to white, glabrous or sparsely pubescent; poisonous . . . .***Toxicodendron***

    24. Buds evident, not imbedded under or surrounded by base of petiole

      30. Spines or thorns present on the stems and often on the midrib of leaves

31. Stipules lacking; leaflets dotted with pellucid glands; wood yellow;
    fruit a capsule . . . .*Zanthoxylum*
31. Stipules present; leaflets lacking pellucid glands; wood brown;
    fruit berrylike (an aggregate of drupelets) or a hip (berry)
    32. Stipules adnate to the petiole half its length or more; leaflets
        evenly serrate . . . .*Rosa*
    32. Stipules free from the petiole; leaflets usually unevenly coarsely
        toothed or often doubly serrate . . . .*Rubus*
30. Spines, thorns, and prickles absent
    33. Leaflets toothed or lobed
        34. Leaflets often more than 20, with rank odor when crushed, with
            glands on the lower surface of the small basal lobes;
            fruit a samara . . . .*Ailanthus*
        34. Leaflets usually fewer than 20, odiferous but not rankly so,
            glabrous, pubescent, or glandular-hairy but lacking
            glandular basal lobes; fruit a follicle, nut, or
            berrylike pome
            35. Petiole base clasping the twig; wood yellow; fruit
                a follicle . . . .*Xanthorhiza*
            35. Petiole base not or only partially enclosing the twig; wood
                brown to nearly black; fruit a nut or berrylike pome
                36. Stipules present, although sometimes ephemeral but then buds
                    red; leaves glabrous or pubescent, not glandular-hairy, lacking
                    resinous odor; branches and twigs light and slender; fruit a
                    berrylike pome . . . .*Sorbus*
                36. Stipules lacking; buds yellow, brown, or black; leaves
                    often glandular-hairy, often with strong resinous odor;
                    branches and twigs heavy and stout; fruit a nut
                    37. Pith chambered; husk of nut indehiscent . . . .*Juglans*
                    37. Pith solid; husk of nut dehiscing into 4 valves . . . .*Carya*
    33. Leaflets entire or with small glandular lobes at the base
        38. Stems twining, vines; leaves with rank odor, the leaflets with glandular
            lobes at the base; fruit a legume . . . .*Wisteria*
        38. Stems erect, shrubs or trees; leaves usually lacking strong odor,
            the leaflets entire, lacking small lobes; fruit a legume, nut,
            drupe, or head of achenes
            39. Leaflets subtended by stipels; fruit a legume . . . .*Amorpha*
            39. Leaflets lacking subtending stipels; fruit a nut, drupe, or
                head of achenes
                40. Leaflets 1–3 cm long; short or tall shrubs . . . .*Potentilla*
                40. Leaflets greater than 4 cm long, glabrous; trees or
                    tall shrubs
                    41. Pith chambered; fruit a large brown nut . . . .*Juglans*
                    41. Pith solid; fruit a drupe . . . .*Rhus*

KEY XIII. PLANTS WITH OPPOSITE OR WHORLED LEAF SCARS

1. Stems climbing or twining; vines
   2. Stems with 6 or more prominent longitudinal ridges, often nearly herbaceous; fruit a cluster of hairy achenes . . . .***Clematis***
   2. Stems smooth, lacking prominent longitudinal ridges; fruit a capsule or berry
      3. Bundle scars in a closed or nearly closed ring; fruit a large fusiform capsule
         4. Stems often with aerial rootlets at the nodes; tendrils absent . . . .***Campsis***
         4. Stems lacking aerial rootlets; leaf tendrils sometimes persisting . . . .***Bignonia***
      3. Bundle scars in a crescent-shaped line, usually 3; fruit a capsule or berry
         5. Stems climbing by aerial rootlets; fruit a capsule; southern . . . .***Decumaria***
         5. Stems twining, lacking aerial rootlets; fruit a berry . . . .***Lonicera***
1. Stems erect; trees or shrubs
   6. Buds and twigs covered with silver to brown scales; shrubs . . . .***Shepherdia***
   6. Buds and twigs glabrous or pubescent, lacking scales; trees
      7. Branchlets bearing short or spur shoots 2–15 mm long, with crowded whorls of leaf scars or persistent leaf sheaths; female reproductive structure a cone . . . .***Larix***
      7. Branchlets lacking spur shoots with whorls of leaf scars or persistent leaf sheaths; female reproductive structure a fruit
         8. Bundle scars 1, or appearing as 1 (sometimes numerous and almost confluent, thus forming a transverse, lunate, or U-shaped line) (alternate step 8, p. 109)
            9. Leaf scars mostly subopposite . . . .***Rhamnus***
            9. Leaf scars mostly opposite
               10. Leaf scars strongly decurrent or raised; fruit a capsule
                  11. Bud scales fleshy; terminal bud usually absent; twigs gray to brown, lacking wings . . . .***Syringa***
                  11. Bud scales thin, not fleshy; terminal bud present; twigs pale green, bright green, or red-brown, sometimes winged . . . .***Euonymus***
               10. Leaf scars scarcely, or not at all, raised or decurrent; fruit various
                  12. Bud scales very loose and open; buds never sunken in the bark; bark exfoliating; low shrubs; fruit a capsule . . . .***Hypericum***
                  12. Bud scales mostly close and firm or else absent; buds occasionally sunken in the bark; bark mostly

Keys to Genera: Abridged

close, not separating; fruit a berry, drupe, capsule, samara,
or nutlets

13. Opposing leaf scars connected by a distinct raised line formed by
the stipular scars

   14. Buds submerged in the bark; branchlets usually light brown or pink-brown
   with close bark, the pith white; leaf scars often whorled, not raised on
   bases of petioles; bundle scars U-shaped; fruit a head
   of nutlets . . . .*Cephalanthus*

   14. Buds axillary; branchlets red-brown, often with shredding bark, pith light
   brown or white; leaf scars strictly opposite, raised on persistent bases of
   petioles; bundle scars nearly circular; fruit a berry . . . . . .*Symphoricarpos*

13. Opposing leaf scars separate, not connected by a distinct stipular
line (an indistinct line present in some species of *Fraxinus*)

   15. Buds often superposed, naked, or the smaller ones covered with a pair of
   nearly valvate scales; twigs usually scurfy . . . .*Callicarpa*

   15. Buds standing alone, not superposed, covered with several scales;
   twigs glabrous or pubescent

      16. Twigs bright green or red-brown, often glaucous, usually 4-angled; trees
      or shrubs . . . .*Euonymus*

      16. Twigs brown, gray, black, or occasionally olive, not
      glaucous, terete

         17. Twigs sometimes short, stiff, horizontal, abruptly
         terminating, or terminating in a spine; buds appressed,
         black or gray, acute; fruit a drupe

            18. Twigs dark gray to black, some terminating in a spine;
            buds black . . . .*Rhamnus*

            18. Twigs mostly light gray or light brown, often abruptly
            terminating; buds gray . . . .*Forestiera*

         17. Twigs variable, not stiff, horizontal, or terminating abruptly
         or in a spine; buds divaricate, variously colored, obtuse or
         acute; fruit a samara, capsule, berry, or drupe

            19. Twigs slender, less than 2 mm in diameter; fruit
            a berry . . . .*Ligustrum*

            19. Twigs stout, greater than 2 mm in diameter; fruit a
            samara, drupe, or capsule

               20. Buds scurfy, brown or black; bundle scars often almost
               separate and numerous, forming a long U-shaped line; trees;
               fruit a samara . . . .*Fraxinus*

               20. Buds glabrous or scarcely pubescent, not scurfy, brown
               or light brown; bundle scars forming a straight or curved
               line; shrubs or small trees; fruit a capsule or drupe

                  21. Bud scales fleshy, green or red-brown; twigs glabrous;
                  fruit a capsule . . . .*Syringa*

                  21. Bud scales thin, not fleshy, brown; twigs usually hirsute;
                  fruit a drupe . . . .*Chionanthus*

8. Bundle scars several, separate
  22. Bundle scars in a closed, or nearly closed, ellipse (*Staphylea* sometimes has as few as 4 bundle scars, which form an ellipse when connected by a line.)
    23. Stipule scars present; ellipse of bundle scars transverse; leaf scars broadly crescent-shaped; older twigs finely white-striped . . . .***Staphylea***
    23. Stipule scars absent; ellipse of bundle scars longitudinal; leaf scars orbicular; older twigs solid brown, gray, or black (*Broussonetia* sometimes has opposite leaf scars but has milky sap.)
      24. Pith chambered or hollow; ellipse of bundle scars not quite closed; leaf scars opposite, fruit a short capsule . . . .***Paulownia***
      24. Pith solid; ellipse closed; leaf scars opposite or whorled; fruit a long capsule . . . .***Catalpa***
  22. Ellipse of bundle scars open, the 3 or more bundle scars quite distinct, in a lunate or U- or V-shaped line, or the leaf scar C-shaped and nearly closed around the bud
    25. Buds sunken in the bark or imbedded under the leaf scars (usually bursting through in late winter); shrubs
      26. Twigs spicy-aromatic . . . .***Calycanthus***
      26. Twigs lacking odor
        27. Buds imbedded in the bark, superposed; twigs branched and striated . . . .***Iva***
        27. Buds imbedded under the leaf scars; twigs unbranched, of uniform color . . . .***Philadelphus***
    25. Buds axillary; shrubs or trees
      28. Stems with hollow pith between the nodes; twigs gray or gray-white; fruit a berry . . . .***Lonicera***
      28. Stems with solid pith; twigs black, brown, red, blue, green, or gray; fruit various
        29. Bud scales of axillary buds solitary or in 0–3 pairs (sometimes 1–2 pairs of extra bracteoles beneath)
          30. Bud scale 1 . . . .***Salix***
          30. Bud scales lacking or in 1–3 pairs
            31. Buds lacking scales densely tomentose, the foliage leaves serving as bud scales; shrubs; fruit drupe-like
              32. Buds small and slender, usually less than 3 mm long . . . .***Rhamnus***
              32. Buds large and stout, usually greater than 3 mm long . . . .***Viburnum***
            31. Buds with scales, pubescent, glabrous, or silky, not tomentose; shrubs or trees; fruit various
              33. Buds scurfy, linear-lanceolate, often curved; fruit drupe-like . . . .***Viburnum***

33. Buds glabrous or pubescent, not scurfy, ovoid to lanceolate;
    fruit a samara, drupe, or drupe-like
    34. Junction of the upper leaf scars forming a raised projection; bud scales
        in 3 pairs; twigs olive, red-brown, or blue, often glaucous or polished;
        fruit a samara . . . . *Acer*
    34. Junction of the leaf scars in plane with the twig, not
        projecting, often notched; bud scales in 1–3 pairs; twigs
        variously colored; fruit a samara, drupe, or drupe-like
        35. First pair of bud scales shorter than the bud; shrubs . . . . . *Viburnum*
        35. First pair of bud scales as long as the bud (until swelling
            begins); trees or shrubs
            36. Bud scales 2; a pair of petiole bases persisting about the terminal
                buds; fruit a drupe . . . . *Cornus*
            36. Bud scales 2 pairs (only 1 pair exposed); lacking persistent
                petiole bases; fruit drupe-like or a samara
                37. Second pair of bud scales hairy; twigs pubescent or older bark
                    white-striped; shrubs or trees; fruit a samara . . . . *Acer*
                37. Second pair of bud scales glabrous or glutinous; twigs glabrous;
                    older bark solid gray or black, not white-striped; shrubs;
                    fruit drupe-like . . . . *Viburnum*
29. Bud scales of axillary buds 4 to many pairs
    38. Bundle scars usually 5, sometimes many more; leaf scars broad;
        twigs stout
        39. Terminal bud present; pith small; trees; fruit a capsule with large
            nut-like seeds . . . . *Aesculus*
        39. Terminal bud absent; pith very large; shrubs;
            fruit berrylike . . . . *Sambucus*
    38. Bundle scars usually 3, rarely 5; leaf scars narrow; twigs stout
        or slender
        40. Opposing leaf scars connected by a decurrent, usually hairy ridge;
            shrubs; fruit a capsule . . . . *Diervilla*
        40. Opposing leaf scars separate, lacking connecting decurrent
            ridge; trees or shrubs; fruit a berry, capsule, or samara
            41. Bud scales of previous year's buds deciduous; trees or shrubs
                42. Upper margin of leaf scar strongly concave; inner bark brown,
                    sweet, at least not bitter; fruit a samara; trees . . . . *Acer*
                42. Upper margin of leaf scar nearly straight; inner bark yellow,
                    bitter; fruit a berry; shrubs or small trees . . . . *Rhamnus*
            41. Bud scales of previous year's buds persisting at the base
                of twigs; shrubs
                43. Leaf scars opposite; buds often superposed; fruit
                    a berry . . . . *Lonicera*
                43. Leaf scars whorled or opposite; buds separate, not superposed;
                    fruit a capsule . . . . *Hydrangea*

KEY XIV. PLANTS WITH ALTERNATE LEAF SCARS

1. Trees with fibrous bark; female reproductive structure a globose cone ....*Taxodium*
1. Trees, shrubs, or vines; bark mostly not fibrous; female reproductive structure a fruit or an ovoid drupe-like cone
   2. Leaf scars, except on young shoots, densely clustered in rings on short spur-like branches; bundle scars 1 or 2; bark often resinous ....*Larix*
   2. Leaf scars mostly scattered along the twigs; bundle scars 1 to many; bark usually not resinous
      3. Stem brier-like, green, often prickly; petiole base persisting; climbing shrubs; vascular bundles scattered in stem cross section; fruit a berry; monocotyledons ....*Smilax*
      3. Stems usually not brier-like, green, or prickly; petiole base usually deciduous; shrubs, trees, or climbing or twining vines; stem vascular bundles in a ring; fruit various; dicotyledons
         4. Stems climbing or twining; vines (alternate step 4, p. 112)
            5. Tendrils present; fruit a berry
               6. Woody partitions through the brown pith usually present at the nodes; outer bark usually forming loose strips ....*Vitis*
               6. Woody partitions absent at the nodes; pith continuous; outer bark solid
                  7. Pith white ....*Ampelopsis*
                  7. Pith brown ....*Parthenocissus*
            5. Tendrils absent; fruit a drupe, berry, capsule, or legume
               8. Stipule scars with short, dense fringe of white hairs at the margin, large, mostly broader than the leaf scar; branchlets with short white and long yellow hairs; pith chambered ....*Pueraria*
               8. Stipule scars, if present, pubescent or glabrous at the margin, small; branchlets glabrous or uniformly pubescent; pith solid or chambered
                  9. Buds in slightly supra-axillary depressions or hidden; leaf scars usually orbicular; fruit a drupe
                     10. Twigs pubescent; fruit red; southern ....*Cocculus*
                     10. Twigs glabrous or nearly so; fruit blue or black
                        11. Stems slender, green, often suffrutescent; drupe less than 1 cm in diameter, blue ....*Menispermum*
                        11. Stems stout, woody; drupe at least 2 cm long, black; southern ....*Calycocarpum*
                  9. Buds axillary, not in depressions; leaf scars variously shaped
                     12. Stems prickly; leaf scars very narrow ....*Rosa*
                     12. Stems smooth or rough, lacking prickles (thorns often present in *Lycium*)

13. Bundle scars 3–7, distinct
    14. Twigs gray, twining; buds glabrous . . . .*Akebia*
    14. Twigs green or brown, twining or not; buds tomentose
        15. Twigs green, twining; buds clustered; fruit a capsule . . . *Aristolochia*
        15. Twigs brown, climbing by rootlets; buds solitary; fruit a white
            drupe; poisonous . . . .*Toxicodendron*
13. Bundle scars 1, or several confluent scars appearing as 1
    16. Pith hollow; stem often almost herbaceous, green or nearly white,
        with a strong odor; twigs angled; leaf scars raised; fruit a
        red berry . . . .*Solanum*
    16. Pith solid or chambered; stem woody, brown, lacking strong
        odor; twigs terete; leaf scars more or less flush with plane of
        twig; fruit a legume, berry, or capsule
        17. Buds silky; leaf scars often 2-horned on the lower side, projecting, not
            decurrent; fruit a legume . . . .*Wisteria*
        17. Buds glabrous or nearly so; leaf scars raised or not, decurrent
            or not, lacking horns or projections; fruit a berry or capsule
            18. Stems ridged, often thorny; buds blunt, often clustered; fruit
                a red berry . . . .*Lycium*
            18. Stems terete, not ridged, not thorny; buds more pointed,
                solitary; fruit a berry or capsule
                19. Pith dark brown . . . .*Schisandra*
                19. Pith white or light brown
                    20. Buds divaricate; leaf scars in plane of twig, not raised;
                        pith white; fruit an orange capsule with red pulp around
                        the seeds . . . .*Celastrus*
                  20. Buds appressed; leaf scars raised; pith light brown; fruit
                      a black berry . . . .*Berchemia*
4. Stems mostly erect trees or shrubs, not climbing or twining,
  rarely prostrate
    21. Branchlets slender, red-brown to purple brown, the most obvious scars
        being twig scars that appear as leaf scars and often subtended by a
        persistent scalelike leaf . . . .*Tamarix*
    21. Branchlets stout or slender, variously colored, the obvious scars
        leaf scars, thus lacking a subtending leaf or leaf scar
        22. Plants procumbent or low shrubs to 1 m tall; leaf scars prominently
            raised, with persistent, hardened stipules; branchlets green, with
            prominent longitudinal ridges; fruit a legume . . . .*Genista*
        22. Trees or shrubs, usually not procumbent; leaf scars raised or
            not, the stipules deciduous or not; branchlets variously
            colored, ridged or not; fruit various
            23. Bundle scar 1 (sometimes spread out into a transverse line)
               (alternate step 23, p. 115)
               24. Buds yellow-brown or rusty, often stalked, often
                  superposed . . . .*Styrax*

24. Buds brown, sessile, superposed, or solitary and distant
  25. Young twigs covered with red-brown scales or minutely
    stellate-pubescent, or buds naked
    26. Bud scales lacking; young twigs minutely stellate-pubescent;
      fruit a capsule . . . .*Clethra*
    26. Bud scales present; young twigs covered with red-brown scales;
      fruit a drupe . . . .*Elaeagnus*
  25. Young twigs pubescent or glabrous, not covered with scales, not
    stellate-pubescent; buds usually with scales
    27. Stipules or stipule scars present; shrubs
      28. Stipules deciduous
        29. Buds superposed; twigs often striate with gray and green or brown
          stripes; fruit a legume . . . .*Amorpha*
        29. Buds solitary, not superposed; twigs yellow or brown,
          lacking striations; fruit of 3 nutlets or multiple in
          a large ball
          30. Low, weak, unarmed shrubs, dying nearly to the ground each
            winter; buds above the leaf scar; fruit of 3 small nutlets, the
            bases usually persisting . . . .*Ceanothus*
          30. Trees, the branches bearing thorns; buds lateral to the thorns;
            fruit multiple, forming a large ball (Compare also with *Celtis,* in
            which bundle scars are fused into 1.) . . . .*Maclura*
      28. Stipules persistent (often very small)
        31. Stipules large, sheathing the stem; bark brown, shredding; fruit of
          many hairy achenes . . . .*Potentilla*
        31. Stipules minute, divergent, not sheathing the stem;
          bark smooth, rough or shredding; fruit a legume
          or berry
          32. Twigs bright green, ridged, angled, or winged, wand-like,
            usually dying at the tip; fruit a legume . . . .*Cytisus*
          32. Twigs gray-brown or gray, mostly smooth or rough,
            lacking ridges, angles or wings, dying at the tip or not;
            fruit a berry . . . .*Ilex*
    27. Stipule scars or stipules absent; shrubs or trees
      33. Bark of branchlets and branches green; aromatic; trees; fruit a
        blue drupe . . . .*Sassafras*
      33. Bark of branchlets gray or brown to black or red, lacking
        strong odor; trees or shrubs; fruit a drupe, capsule, samara,
        follicle, berry, or drupe-like
        34. Twigs finely white-speckled or granulose, green or
          red-brown; shrubs; fruit a berry, drupe, or drupe-like
          35. Twigs and branchlets brittle or stiff, not flexible or leathery; fruit
            a red, blue, or black berry . . . .*Vaccinium*
          35. Twigs and branchlets leathery, very flexible; fruit a red or yellow
            drupe or dry and drupe-like . . . .*Daphne*

34. Twigs uniformly colored, neither white-speckled nor granulose;
      shrubs or trees; fruit a berry, follicle, samara, or capsule
  36. Trees
    37. Pith continuous
      38. Visible bud scales 4–6; fruit a 5-celled capsule borne in
          spreading racemes . . . . ***Oxydendrum***
      38. Visible bud scales 2; fruit a large berry . . . . ***Diospyros***
    37. Pith chambered
      39. Buds superposed; fruit a dry, 4-winged capsule . . . . ***Halesia***
      39. Buds solitary; fruit a samara or large berry . . . . ***Diospyros***
  36. Shrubs
    40. Buds, except on strong, young, long shoots, clustered at the end of
        twigs; terminal bud often very large; fruit a capsule . . . . ***Rhododendron***
    40. Buds scattered along twig, not clustered; terminal bud small;
        fruit a berry, follicle, or capsule
      41. Winter twigs often with red or red-brown catkin-like racemes of
          flower buds . . . . ***Leucothoe***
      41. Winter twigs lacking catkin-like racemes; fruits or flower
          buds, if present, in corymbs, fascicles, racemes, or solitary
        42. Twigs with 2 kinds of buds, the larger flower buds with several
            visible scales, the shorter buds with 2 visible scales, all red-yellow
            to yellow-brown; twigs red to red-brown, mostly pubescent; fruit
            a berry . . . . ***Gaylussacia***
        42. Twigs with only 1 kind of bud; twigs red-brown to black,
            pubescent or glabrous; fruit a follicle, berry, or capsule
          43. Visible bud scales mostly 2 or 3; fruit a capsule
            44. Shrubs 2–3 m tall; branchlets red-brown to light gray; flowers
                clustered or in axillary racemes . . . . ***Lyonia***
            44. Shrubs less than 2 m tall; branchlets light brown or
                gray to nearly black; flowers solitary, clustered, or
                in terminal racemes
              45. Branchlets gray to black; flowers in few-flowered
                  terminal corymbs . . . . ***Menziesia***
              45. Branchlets light brown; flowers solitary or in
                  clusters; southern . . . . ***Securinega***
          43. Visible bud scales several; fruit a capsule, follicle, or berry
            46. Leaf scars mostly flush with the twig; fruit a
                capsule or berry
              47. Twigs glabrous or pubescent, sometimes glaucous; shrub
                  less than 5 m tall; fruit a berry, solitary, or in clusters or
                  racemes above the leaf scar . . . . ***Vaccinium***
              47. Twigs glabrous, often glaucous; shrubs about 2 m tall;
                  fruit a capsule, in racemes of corymbs on stems lacking
                  leaf scars; southern . . . . ***Zenobia***

46. Leaf scars, at least the upper edge, projecting slightly from the
surface of the twig; fruit a berry or follicle
    48. Leaf scars very rough and uneven; bark usually bronze-brown, tending
to exfoliate; stems wand-like or much branched and recurved with
slender twigs; fruit a follicle . . . .***Spiraea***
    48. Leaf scars nearly smooth and even; bark on older branches gray,
smooth, often with a white exfoliating crust; stems much branched,
neither wand-like nor recurved; fruit a dark red berry . . .***Nemopanthus***
23. Bundle scars more than 1
    49. Bundle scars more than 3, in any arrangement except a single
lunate line (alternate step 49, p. 116)
        50. Branches and stems with thorns or prickles
            51. Branches with thick, long, sharp-pointed thorns; buds flat-topped
and depressed; branches light olive-gray; fruit multiple, large
and fleshy . . . .***Maclura***
            51. Branches with prickles, usually broad-based, short, not long
and thorn-like; buds pointed or rounded, neither flat-topped
nor depressed; branches variously colored; fruit drupe-like
                52. Twigs and branchlets with well-spaced, short, triangular spines;
trees or shrubs to 10 m tall . . . .***Aralia***
                52. Twigs and branchlets densely covered with many
lanceolate spines; shrubs or trees
                    53. Bundle scars about 20; leaf scars nearly completely encircling
the twig . . . .***Aralia***
                    53. Bundle scars about 15; leaf scars extending $^2/_3$ of the way around
the twig . . . .***Oplopanax***
        50. Branches and stems lacking thorns or prickles
            54. Leaves usually deciduous above the base, the enlarged leaf base
persisting, forming a raised leaf scar; shrubs 1–3 m tall; fruit a
legume; southern . . . . . . , ...***Sesbania***
            54. Leaves deciduous at the base, forming a simple flushed or
raised leaf scar; trees or shrubs; fruit various
                55. Stipule scars and stipules absent
                    56. Leaf scars nearly circular; twigs slender; axillary buds not visible;
flower buds terminal, clustered, similar to young catkins; low or
decumbent shrubs; fruit a drupe . . . .***Rhus***
                    56. Leaf scars inversely triangular or oblong; twigs stout;
axillary buds usually visible; flower buds scattered or
not conspicuously terminal; shrubs or trees; fruit a
drupe or nut
                        57. Bark of twigs mottled; lateral buds small, solitary; poisonous
shrubs; fruit a white or pale drupe . . . .***Toxicodendron***
                        57. Bark of twigs a uniform color, not mottled; lateral buds large,
superposed; trees; fruit a nut . . . .***Carya***

55. Stipule scars or stipules present
    58. Bud scales 1 or 2, united into a cap; fruit cone-shaped
        59. Bud scales 1, with a scar on the back; leaf scars mostly lunate; fruit
            a fleshy "cone" . . . .*Magnolia*
        59. Bud scales 2, valvate; leaf scars mostly orbicular; fruit a dry "cone,"
            at least its axis persisting . . . .*Liriodendron*
    58. Bud scales several, imbricated; fruit an acorn, nut, drupe-like,
        or multiple
        60. Terminal bud present; fruit a nut or acorn
            61. Buds linear-lanceolate, sharp-pointed, scattered along the twig;
                trees; fruit a 3-angled nut, 2 together in a spiny,
                4-valved involucre . . . .*Fagus*
            61. Buds ovoid, obtuse, or rounded, clustered near the end of the
                twig; twigs often angled or fluted; trees or large shrubs;
                fruit an acorn . . . .*Quercus*
        60. Terminal bud absent; fruit a nut, drupe-like, or multiple
            62. Visible bud scales 4 or more or catkins present; trees
                or shrubs
                63. Buds deltoid, appressed or slightly spreading; trees; fruit
                    multiple, berrylike . . . .*Morus*
                63. Buds ovoid, spreading; tall shrubs; fruit a nut with
                    an involucre . . . .*Corylus*
            62. Visible bud scales 2 or 3; catkins absent in winter; trees
                64. Twigs scabrous, often mottled; sap milky; fruit
                  multiple, berrylike . . . .*Broussonetia*
                64. Twigs glabrous or puberulent, not mottled; sap clear;
                  fruit drupe-like or a nut
                  65. Buds and twigs red-brown or olive; twigs usually zigzagging;
                    buds with 1 scale to the side at the base, making it appear
                    lopsided; pith terete; fruit drupe-like, attached to a
                    leafy bract . . . .*Tilia*
                  65. Buds and twigs light olive-brown; twigs nearly straight; buds
                    lacking extra side scale, not lopsided; pith 5-sided or angled;
                    fruit a nut with prickly involucre . . . .*Castanea*
49. Bundle scars 3 or more in a single lunate line, or, in *Juglans*, often
   3 U-shaped groups that form a lunate line (If the leaves are not
   deciduous at the base, then the bundle scars must be counted in a
   section cut through the base of the petiole.)
    66. Pith minute, less than $^1/_{10}$ of the twig diameter; stipule scars
      so small as to be nearly invisible most of the year; leaf scars
      2-ranked; branchlets zigzagging; fruit a samara, winged nut,
      or drupe
        67. Pith chambered . . . .*Celtis*
        67. Pith solid

68. Buds globose or subovoid; twigs dark red-brown to
     light green . . . .*Planera*
68. Buds ovoid; twigs brown to black . . . .*Ulmus*
66. Pith much larger than $1/10$ the diameter of the twig; stipule scars
  present or not; leaf scars alternate or spiral but rarely 2-ranked;
  branchlets linear, not zigzagging; fruit various
   69. Stipule scars or stipules present (alternate step 69, p. 120)
     70. Branches with spines
       71. Branchlets bristly hairy; fruit a legume; southern . . . .*Mimosa*
       71. Branchlets glabrous or pubescent, not bristly;
         fruit a pome . . . .*Chaenomeles*
     70. Branches lacking spines
       72. Stipule scars extending entirely around the stem; leaf scar
         nearly encircling the bud; bud with 1 scale; bark mottled,
         flaking, pale to brown; fruit a stalked globular head of
         hairy achenes . . . .*Platanus*
       72. Stipule scars not extended or extending only partway
         around the stem; leaf scar subtending but not encircling
         the bud; buds naked or with 1 or more scales; bark usually
         not mottled, variously colored; fruit various
         73. Buds stalked
           74. Buds scurfy or glutinous; pith 3-sided; fruit
             cone-like . . . .*Alnus*
           74. Buds densely tomentose or glabrous; pith terete:
             fruit a woody capsule
             75. Buds with 2 scales, white-tomentose . . . .*Fothergilla*
             75. Buds naked, yellow-tomentose . . . .*Hamamelis*
         73. Buds sessile
           76. Buds naked . . . .*Rhamnus*
           76. Buds with 1 or more scales
             77. Bud scale 1, hood-like, the suture down the inner face of
               the bud . . . .*Salix*
             77. Bud scales more than 1
               78. Buds asymmetrical or lopsided, the first scale short and
                 bulging on 1 side . . . .*Tilia*
               78. Buds symmetrical, lacking 1 short scale to 1 side
                 79. Exposed bud scales 2 or 3 (rarely 4)
                   80. Exposed bud scales 3, the 2 lateral ones large, their
                     margins nearly meeting, the third scale carinate, the
                     swollen or keeled portion protruding between the
                     2 lateral scales; buds broadly deltoid, appressed;
                     leaf scars sometimes with a fourth or fifth bundle
                     scar above the others; fruit a capsule with 3 large
                     seeds; southern . . . .*Sapium*

Keys to Genera: Abridged

80. Exposed bud scales 2–3, of varying shapes and arrangements; bundle scars usually 3; buds globose to deltoid, appressed or divergent; fruit a legume, cone-like, or drupe-like
    81. Twigs usually with evident longitudinal ridges, with or without inconspicuous lenticels; shrubs; fruit a legume
        82. Buds superposed . . . .*Amorpha*
        82. Buds solitary, separate . . . .*Lespedeza*
    81. Twigs smooth, lacking longitudinal ridges, often with prominent lenticels; trees or shrubs; fruit a legume, cone-like, or drupe-like
        83. Buds less than 1 mm long, often wider than long; twigs light red-brown; fruit a legume . . . .*Albizia*
        83. Buds usually greater than 1 mm long, longer than wide; twigs dark brown, black, yellow, or deep purple-brown; fruit cone-like or dry and drupe-like
            84. Buds globose or ovoid, lopsided due to a short scale on 1 side; twigs zigzagging; lacking wintergreen odor or flavor; fruit drupe-like, dry, attached to a leafy bract . . . .*Tilia*
            84. Buds elongated, acute, symmetrical; twigs straight, not zigzagging (spurs with a terminal bud); often with wintergreen flavor; fruit cone-like . . . .*Betula*
79. Exposed bud scales 4 to many
    85. Shrubs, usually with sharp prickles; stems biennial; fruit an aggregate of drupelets . . . .*Rubus*
    85. Shrubs or trees with smooth or rough stems and trunks, lacking prickles; stems perennial; fruit simple
        86. Bud tips appressed, brown; pith mostly chambered; trees; fruit a drupe . . . .*Celtis*
        86. Bud tips divaricate or at least not appressed; pith continuous, not chambered; trees or shrubs; fruit various
            87. First scale of axillary bud anterior; buds often sticky; trees . . . .*Populus*
            87. First scale of axillary bud lateral; buds usually not sticky; trees or shrubs
                88. Leaf scars strongly decurrent; bark light brown, shreddy; fruit a follicle . . . .*Physocarpus*
                88. Leaf scars slightly or not at all decurrent; bark gray, black, brown, or yellow-brown, tight or shedding; fruit various
                    89. Bud scales evident in 2 vertical rows; bundle scars typically sunken in a smooth, corky layer that covers the leaf scars; catkins and spurs absent
                        90. Buds globose or short conical; twigs dark red; fruit drupe-like; growing in swamps . . . .*Planera*

90. Buds ovoid; twigs brown, not dark red; fruit a samara; growing in
diverse habitats . . . . *Ulmus*

89. Bud scales spiral, not in 2 rows; bundle scars evident at leaf scar
surface, not sunken in a corky layer; bark smooth, thin, papery,
shedding or curly, often with horizontal lines, or sinewy; catkins
and spurs present or not

  91. Growth of previous season densely covered at base with leaf scars,
the numerous short spurs on older wood with many leaf scars and
a terminal bud; bark of twigs often with wintergreen flavor;
lenticels elongated . . . . *Betula*

  91. Growth of previous season with leaf scars scattered, not
clustered at base, lacking short spurs; bark of twigs lacking
wintergreen flavor; lenticels small or lacking
(except in *Prunus*)

    92. Low weak shrubs; terminal bud present; fruit a 3-lobed capsule, the base
usually persistent . . . . *Ceanothus*

    92. Trees or shrubs; terminal bud absent; fruit various

      93. Buds globose; plants aromatic or resinous

        94. Twigs glandular, pubescent; fruit symmetrical . . . . *Comptonia*

        94. Twigs lacking glands, pubescent or glabrous;
fruit asymmetrical . . . . *Myrica*

      93. Buds ovoid; twigs glabrous or pubescent, lacking glands,
resin, and aroma

        95. Stipule scars positioned nearly between the upper edge of the leaf
scar and the twig; fruit a drupe . . . . *Prunus*

        95. Stipule scars lateral to the leaf scars; fruit various

          96. Large or small trees; buds 4-angled, striate or not,
pale brown to red-brown; bark gray or pale brown,
smooth or scaly; fruit winged nuts or samaras
in catkins

            97. Bark of trunk scaly, dark gray or red-brown; buds usually
3–7 mm long, the scales often striate; staminate catkins
usually present in winter . . . . *Ostrya*

            97. Bark of trunk smooth, sinewy-fluted, light gray; buds usually
2–4 mm long, the scales smooth, continuous, not striate;
staminate catkins usually present only in spring, not
in winter . . . . *Carpinus*

          96. Shrubs or small trees; buds terete, uniform, pale
brown to black; bark dark brown or black, smooth
or rough; fruit a pome or berrylike

            98. Bud scales dark brown or black; twigs gray or green;
fruit berrylike . . . . *Rhamnus*

            98. Bud scales red, gray, or light brown; twigs olive or
red-brown; fruit a pome . . . . *Cydonia*

69. Stipule scars and stipules absent, or, rarely, modified into spines
 99. Pith chambered or with woody partitions in solid pith
  100. Pith solid with transverse woody partitions
   101. Bundle scars 5–7; lateral buds globose, densely dark-tomentose; terminal bud naked; fruit large, berrylike . . . .***Asimina***
   101. Bundle scars 3; lateral buds ovoid, nearly or quite glabrous; terminal bud scaly; fruit a small drupe . . . .***Nyssa***
  100. Pith chambered
   102. Wood bright yellow; leaf scars almost encircling the twigs; buds solitary, not superposed; bud scales thin; low shrubs; fruit a follicle . . . .***Xanthorhiza***
   102. Wood brown to white, not bright yellow; leaf scars extending only partway around twigs; buds superposed; bud scales thick or thin; trees or shrubs; fruit a nut or capsule
    103. Buds large, at least 5 mm long, with thick scales; trees; fruit a nut . . . .***Juglans***
    103. Buds small, much less than 5 mm long, with thin scales; shrubs; fruit a 2-valved capsule . . . .***Itea***
 99. Pith continuous and homogenous
  104. Bundle scars more than 3
   105. Stems or branchlets prickly, often very stout; fruit drupe-like
    106. Bundle scars about 20; leaf scars nearly completely encircling the twig . . . .***Aralia***
    106. Bundle scars about 15; leaf scars extending $^2/_3$ of the way around the twig . . . .***Oplopanax***
   105. Stem and branchlets smooth, rough, or ridged, not prickly, slender or stout; fruit various
    107. Pith with transverse woody partitions; lateral buds globose, densely dark-tomentose, the terminal bud naked; fruit large, berrylike . . . .***Asimina***
    107. Pith solid but lacking transverse woody partitions; lateral buds variously shaped and colored, the terminal bud, if present, naked or scaly; fruit various
     108. Leaf scars deeply V-shaped, extending nearly around the bud; buds often tomentose or silky; twigs slender or stout; fruit a drupe, follicle, legume, or berry
      109. Bundle scars projecting out of nearly white leaf scar; buds superposed; fruit a legume . . . .***Cladrastis***
      109. Bundle scars flush with other leaf scar tissue, not projecting; buds solitary, not superposed; fruit a drupe, berry, or follicle

110. Wood bright yellow; terminal bud present; fruit
a follicle .... *Xanthorhiza*
110. Wood gray to light brown; terminal bud absent; fruit a
drupe or berry
    111. Twigs pliable, slender; bark tough and fibrous, not resinous;
fruit a drupe .... *Dirca*
    111. Twigs rigid, stout, or slender; fruit a red drupe
or berry .... *Rhus*
108. Leaf scars semicircular, deltoid, or lunate, extending not more
than halfway around the depressed bud; buds pubescent or
glabrous; twigs mostly very stout; fruit a berrylike pome,
legume, drupe, or samara
    112. Buds dark red; bark lacking resinous juice or strong odor; base of
petiole often persistent; terminal bud present; fruit a berrylike
pome, red .... *Sorbus*
    112. Buds brown; bark often with resinous juices or foul odor;
base of the petiole deciduous; terminal bud present or absent;
fruit a samara, drupe, or legume
        113. Buds superposed, imbedded in the bark; pith salmon-colored;
terminal bud absent; twigs pale gray or brown; fruit
a legume .... *Gymnocladus*
        113. Buds solitary, exposed; pith brown; terminal bud present
or absent; twigs usually yellow-brown to brown; fruit a
samara or drupe
            114. Bark lacking resinous juices, not toxic, but with foul odor;
terminal bud absent; fruit a samara .... *Ailanthus*
            114. Bark with resinous, poisonous juices, with little or no odor;
terminal bud present; shrub; fruit a drupe ...... *Toxicodendron*
104. Bundle scars 3
  115. Leaf scar a narrow line extending about halfway around the stem, not
decurrent; plant often prickly; fruit red to red-brown, a
berrylike pome .... *Rosa*
  115. Leaf scar broad or narrow, but not forming a line, often
decurrent; plants prickly or not; fruit various
    116. Twigs prickly or buds red-tomentose
      117. Bud scales blunt, terminated by a scar; inner bark yellow ... *Berberis*
      117. Bud scales acute or blunt, not terminated by a scar; inner
bark dull, not yellow
        118. Stems often recurved or pendulous; fruit a berrylike cluster
of drupelets .... *Rubus*
        118. Stems and twigs straight, not conspicuously recurved
or pendulous; fruit a berry or follicle
          119. Prickles 1 to 3 below each leaf scar, the latter decurrent; buds
lanceolate, brown with thin scales; fruit a berry .... *Ribes*

119. Prickles 2 at each leaf scar or scattered; buds depressed, red-tomentose
or nearly black, with thick scales; fruit a follicle . . . .***Zanthoxylum***

116. Twigs smooth or rough, not prickly, sometimes with thorns;
buds yellow to brown or black, glabrous or pubescent, not
red-tomentose

  120. Leaf scars semicircular or broadly lunate, large, greater than
2 mm in diameter

    121. Bark with resinous juice; fruit a drupe

      122. Buds acute; fruit on a plumose stalk . . . .***Cotinus***

      122. Buds rounded; fruit sessile or on a glabrous or
pubescent stalk . . . .***Rhus***

    121. Bark lacking resinous juice; fruit a capsule or drupe, but not
borne on a plumose stalk

      123. First bud scale of lateral buds anterior; bud
scales puberulent . . . .***Populus***

      123. First bud scale of lateral buds lateral, at least not anterior;
bud scales pubescent or glabrous

        124. Bud scales ciliate, polished; buds, when broken, resinous
and aromatic; leaf scars raised; branches often with corky
winged ridges of bark; pith large, angular; fruit a head of
many capsules . . . .***Liquidambar***

        124. Bud scales glabrous; buds lacking resin or strong odor;
leaf scars mostly in plane of twig, not prominently raised;
branches lacking corky ridges; pith terete, not angular;
fruit a drupe . . . .***Prunus***

  120. Leaf scars narrowly lunate, or small, less than 2 mm in diameter

    125. Internodes very unequal, branches much exceeding the central axis;
fruit a drupe . . . .***Cornus***

    125. Internodes nearly equal, branches shorter than the central
axis; fruit various

      126. Buds superposed

        127. Leaf scars deeply V-shaped, partly surrounding the
bud, or buds bursting through the leaf scars

          128. Buds in the upper angle of V-shaped leaf scars, pubescent,
mostly superposed; twigs lacking thorns, spines, and prickles;
fruit a samara . . . .***Ptelea***

          128. Buds bursting through the leaf scars, pubescent or
glabrous, superposed or solitary; twigs usually
bearing thorns, spines, or prickles; fruit a legume

            129. Buds glabrous; branchlets often bearing stout branched
thorns; twigs mostly terete . . . .***Gleditsia***

            129. Buds pubescent; branchlets usually bearing unbranched
nodal spines or prickles; twigs often strongly
angled . . . .***Robinia***

127. Leaf scars variously shaped but not deeply V-shaped, free
from the buds
   130. Twigs furrowed, green or gray-brown; fruit a head of
hairy achenes . . . .*Baccharis*
   130. Twigs even, not furrowed, finely white-speckled; fruit a
drupe or legume
      131. Twigs spicy-aromatic; most branchlets bearing globose collateral
flower buds; shrubs; fruit a red drupe . . . .*Lindera*
      131. Twigs lacking aroma; flower buds axillary, not easily distinguished
from vegetative buds; small trees; fruit a legume . . . .*Cercis*
126. Buds solitary, not superposed
   132. Bark, at least the older, shredding; leaf scars often strongly decurrent; bud
scales very thin, often glandular; shrubs; fruit a berry . . . .*Ribes*
   132. Bark close, not shredding; leaf scars not or only slightly decurrent;
bud scales thick or thin, usually lacking glands; fruit various
      133. Second bud scale usually less than half the length of the bud, bud scales
lacking glands on the teeth, thin, closely appressed, mostly brown with
a black tip; shrubs or small trees . . . . *Amelanchier*
      133. Second bud-scale about half the length of bud or more, bud
scales glandular-toothed, rather thick, somewhat spreading
or divaricate, red to red-brown, lacking a black tip; shrubs
         134. Branchlets slender, less than 3 mm in diameter . . . .*Aronia*
         134. Branchlets stout, usually greater than 5 mm in diameter
            135. Buds small (less than 1 mm long), densely rusty-tomentose; twigs
densely rusty-tomentose; branchlets dark gray to black; leaf scars
clustered on knobs, spurs, or very short branchlets; southern
plants of sandy or wet areas . . . .*Bumelia*
            135. Buds usually greater than 1 mm long, pubescent or
glabrous; twigs glabrous or pubescent; branchlets
various shades of color; leaf scars usually not clustered
on knobs or spurs (sometimes in *Pyrus*)
               136. Branchlets clustered near the end of each season's growth,
often with minute golden resin-granules; scales of the terminal
bud acute, the others rounded; shrubs; fruit dry or a small
waxy drupe . . . .*Myrica*
               136. Branchlets more evenly distributed, lacking
resin-granules; scales of terminal and lateral buds
similar, usually rounded; shrubs or trees; fruit a
pome, berry, or aggregate of berries
                  137. Leaves deciduous above the base of petiole . . . .*Rubus*
                  137. Leaves deciduous at the base of petiole
                     138. Shrubs less than 3 m tall
                        139. Branches many, short; rare native of
southern swamps . . . .*Litsea*

139. Branches few, longer; common, of various habitats . . . . *Ribes*
138. Shrubs, or trees greater than 3 m tall
140. Plant a low or creeping shrub . . . . *Ribes*
140. Plant a tree or tall shrub
    141. Scales of terminal bud narrowly ovate, thick; buds narrowly ovoid;
        twigs dark red or bronze-gray; lateral twigs mostly sharp and
        thorn-like; small trees . . . . *Malus*
    141. Scales of terminal bud broadly ovate, thinner, appressed,
        dentate or entire; buds usually broadly ovoid; twigs gray or
        olive to dark brown; lateral twigs lacking or bearing thorns,
        sharp or dull; trees or shrubs
        142. Axillary buds flattened and closely appressed, broadly ovoid, mostly
            pubescent; twigs mostly dark, rarely olive; trees with a broad,
            spreading crown; fruit a pome with papery carpel walls . . . . . *Malus*
        142. Axillary buds plump and divaricate, sharp-pointed or
            rounded, pubescent or glabrous; twigs gray to dark
            brown; shrubs or trees with narrow or spreading crown;
            fruit a pome with stony or papery carpel walls
            143. Buds acute, mostly glabrous, conical, sharp; branchlets sometimes
                with short blunt spurs; small trees . . . . *Pyrus*
            143. Buds obtuse, rarely acute, pubescent or glabrous, globose to
                ovoid; branchlets often with long stiff, sharp thorns; small trees
                or shrubs . . . . *Crataegus*

KEYS TO SPECIES

Note: Native and naturalized species appear in bold type; cultivated species are not bold. The following abbreviations are used: N = naturalized; ON = occasionally or infrequently naturalized; S = rarely native north of North Carolina and Tennessee; CS = rarely cultivated north of North Carolina and Tennessee; OC = occasionally or rarely cultivated.

Abelia ×grandiflora   glossy abelia

Abeliophyllum distichum   abelialeaf, abeliophyllum (OC)

Abies   fir

1. Leaves silver-blue, glaucous, mostly greater than 4 cm long . . . . . . *A. concolor*
1. Leaves dark green, except for white stomatal bands on lower
surface, less than 3 cm long
   2. Twigs with dark gray hairs; cone bracts hidden (protruding in the
   var. **phanerolepis**) . . . . **A. balsamea**
   2. Twigs with red hairs, often appearing gray due to dust particles;
   cone bracts protruding . . . . **A. fraseri** (S, CULTIVATED NORTH)

Acer   maple

Summer key

1. Leaves pinnately compound
   2. Leaflets densely pubescent, like velvet, on lower surface, mostly
   undulate to coarsely crenate . . . . *A. maximowiczianum* (OC)
   2. Leaflets glabrous or sparsely pubescent on lower surface, mostly
   only on petiolules and main veins, mostly coarsely dentate to lobed

     3. Leaflets mostly less than 4 cm long; bark curling, peeling,
       and flaking . . . .*A. griseum*
     3. Leaflets mostly greater than 6 cm long; bark tight, not peeling
       or flaking . . . .**A. negundo**
1. Leaves simple, mostly palmately lobed
    4. Leaves with cup-shaped, obtuse to cuneate base and 2-lobed,
      the margins entire . . . .*A. buergeranum* (OC)
    4. Leaves with truncate to cordate base and 2- to many-lobed,
      the margins entire or serrate
      5. Leaves silvery-white or glaucous on lower surface
        6. Leaves sometimes greater than 15 cm wide, the lobes
          crenate-serrate, the secondary veins along each of the 35
          main veins prominent and numerous, evident on
          both surfaces . . . .**A. pseudoplatanus** (ON)
        6. Leaves less than 15 cm wide, the lobes remotely, crenately,
          or sharply serrate, the secondary veins along each of the 3–5
          main veins inconspicuous and few
          7. Lobes of leaves with 1–3 large teeth on each margin; buds
            pointed, brown or yellow-brown, not red; collateral
            buds absent . . . .**A. saccharum**
          7. Lobes of leaves with 5 or more small and large teeth on
            each margin; buds blunt, mostly red; collateral buds
            often present
            8. Leaves with middle, main lobe about half the length of the
              blade, the sinus between it and the side lobes V-shaped;
              branchlets straight; samaras glabrous, less than
              3 cm long . . . .**A. rubrum** and *A.* ×*freemanii*
            8. Leaves with middle, main lobe about $2/3$ the length of
              the blade, the sinus between it and the side lobes U-shaped or
              narrower at the opening than at the base; branchlets arching
              upward; samaras pubescent when young, greater than
              3 cm long . . . . **A. saccharinum** and *A.* ×*freemanii*
      5. Leaves green or pale green, not silvery-white on lower surface
        9. Leaves with obtuse or rounded lobes and
          entire margins . . . .**A. campestre** (ON)
        9. Leaves with acute or acuminate lobes or leaves with
          serrate margins
          10. Leaves not or slightly lobed, the margins serrate
            11. Leaves irregularly doubly toothed, often slightly lobed,
              the lateral veins 6–8 on each side of
              the midrib . . . .*A. tataricum* (OC)
           11. Leaves regularly at least doubly or triply toothed,
              never lobed, the lateral veins at least 20 on each side
              of the midrib . . . .*A. carpinifolium* (OC)

10. Leaves lobed, the margins entire or serrate

   12. Leaves mostly with 3 lobes; buds sessile or stalked, the scales
several or only a pair visible

      13. Buds with several pairs of imbricated
scales, sessile . . . . **.*A. ginnala*** (ON)

      13. Buds with 1 pair of valvate scales visible, stalked

         14. Twigs and buds pubescent; leaves glabrous or with pale or
white pubescence on lower surface; bark brown, uniform,
not striped . . . . **.*A. spicatum***

         14. Twigs and buds glabrous; leaves rusty-pubescent on
lower surface; bark on young trees uniformly brown
or green and white-striped

            15. Bark on young trees green, striped white; leaves with
lower surface sparsely pubescent, only a few
hairs clustering . . . . **.*A. pensylvanicum***

            15. Bark uniform, brown; leaves with lower surface often
densely pubescent, usually with large, prominent clusters
of hairs . . . . **.*A. rufinerve*** (OC)

   12. Leaves mostly with 5–11 lobes; buds sessile, with several
pairs of imbricated scales

      16. Leaves mostly with 5 lobes, the margins entire or
coarsely dentate

         17. Sap from broken petioles milky; buds green or red,
stout and blunt . . . . **.*A. platanoides*** (N)

         17. Sap from broken petioles clear, not milky; buds brown,
slender and acute

            18. Leaf lobes entire, the basal lobes reduced but prominent

               19. Leaves glabrous or sparsely pubescent on the lower surface
just above the petiole; young twigs usually lacking
bloom; samara wings mostly diverging, forming a
V or U . . . .*A. truncatum* (OC)

               19. Leaves prominently woolly on the lower surface
just above the petiole; young twigs often glaucous;
samara wings diverging, forming a broad U or nearly
horizontal . . . .*A. cappadocicum* (OC)

            18. Leaf lobes with large teeth or small lobes, the basal lobes
greatly reduced, often inconspicuous

               20. Leaves pale green, glabrous or occasionally pubescent
on lower surfaces, the margins and tips horizontal,
not drooping; stipules absent . . . . **.*A. saccharum***

               20. Leaves yellow-green to dark green, pubescent on
lower surface, the margins and tips drooping; stipules
often present . . . . **.*A. nigrum***

      16. Leaves mostly with 7–11 lobes, the margins serrate

21. Leaves mostly with 7 lobes, the lobes separated
more than halfway toward
the base . . . . ***A. palmatum*** (ON) and *A. japonicum* "Aconitifolium"
21. Leaves mostly with 9–11 lobes, the lobes separated less than
halfway toward the base . . . . *A. japonicum* except "Aconitifolium"

Winter key

1. Buds with 2 exposed valvate scales, strictly stalked; shrubs or
small trees
   2. Twigs and buds pubescent; bark brown to gray, not striped . . ***A. spicatum***
   2. Twigs and buds glabrous; bark on young trees green and
      white-striped or uniformly brown
      3. Bark on young trees green, white-striped . . . . ***A. pensylvanicum***
      3. Bark uniform, brown . . . . *A. rufinerve* (OC)
1. Buds with several exposed scales, essentially sessile; mostly trees
   4. Buds white-downy; twigs often green or purple, glaucous; opposite leaf scars
      meeting in a sharp-pointed raised projection . . . . ***A. negundo***
   4. Buds brown, gray, yellow-brown, or red, glabrous or pubescent,
      not white-downy; twigs usually lacking a bloom; opposite leaf
      scars not meeting or meeting in a raised or flat line
      5. Bark thin, red-brown or cinnamon, flaking, curling,
         and peeling . . . . *A. griseum*
      5. Bark mostly thick, gray to brown, not curling or peeling
         6. Buds slender, sharp-pointed, with 4–8 pairs of visible
            brown to yellow-brown scales
            7. Twigs slender, red-brown, glossy; bark of
               trunk gray . . . . ***A. saccharum***
            7. Twigs stout, orange-brown, dull, often mottled with gray; bark of
               trunk dark gray . . . . ***A. nigrum***
         6. Buds stout, blunt, with fewer pairs of visible brown,
            red-brown, or gray scales
             8. Collateral buds often present; bud scales bright red
               9. Bark flaking on old trunk, leaving brown areas;
                  twigs with rank odor
                  when broken . . . . ***A. saccharinum*** and *A. ×freemanii*
               9. Bark of trunk rough but generally not flaking;
                  twigs lacking rank odor
                  when broken . . . . ***A. rubrum*** and *A. ×freemanii*
            8. Collateral buds absent; bud scales green, brown,
            or red-brown (red only in ***A. platanoides***)
              10. Terminal bud greater than 5 mm long; opposite leaf
                scars joined by a line or not; twigs stout; trees
                 11. Buds red, appressed; edges of leaf
                    scars meeting . . . . ***A. platanoides*** (N)

11. Buds green, spreading; edges of leaf scars
     not meeting ....***A. pseudoplatanus*** (ON)
10. Terminal bud less than 5 mm long or absent; opposite leaf scars
    joined by a line; twigs slender or medium; small trees or shrubs
   12. Branchlets green, stipule lines and leaf scar margins raised
      13. Branchlets pubescent, at least when young; fruit about
          25 mm long ....*A. japonicum*
      13. Branchlets glabrous; fruit usually less than
          25 mm long ....***A. palmatum*** (ON)
   12. Branchlets brown, stipule lines and leaf scar margins flat,
       not raised
      14. Bud scales dark red-brown, rusty-pubescent at the margins;
          twigs pubescent, often woolly toward
          the tips ....*A. maximowiczianum* (OC)
      14. Bud scales gray to red-brown, gray-pubescent or glabrous;
          twigs glabrous or pubescent
         15. Bud scales sharply keeled, often divergent, ciliate or with
             prominent tufts of hairs toward the apex
            16. Bud scales with prominent tufts of hair toward
                the apex ....*A. buergeranum* (OC)
            16. Bud scales ciliate toward the apex ....*A. carpinifolium* (OC)
         15. Bud scales rounded or broadly keeled, tight, glabrous
            17. Buds gray, pubescent; twigs mostly pubescent;
                branchlets often corky ....***A. campestre*** (ON)
            17. Buds brown, glabrous; twigs glabrous; branchlets
                hard, not corky
               18. Twigs and branchlets mostly light gray ......***A. ginnala*** (ON)
               18. Twigs and branchlets brown to dark red-brown
                  19. Samara wings mostly diverging at less than
                      60-degree angles to each other ....*A. tataricum* (OC)
                  19. Samara wings mostly diverging at greater than
                      60-degree angles to each other
                     20. Young twigs often glaucous; samara wings
                         diverging in a broad U or often
                         nearly horizontal ....*A. cappadocicum* (OC)
                     20. Twigs glabrous; samara wings usually
                         diverging into a U or V ....*A. truncatum* (OC)

Actinidia   bower actinidia, kiwi, tara vine

1. Branchlets and lower leaf surface glabrous or pubescent only on the veins;
   fruit glabrous, 2–3 cm long ....*A. arguta*
1. Branchlets and lower leaf surface tomentose; fruit pubescent,
   3–5 cm long ....*A. chinensis*

Aesculus   buckeye, horse-chestnut

1. Buds gummy; leaflets 5 or 7
    2. Leaflets usually 7, sessile or with petiolules
       1–5 mm long . . . . **A. hippocastanum** (N)
    2. Leaflets usually 5, mostly with petiolules 12–25 mm long . . . . . . *A. ×carnea*
1. Buds smooth, dry, not gummy; leaflets usually 5
    3. Buds elongated, mucronate; injured twigs with a strongly fetid
       odor; branchlets usually glossy; fruit prickly, at least
       when young . . . . **A. glabra**
    3. Buds ovate, acute; injured twigs with only a slightly fetid odor;
       branchlets usually dull; fruit smooth, not prickly
        4. Leaflets lacking clusters of red hairs in the axils of the veins on
           lower surface; fruit less than 4 cm long . . . . **A. parviflora** (s)
        4. Leaflets with sparse or dense clusters of red hairs in the axils of
           the veins on lower surface; fruit greater or less than 4 cm long
            5. Leaflets often irregularly toothed, sometimes sharply serrate;
               fruit mostly greater than 4 cm long . . . . **A. pavia** (s)
            5. Leaflets usually finely crenate or crenate-serrate; fruit less
               than 4 cm long or greater than 5 cm long
                6. Petiolules mostly greater than 3 mm long; leaflets
                   minutely velutinous on lower surface; fruit mostly less
                   than 4 cm long . . . . **A. sylvatica** (s)
                6. Petiolules mostly less than 3 mm long; leaflets glabrate on
                   lower surface; fruit mostly greater than 5 cm long . . . . **A. flava**

Ailanthus altissima   tree-of-heaven, ailanthus (N)

Akebia quinata   five-leaf akebia (ON)

Albizia julibrissin   silk tree, mimosa (N)

Alnus   alder

1. Buds sessile, glutinous; leaves sharply serrulate, green on both surfaces;
   nutlets strongly winged . . . . **A. viridis**
1. Buds stalked, scurfy; leaves dentate, finely, doubly, or coarsely
   serrate; nutlets wingless or nearly so
    2. Leaves with wedge-shaped base; flowering in fall; fertile catkins
       usually solitary . . . . **A. maritima**
    2. Leaves with acute, obtuse, cordate, or rounded base; flowering in
       spring; fertile catkins clustered
        3. Leaves obovate to orbicular, the apex notched or truncate, the margin
           dentate or serrate; lateral buds with 1 or more scars on the
           stalk; tree . . . . **A. glutinosa** (N)

3. Leaves elliptical, ovate, or obovate, acute, the apex neither
notched nor truncate, the margin finely, doubly, or coarsely
serrate; lateral buds lacking scars; shrubs or small trees
    4. Leaves obovate, base acute, margin finely serrate; staminate catkins nearly
at right angles to the pistillate catkins; nutlets ovate . . . . *A. serrulata*
    4. Leaves broadly elliptical to ovate, base broad and rounded,
margin doubly or coarsely serrate; staminate and pistillate
catkins both drooping; nutlets orbicular . . . . *A. incana* **ssp.** *rugosa*

Amelanchier   juneberry, shadbush, serviceberry

Only a summer key is offered here and even it may be inadequate to determine
species in this taxonomically trying genus. I hold as much as is feasible to Muenscher's treatment (which is based on K. M. Wiegand's 1912 treatment in *Rhodora*
14:117–161), while recognizing that there is much hybridization among species,
resulting in many named and unnamed entities.

1. Leaves imbricate in the bud, petioles stout, 2–10 mm long, the leaf base acute,
tapering into petiole with raised margins; flowers solitary
(1–3 in a cluster) . . . . *A. bartramiana*
1. Leaves conduplicate in the bud, petioles slender, 8–25 mm long,
the base rounded, cordate, or subcordate, not tapered; flowers
in racemes
    2. Leaves acuminate, finely serrate; ovary summit usually glabrous
        3. Leaves pubescent, at least along their midribs or petioles . . . *A. arborea*
        3. Leaves glabrous even when young . . . . *A. laevis*
    2. Leaves blunt, rounded, acute, or mucronate, finely or coarsely
serrate; ovary summit glabrous to tomentose
        4. Leaves finely toothed, often greater than 7 teeth per cm of
margin; ovary summit glabrous
            5. Shrub less than 2 m tall; fruit stalks mostly less than
10 mm long . . . . *A. obovalis*
            5. Shrub greater than 2 m tall; fruit stalks often greater than
12 mm long . . . . *A. canadensis*
        4. Leaves coarsely toothed, about 2–6 per cm of margin; ovary
summit glabrous to tomentose
            6. Leaves toothed to the base, the veins prominent to the tip
of the teeth; buds 4–8 mm long
                7. Leaves truncate, 2–5 teeth per cm of margin . . . . *A. alnifolia*
                7. Leaves subcordate or with rounded bases, 4–5
teeth per cm of margin . . . . *A. sanguinea*
            6. Leaves entire or toothed only midway to the base, the veins
less prominent, often not entering the teeth; buds
4–13 mm long

8. Leaves pubescent or tardily glabrate, veins 16–20, prominent, the upper often entering the teeth; buds 4–9 mm long; stoloniferous shrub, usually strongly surculose; ovary summit tomentose . . . . *A. humilis*
8. Leaves glabrate, veins often greater than 20, usually anastomosing before the margin; buds 6–13 mm long; tree, or stoloniferous or non-stoloniferous shrub; ovary summit glabrous to sparsely pubescent
   9. Tree or non-stoloniferous shrub . . . . *A.* ×*interior*
   9. Stoloniferous shrub
      10. Young leaves glabrous on lower surface; Canada, Maine, northern Michigan, and Minnesota . . . . *A. fernaldii*
      10. Young leaves tomentose on lower surface; widely distributed . . . . *A. spicata*

Amorpha   false-indigo

1. Leaflets less than 1 cm long . . . . *A. nana*
1. Leaflets greater than 1 cm long
   2. Leaves and twigs short, white-tomentose (canescent) . . . . *A. canescens*
   2. Leaves and twigs glabrous to sparsely or shortly pubescent, not tomentose or canescent
      3. Shrubs often greater than 2 m tall; leaflets with obvious mucro; widespread . . . . *A. fruticosa*
      3. Shrubs less than 2 m tall; leaflets with mucro absent, inconspicuous, glandular, or obvious; southern
         4. Leaflets emarginate, the mucro glandular or absent . . . . *A. glabra* (s)
         4. Leaflets obtuse or rounded, usually with inconspicuous glandular mucro
            5. Petiole longer than the width of the lowest leaflet; shrub 1–2 m tall . . . . *A. schwerinii* (s)
            5. Petiole shorter than the width of the lowest leaflet; shrub less than 1 m tall
               6. Fruit glabrous; leaflets glabrous or sparsely pubescent, with inconspicuous punctate glands or dots . . . . *A. georgiana* (s)
               6. Fruit pubescent; leaflets densely short-pubescent, with conspicuous punctate glands or dots . . . . *A. herbacea* (s)

Ampelopsis   ampelopsis

1. Leaves compound . . . . *A. arborea* (s)
1. Leaves simple, with or without lobes
   2. Twigs pubescent; leaves with or without lobes . . . . *A. brevipedunculata*
   2. Twigs glabrous; leaves toothed, mostly lacking lobes . . . . *A. cordata*

Andromeda   bog-rosemary

1. Leaves greater than 3 cm long, the lower surface pubescent; fruit erect . . . . *A. glaucophylla*
1. Leaves less than 3 cm long, the lower surface glabrous; fruit pendulous . . . . *A. polifolia*

Aphananthe aspera   muku tree (OC)

Aralia   Hercules'-club, sarsaparilla

1. Stems with slender bristles; shrubs to 1 m tall; umbels in a corymb . . . . *A. hispida*
1. Stems spiny, forming tall shrubs or trees; umbels in a large compound panicle
    2. Leaflets with stalks at least 2 mm long; twigs light brown . . . . . *A. spinosa*
    2. Leaflets subsessile; twigs light gray . . . . *A. elata* (CS)

Araucaria araucana   monkey-puzzle tree (CS)

Arceuthobium pusillum   eastern dwarf-mistletoe

Arctostaphylos   bearberry

1. Leaves entire; fruit red . . . . *A. uva-ursi*
1. Leaves serrulate; fruit black . . . . *A. alpina*

Ardisia crenata   coralberry

Aristolochia   Dutchman's pipe

1. Leaves glabrate; winter twigs glabrous, stout, green; large vessels not clearly visible in cross section of stem . . . . *A. macrophylla*
1. Leaves tomentose; winter twigs finely pubescent, slender, glaucous, blue-green; large vessels clearly visible in cross section of stem . . . . *A. tomentosa*

Aronia   chokeberry

1. Twigs, buds, and leaves glabrous or glabrate; fruit dark purple or black . . . . *A. melanocarpa*
1. Twigs, buds, and lower surface of leaves canescent-tomentose, sometimes glabrate; fruit red or purple-black
    2. Leaves turning bright red in fall; fruit red . . . . *A. arbutifolia*
    2. Leaves turning brown or red-brown or orange-brown in fall; fruit purple-black . . . . *A. ×prunifolia (A. arbutifolia × A. melanocarpa)*

Artemisia abrotanum   wormwood, southernwood, old man (N)

Arundinaria gigantea   giant cane

Asimina   pawpaw

1. Shrub to 2 m tall; larger leaves less than 18 cm long and 10 cm wide; fruit 1–3 cm long; seeds 12–15 mm long . . . .*A. parviflora* (S)
1. Tree to 10 m tall; larger leaves to 25 cm long and 15 cm wide; fruit mostly greater than 3 cm long; seeds greater than 15 mm long . . . .*A. triloba*

Aucuba japonica   Japanese aucuba

Baccharis   groundsel tree

1. Leaves linear, less than 5 mm wide . . . .*B. angustifolia* (S)
1. Leaves usually elliptic or obovate, usually greater than 7 mm wide
    2. Petioles 5–12 mm long; flowering heads usually borne on peduncles . . . .*B. halimifolia*
    2. Petioles absent or to 7 mm long; flowering heads usually sessile . . . .*B. glomeruliflora* (S)

Berberis   barberry

1. Branchlet spines greater than 1 cm long; leaves evergreen, the spiny teeth greater than 1 mm long . . . .*B. julianae*
1. Branchlet spines less than 1 cm long; leaves deciduous, the spiny teeth, if present, less than 1 cm long
    2. Leaves entire; twigs red-brown to purple; spines of the branchlets simple or with 2 small lateral branches; compact low shrubs; flowers and fruits solitary or in small clusters . . . .*B. thunbergii* (N)
    2. Leaves serrate or dentate; twigs red-brown, brown, gray, orange-brown, or buff; spines of the branchlets branched, of three nearly equal parts; tall shrubs; flowers and fruits in drooping racemes
        3. Twigs gray, orange-brown, or buff; leaves with lateral veins prominent on lower surface . . . .*B. vulgaris* (N)
        3. Twigs red-brown to brown; leaves with lateral veins inconspicuous on lower surface . . . .*B. canadensis*

Berchemia scandens   supplejack (S)

Betula   birch

Summer key

1. Leaves obovate, orbicular, or reniform, with rounded apex, margin crenate-dentate; shrubs

2. Plants less than 60 cm tall; leaves less than 1 cm long; nut
   lacking wings . . . .*B. michauxii* (s)
2. Plants greater than 1 m tall or sometimes depressed and shorter;
   leaves mostly greater than 1 cm long; nut narrowly winged
      3. Twigs pubescent, smooth, not glandular . . . .*B. pumila*
      3. Twigs glabrous or puberulent, glandular-warty . . . .*B. glandulosa*
1. Leaves ovate or at least with acute or tapering apex, margin usually
   doubly serrate; trees or shrubs
      4. Shrubs; larger leaves usually less than 4 cm long, with 2–4 pairs
      of lateral veins . . . .*B. minor*
      4. Trees; larger leaves usually greater than 4 cm long,
      with 5–12 pairs of lateral veins
         5. Bark of trunk white; twigs usually with prominent lenticels
            6. Apex of leaves long-tapering, the petiole 2–3 cm long; bark
            of trunk tight, not peeling in layers; twigs rough
            glandular-warty . . . .*B. populifolia*
            6. Apex of leaves acuminate, the petiole 1–3 cm long; bark of
            trunk peeling in thin papery layers; twigs with
            scattered glands
               7. Leaves pubescent on lower surface . . . .*B. papyrifera*
               7. Leaves glabrous on lower surface
                  8. Branchlets pendulous on older trees . . . .*B. pendula* (n)
                  8. Branchlets horizontal . . . .*B. platyphylla* (on)
         5. Bark of trunk black, gray-black, bronze, red-brown, gray-
         brown, tan, or white tinged with pink or bronze; twigs usually
         with inconspicuous (prominent in *B. cordifolia*) lenticels
            9. Bark of twigs with wintergreen flavor; bark of trunk black,
            gray, or bronze
               10. Bark of trunk bronze, curling and peeling in thin layers;
               leaves with lateral veins lacking
               prominent branches . . . .*B. alleghaniensis*
               10. Bark of trunk gray or black, smooth, not peeling; leaves
               with lateral veins often having prominent branches
                  11. Bark black; twigs glabrous; leaves thin . . . .*B. lenta*
                  11. Bark gray; twigs pubescent; leaves thick . . . .*B. grossa*
            9. Bark of twigs lacking wintergreen flavor; bark of trunk
            red-brown, gray-brown, light brown, tan, or white tinged
            with brown
               12. Leaves with 9 or more pairs of lateral veins, the
               petiole glabrous
                  13. Leaves singly toothed, the teeth less than
                 1 mm long . . . .*B. cordifolia*
                 13. Leaves often with 3 or more teeth between the teeth
                 that are the extensions of the veins, the teeth 1 mm
                 or more long . . . .*B. maximowicziana* (oc)

12. Leaves with fewer than 9 pairs of lateral veins, the petiole pubescent
   14. Mature bark of trunk light brown to white tinged with brown,
      tight or exfoliating in thin sheets; lower surface of
      leaves glabrate . . . .*B. pubescens*
   14. Mature bark of trunk red-brown to gray-brown, shaggy,
      exfoliating irregularly; the lower surface of leaves pubescent at
      least on the midrib
      15. Leaves cuneate, rarely truncate at the base, the lower surface often
         glaucous; twigs sparsely glandular; nut 3 mm long . . . .**B. nigra**
      15. Leaves mostly rounded or subcordate at the base, the
         lower surface lacking bloom; twigs sparsely or densely
         glandular; nut 1–2 mm long
         16. Bark on young trees shiny, smooth, not peeling; leaves sometimes
            greater than 8 cm long . . . .*B. grossa*
         16. Bark flaking, peeling, or curling back; leaves less than
            8 cm long
            17. Twigs densely glandular; bark curling back but
               persisting; nut 2 mm long . . . .*B. davurica* (oc)
            17. Twigs sparsely glandular; bark peeling in long sheets;
               nut 1 mm long . . . .*B. albosinensis* (oc)

Winter key

1. Shrubs; bark dark brown or dark red-brown; buds 2–6 mm long
   2. Plants less than 60 cm tall; nut lacking wings . . . .**B. michauxii** (s)
   2. Plants mostly greater than 1 m tall or sometimes depressed and
      matted; nut winged
      3. Bark dark red-brown; buds 2–4 mm long, less than
         3 mm wide . . . .*B. minor*
      3. Bark dark brown; buds 5–6 mm long, greater than 3 mm wide
         4. Twigs very rough, glandular-warty, with
            lenticels prominent . . . .*B. glandulosa*
         4. Twigs smoother, with scattered glands, with lenticels
            inconspicuous . . . .*B. pumila*
1. Trees; bark variously colored; buds 5–13 mm long
   5. Bark of twigs lacking wintergreen flavor; bark of trunk white,
      tan, brown, or red-brown
      6. Bark of trunk red-brown, gray-brown, light brown, tan, or
         white tinged with brown or pink
         7. Twigs with prominent lenticels . . . .*B. cordifolia*
         7. Twigs with inconspicuous lenticels
            8. Mature bark of trunk red-brown or gray-brown, shaggy,
               shedding irregularly
               9. Twigs sparsely glandular; nut 3 mm long . . . .**B. nigra**

9. Twigs usually densely glandular; nut
   2 mm long . . . .*B. davurica* (oc)
8. Mature bark of trunk light brown, tan, or white tinged
   with brown, tight or exfoliating in thin sheets
   10. Twigs lacking glands; nut 1 mm long . . . .*B. pubescens*
   10. Twigs usually sparsely glandular; nut
       2 mm long . . . .*B. albosinensis* (oc)
6. Bark of trunk white
   11. Branchlets pendulous . . . .***B. pendula*** (n)
   11. Branchlets horizontal
       12. Bark with conspicuous white lenticels to 2 mm long;
           twigs glandular . . . .*B. maximowicziana* (oc)
       12. Bark of trunk with small variously colored lenticels usually
           less than 1 mm long; twigs glandular or not
           13. Twigs glabrous, rough glandular-warty and bark
               peeling freely . . . .***B. platyphylla*** (on)
           13. Twigs glabrous, rough glandular-warty and bark tight,
               or twigs pubescent, not glandular-warty and bark
               peeling freely
               14. Twigs and branchlets glabrous, very rough glandular-warty;
                   staminate catkins usually solitary; bark dull chalky-white,
                   tight, not peeling freely or separating into film-like layers;
                   stems usually in clumps, the lateral branches on
                   each stem from near the ground, with black
                   triangular spots below their insertion on
                   the trunk . . . .***B. populifolia***
               14. Twigs and branchlets mostly pubescent but not rough
                   glandular-warty; staminate catkins usually clustered;
                   bark lustrous, separating into thin, film-like papery
                   layers; stems usually solitary, lacking
                   triangular spots . . . .***B. papyrifera***
5. Bark of twigs with wintergreen flavor; bark of trunk bronze,
   gray-black, or black
   15. Twigs pubescent and red-brown . . . .*B. grossa*
   15. Twigs glabrous and red-brown or pubescent and
       yellow-brown
       16. Buds glabrous, the lower scale occasionally ciliate, sharply divergent
           or recurved falcate; twigs glabrous, glossy reddish-brown; staminate
           catkins narrowly cylindrical, 3.0–3.5 mm in diameter; pistillate
           catkins ovoid, the scales glabrous; bark close, dark red-brown,
           gray-black, or black, broken into irregular plates on
           older trunks . . . .*B. lenta*
       16. Buds pubescent or at least with heavily ciliated scales,
           mostly ascending or slightly divergent; twigs somewhat

pubescent, buff or yellow-brown; staminate catkins rather broadly
cylindrical, 4–5 mm in diameter; pistillate catkins oblong, the scales
pubescent; bark of trunk buff or bronze, separating into thin, film-like
papery layers . . . .*B. alleghaniensis*

Bignonia capreolata   trumpet flower

Borrichia frutescens   sea ox-eye (s)

Broussonetia papyrifera   paper-mulberry (ON)

Buckleya distichophylla   buckleya (s)

Buddleia   butterfly bush, summer-lilac

1. Leaves serrate or serrate-crenate, 10–15 cm long . . . .*B. davidii* (N)
1. Leaves entire or denticulate, 4–8 cm long . . . .*B. lindleyana* (s)

Bumelia   bumelia

1. Leaves glabrous or sparsely pubescent on lower surface . . . .*B. lycioides* (s)
1. Leaves woolly pubescent on lower surface
    2. Lower leaf surface with dense, matted hairs, often with
       a sheen . . . .*B. tenax* (s)
    2. Lower leaf surface with dense, but not matted hairs, lacking
       a sheen . . . .*B. lanuginosa* (s)

Buxus   boxwood

1. Leaves mostly widest at or below the middle . . . .*B. sempervirens* (ON)
1. Leaves mostly widest above the middle . . . .*B. microphylla*

Callicarpa   beautyberry

1. Petioles usually less than 15 mm long . . . .*C. dichotoma* (ON)
1. Petioles usually 15–30 mm long . . . .*C. americana*

Calluna vulgaris   heather

Calocedrus decurrens   incense-cedar (OC)

Calycanthus   Carolina allspice, sweetshrub

1. Lower surface of leaves pubescent; twigs with scattered hairs; leaf scars
   somewhat sloping . . . .*C. floridus* var. *floridus* (s)
1. Lower surface of leaves glabrous or nearly so except along the veins; twigs
   minutely puberulent; leaf scars shelf-like . . . .*C. floridus* var. *glaucus*

Calycocarpum lyonii   cupseed

Camellia   camellia

1. Leaves up to 10 cm long; young twigs glabrous . . . .*C. japonica*
1. Leaves about 5 cm long; twigs pubescent . . . .*C. sasanqua*

Campsis   trumpet vine, trumpet creeper

1. Leaves pubescent on lower surface . . . .**C. radicans** and *C.* ×*tagliabuana*
1. Leaves glabrous or subglabrous
   on lower surface . . . .*C. grandiflora* and *C.* ×*tagliabuana*

Caragana   peashrub

1. Leaflets 8–12; stipules stiff, becoming spines . . . .*C. arborescens*
1. Leaflets 4; stipular spines lacking . . . .*C. frutex*

Carpinus   hornbeam, blue-beech

1. Buds nearly glabrous, usually greater than 4 mm long . . . .*C. betulus* (oc)
1. Buds pubescent, less than 4 mm long . . . .**C. caroliniana**

Carya   hickory

Summer key

1. Leaves mostly with more than 9 leaflets; husk of fruit thin
   (less than 2 mm thick); shell of nut thin
   2. Leaflets usually 9 or 11; shell of nut uneven or wrinkled;
      swamps and floodplains . . . .*C. aquatica* (s)
   2. Leaflets 11–17; shell of nut smooth; wet forests . . . .*C. illinoinensis*
1. Leaves with 9 or fewer leaflets; husk of fruit thick (mostly greater
   than 3 mm thick) or thin
   3. Leaflets 5–7
      4. Buds yellow, the scales valvate; shell of
         nut smooth . . . .*C. myristicaeformis* (s)
      4. Buds brown or gray, the scales imbricate; shell of nut
         uneven or ridged
         5. Leaflets densely ciliate when young, with tufts of hairs
            remaining on the teeth; bark shaggy or shedding
            6. Terminal leaflet obovate, fruit usually greater than
               4 cm long . . . .*C. ovata*
            6. Terminal leaflet lanceolate or oblanceolate; fruit less
               than 4 mm long . . . .*C. carolinae-septentrionalis* (s)
         5. Leaflets serrate but not densely ciliate, lacking tufts of hairs;
            bark smooth or tight, not shaggy or shedding

7. Rachis and lower surface of midrib covered with clusters of hairs and silvery scales . . . .*C. pallida*

7. Rachis and midrib glabrous or pubescent, but lacking clusters of hairs and silvery scales . . . .*C. glabra*

3. Leaflets 7–9

   8. Buds yellow; husk of fruit thin; shell of nut thin

      9. Shell of nut smooth; husk of fruit splitting its entire length; leaflets 9 or fewer . . . .*C. myristicaeformis* (s)

      9. Shell of nut uneven; husk of fruit splitting only about halfway; leaflets 7–11 . . . .*C. cordiformis*

   8. Buds tan, gray, or brown; husk of fruit thick or thin; shell of nut thick

      10. Leaves densely stellate-tomentose on the lower surface; husk of fruit thick . . . .*C. tomentosa*

      10. Leaves pubescent to glabrous on the lower surface with stellate hairs scattered or more dense only in the axils of the veins; husk of fruit thick or thin

         11. Mature leaves pubescent; husk of fruit thick . . . .*C. laciniosa*

         11. Mature leaves glabrous or nearly so; husk of fruit thin . . . . .*C. ovalis*

Winter key

1. Bud scales 4–6, in pairs, valvate

   2. Bud scales yellow-scurfy, glandular

      3. Shell of nut smooth; husk of fruit splitting its entire length . . . .*C. myristicaeformis* (s)

      3. Shell of nut uneven; husk of fruit splitting only about halfway . . . .*C. cordiformis*

   2. Bud scales brown

      4. Shell of nut smooth; bud scales with clusters of bright yellow hairs . . . .*C. illinoinensis*

      4. Shell of nut uneven; bud scales lacking clusters of hairs . . . .*C. aquatica* (s)

1. Bud scales numerous, imbricated

   5. Terminal bud small, usually less than 15 mm long; twigs slender

      6. Buds puberulous; bud scales silvery; bark of trunk rough-furrowed . . . .*C. pallida*

      6. Buds glabrous or puberulent; bud scales brown or red-brown; bark of trunk smooth and close-furrowed or separating into plate-like scales

         7. Bark of trunk close-furrowed; buds glabrous . . . .*C. glabra*

         7. Bark of trunk ridged, separating into plate-like scales; buds glabrous to pubescent . . . .*C. ovalis*

   5. Terminal bud large, 10–25 mm long; twigs stout

8. Bark of trunk rough and close, not separating in long strips or plates; outer bud scales early deciduous; terminal bud broadly ovate . . . . *C. tomentosa*
8. Bark of trunk separating in long strips or plates; outer bud scales persistent; terminal bud elongate-ovate
    9. Twigs buff or orange . . . . *C. laciniosa*
    9. Twigs gray, olive, or red-brown
        10. Fruit mostly greater than 4 mm long; widely distributed . . . . . *C. ovata*
        10. Fruit mostly less than 4 mm long; Georgia, North Carolina, South Carolina, Tennessee, and Virginia . . . . *C. carolinae-septentrionalis* (s)

Castanea   chestnut

1. Large or small, low shrubs forming thickets; twigs and branchlets glabrous; leaves 5–10 cm long, the lower surface glabrous . . . . . *C. alnifolia* (s)
1. Large shrubs or trees; twigs usually pubescent; leaves 6–20 cm long, the lower surface often pubescent
    2. Leaves ovate, ovate-lanceolate, elliptic, or obovate, the base cuneate
        3. Young twigs glabrous or with appressed, straight, unbranched white (clear) hairs; lower leaf surface glabrous or with minute glandular hairs and occasional unbranched hairs along the veins, the veins in 18 or more pairs . . . . *C. dentata*
        3. Young twigs with stellate yellow pubescence, usually dense, glandular hairs also present or not; lower leaf surface densely stellate-pubescent or nearly glabrous, the veins in fewer than 18 pairs . . . . *C. mollissima* (ON)
    2. Leaves oblong or oblong-lanceolate, the base oblique, rounded, truncate, or cordate
        4. Petioles 13–20 mm long; leaves with the basal 3 pairs of veins lacking teeth, the lower surface often with 4-rayed usually scattered stellate hairs . . . . *C. sativa*
        4. Petioles usually less than 10 mm long; leaves with the basal 3 pairs of veins toothed, the lower surface with dense or sparse stellate hairs mostly 8-rayed but always at least 6-rayed
            5. Leaves greater than 17 cm long, less than 3 times as long as wide; small to large tree . . . . *C. mollissima* (ON)
            5. Leaves less than 15 cm long, at least 3 times as long as wide; shrub or small tree
                6. Leaves less than 4 cm wide, the teeth merely straight spines with little or no broadening at the base, the sinus between them reduced or lacking; nuts 2–3 . . . . *C. crenata*

6. Leaves 4–6 cm wide, the teeth broad at the base, the sinus pronounced;
   nut usually solitary, rarely 2 . . . .*C. pumila*

Catalpa   catalpa, Indian-bean

1. Trees, usually large; leaves pubescent; flowers 2–4 cm wide, white
   spotted with yellow and brown; capsules 8–15 mm thick
     2. Bark of trunk thin, scaly; leaves usually abruptly acuminate;
        trees with low spreading crown; capsules sometimes less than
        1 cm thick . . . .*C. bignonioides*
     2. Bark of trunk thick, ridged; leaves long-acuminate; trees with tall slender
        crown; capsules often greater than 1 cm thick . . . .*C. speciosa*
1. Small trees or large shrubs; leaves glabrous or glabrate;
   flowers 1–2 cm wide, yellow with orange and violet markings;
   capsules 5–8 mm thick
     3. Leaves green when young, the petiole green . . . .*C. ovata* (ON)
     3. Leaves purple or purple-green when young, the petiole
        remaining purple . . . .*C. ×hybrida*

Ceanothus   New Jersey–tea, redroot

1. Large shrub to 4 m tall; inflorescences arising from branches of the
   previous year; Michigan, North Dakota, and west . . . .*C. sanguineus*
1. Low shrubs to 1 m tall; inflorescences born on shoots of the season;
   Montana, North Dakota, Minnesota, and east
     2. Leaves usually ovate, blades usually 2.5–5.0 cm long, the apex
        acute or acuminate; twigs covered with orange, glandular
        excretions; peduncles axillary, equaling or exceeding
        the leaves . . . .*C. americanus*
     2. Leaves usually elliptic-lanceolate, blades usually 1.0–2.2 cm long, the apex
        obtuse or rounded; twigs warty; peduncles terminal, shorter than
        the leaves . . . .*C. herbaceus*

Cedrus   cedar

1. Leading shoot and branch tips pendulous . . . .*C. deodara*
1. Leading shoot upright, stiff, branch tips rarely pendulous
     2. Branchlets densely pubescent; leaves blue . . . .*C. atlantica*
     2. Branchlets glabrous or sparsely pubescent; leaves green . . . .*C. libani*

Celastrus   bittersweet

1. Leaves rounded to obovate, usually twice as long as wide; fruit greater than 3,
   in terminal panicles, the panicles 5–10 cm long . . . .*C. scandens*

1. Leaves ovate-oblong to obovate, usually less than twice as long as wide; fruit, usually 2–3, in axillary clusters, the clusters less than 5 cm long ....*C. orbiculatus* (N)

Celtis   hackberry, sugarberry

1. Leaves yellow-green on both surfaces, lanceolate, usually greater than twice as long as wide, the margin entire or with a few teeth; bud scales puberulent ....*C. laevigata* (S)
1. Leaves dark green on upper surface, pale green on lower surface, lanceolate-ovate to ovate or deltoid, usually twice as long as wide or less, the margin entire or with many, prominent teeth; bud scales pubescent or glabrous
    2. Leaves mostly 6–8 cm long, 4–5 cm wide, chartaceous, strongly asymmetrical at the base, margin sharply toothed to near the base; fruits bitter, wrinkled when dry, the pedicels mostly 10–20 mm long; seeds pitted or angled with minute ridges ....*C. occidentalis*
    2. Leaves mostly 4–6 cm long, 2–4 cm wide, coriaceous, equal or slightly asymmetrical at the base; margins toothed completely, partly, or entire; fruits sweet, smooth when dry, the pedicels mostly 3–6 mm long; seeds smooth or obscurely pitted
        3. Branchlets red-brown, light brown to gray ....*C. bungeana* (OC)
        3. Branchlets dark brown to black ....*C. tenuifolia*

Cephalanthus occidentalis   buttonbush
Cephalotaxus   plum-yew

1. Leaves acuminate, mostly less than 4 cm long ....*C. harringtonia* (OC)
1. Leaves gradually tapering to a point, mostly greater than 4 cm long ....*C. fortunei* (OC)

Ceratiola ericoides   rosemary (S)
Cercidiphyllum japonicum   katsura, cotton candy tree (ON)
Cercis   redbud

1. Leaves rounded or emarginate at the apex ....*C. siliquastrum* (OC)
1. Leaves abruptly short-acuminate
    2. Leaves with margins green, the petioles usually 3 cm long ...*C. canadensis*
    2. Leaves with margins transparent, the petioles usually 2 cm long ....*C. chinensis* (OC)

Chaenomeles   flowering-quince

1. Leaves coarsely crenate, 2–5 cm long; twigs scabrous-tomentose; fruit less
   than 4 cm long ....*C. japonica* (N)
1. Leaves sharply serrate, 4–10 cm long; twigs glabrous or slightly pubescent;
   fruit 5–6 cm long ....*C. speciosa* (N)

Chamaecrista fasciculata   partridge-pea, locust weed
Chamaecyparis   false-cypress

1. Seam where lateral leaves meet usually visible; leaves not sharply
   keeled, the tips obtuse to acute, pressed close to the twig
   2. Visible portion of lateral leaves at least 2 times as long as that of the facial
      leaf; twigs 2 mm in diameter ....*C. obtusa*
   2. Visible portion of lateral leaves about equal in length or slightly
      longer than that of the facial leaf; twigs about 1 mm in diameter
      3. Leaves blue to blue-green, with white markings on lower surface of
         branchlet; male cones red ....*C. lawsoniana*
      3. Leaves green, pale green on lower surface of branchlet; male
         cones yellow ....**C. thyoides**
1. Seam where lateral leaves meet hidden; leaves sharply keeled, the
   tips sharply acute, spreading away from the twig
   4. Branchlets terete or nearly so; leaves less keeled, with white markings
      on lower surface; visible portion of lateral leaves often shorter than that
      of the facial leaf ....*C. pisifera* (OC)
   4. Branchlets appearing 4-angled; leaves sharply keeled, lacking
      white markings; visible portion of lateral leaves often equal in
      length or longer than that of the facial leaf
      5. Cone scales 4–6, each bearing 2 seeds ....*C. nootkatensis* (OC)
      5. Cone scales 8, each bearing 5 seeds ....×*Cupressocyparis leylandii*

Chamaedaphne calyculata   leatherleaf
Chimaphila   pipsissewa, prince's-pine, wintergreen

1. Leaf base obtuse; upper surface of leaf marked with white veins
   or spots ....*C. maculata*
1. Leaf base acute; upper surface of leaf green, lacking white veins
   or spots ....*C. umbellata*

Chimonanthus praecox   wintersweet (OC)
Chionanthus   fringetree

1. Leaves mostly oblong to obovate, mostly greater than
   10 cm long ....*C. virginicus*

1. Leaves mostly ovate or elliptic, less than
   10 cm long . . . .*C. retusus* (ON)

Cladrastis kentuckea   yellowwood
Clematis   clematis, virgin's bower, woodbine

This genus of climbing, half-woody to herbaceous vines contains many native species as well as popular cultivated species and hybrids. Vegetative characters are inadequate to provide a winter key and the following summer key, which is adapted from Blackburn, may be of limited usefulness.

1. Leaves or leaflets entire or lobed, not toothed
    2. Leaves densely, softly hairy on lower surface, simple
       or trifoliate . . . .*C. lanuginosa*
    2. Leaves glabrous or finely hairy on lower surface, the leaflets
       3 or more
        3. Leaflets usually more than 5, the terminal leaflet smaller
           than the basal ones
            4. Leaflets often sessile; leaves simple and sessile at base of
               the shoot . . . .*C. addisonii*
            4. Leaflets stalked; simple leaves, if present, stalked
                5. Leaflets glaucous or at least pale blue-green on both surfaces,
                   apiculate at the otherwise rounded apex . . . .*C. texensis* (CS)
                5. Leaflets green, lacking bloom, obtuse to acute or acuminate
                    6. Young twigs densely pubescent; leaflets finely and densely
                       pubescent on lower surface . . . .*C. viorna*
                    6. Young twigs sparsely pubescent or glabrous; leaflets
                       sparsely pubescent to glabrous
                        7. Leaflets mostly acute . . . .*C. crispa*
                        7. Leaflets mostly obtuse . . . .*C. viticella* (ON)
        3. Leaflets usually 3 or 5, the terminal leaflet equal in size to the
           basal ones
            8. Leaflets glabrous or subglabrous on lower surface
                9. Leaves evergreen, leathery; leaflets 3 . . . .*C. armandii*
                9. Leaves deciduous, thin; leaflets 5 . . . .*C. terniflora* (ON)
            8. Leaflets finely hairy on lower surface
                10. Leaflets 2–5 cm long . . . .*C. flammula* (ON)
                10. Leaflets mostly greater than 5 cm long
                    11. Lower leaves with 2 distant pairs of leaflets; upper leaves
                        sometimes simple . . . .*C. patens*
                    11. Lower leaves with leaflets and leaflet pairs closer
                        together; upper leaves with 3 or more leaflets
                        12. Leaflets mostly 5; flowers mostly in groups
                            of 5 . . . .*C. ×lawsoniana*

    12. Leaflets mostly 3; flowers mostly in groups of 3 or solitary
        13. Leaflets acute, abruptly acuminate, or sometimes acuminate; flowers
            blue or purple to red . . . .***C. ×jackmannii*** (N)
        13. Leaflets acuminate to long-acuminate; flowers white . . .***C. florida*** (ON)
1. Leaves or leaflets toothed or some leaves lobed and partly toothed,
  a few leaves entire
  14. Leaflets mostly 3
      15. Leaves rounded to cordate at the base; leaves and young
          shoots scattered along the branchlet
        16. Leaflets usually greater than 6 cm long, 3, sometimes 5,
            the basal ones lacking lobes, not divided
          17. Leaflets mostly with fewer than 8 large teeth . . . . . .***C. virginiana***
          17. Leaflets mostly with 10 or more large teeth . . . .*C. ×jouiniana*
        16. Leaflets usually less than 6 cm long, 3, sometimes the basal ones
            3-lobed or with 3 additional leaflets . . . .*C. apiifolia*
      15. Leaves cuneate to rounded at the base; leaves and young
          shoots clustered at nodes on older growth
        18. Stalks of the leaflets mostly greater than 2 cm long . . . .***C. occidentalis***
        18. Stalks of the leaflets mostly less than 2 cm long . . . .*C. montana*
  14. Leaflets 5 or more
      19. Leaflets with a few irregular teeth
        20. Leaflets thin, mostly more than 3 cm wide . . . .*C. vitalba*
        20. Leaflets thick, mostly less than 3 cm wide . . . .***C. catesbyana*** (s)
      19. Leaflets evenly toothed
        21. Stalks of leaflets mostly greater than 2 cm long; twigs glabrous
          or subglabrous . . . .***C. orientalis*** (ON)
        21. Stalks of leaflets less than 2 cm long; twigs with long hairs
          22. Leaflets with teeth spreading horizontally
            or outward . . . .***C. tangutica*** (ON)
          22. Leaflets with teeth ascending or pointing forward
            toward the apex . . . .*C. serratifolia* (oc)

Clerodendrum   glory-bower, clerodendrum

1. Leaves lanceolate, the base cuneate . . . .*C. indicum* (cs)
1. Leaves ovate, the base truncate
  2. Leaf margins with glandular teeth . . . .***C. japonicum*** (ON)
  2. Leaf margins entire or toothed . . . .*C. trichotomum*

Clethra   sweet pepperbush

1. Leaf blades 3–10 cm long, obovate, sharply serrate, obtuse to nearly acute;
  racemes of capsules usually in erect panicles; bark of older stems gray or black-
  brown, often flaking; pubescence of twigs appressed . . . .***C. alnifolia***

1. Leaf blades 7–15 cm long, oval, oblong, or elliptic, acuminate, finely serrate; racemes of capsules usually solitary, flexible, drooping; bark of older stems red-brown, exfoliating in thin strips; pubescence of twigs spreading . . . . *C. acuminata*

Cleyera japonica   Japanese cleyera (cs)

Cocculus carolinus   Carolina-moonseed (s)

Colutea arborescens   bladder-senna (on)

Comptonia peregrina   sweetfern

Corema conradii   broom-crowberry

Cornus   dogwood

Summer key

1. Leaves alternate . . . .*C. alternifolia*
1. Leaves opposite
    2. Leaves scabrous-pubescent on upper and lower surface; twigs and branchlets scabrous
        3. Leaves minutely or barely scabrous on lower surface; pith white or beige; widely distributed . . . .*C. drummondii*
        3. Leaves strongly scabrous on lower surface; pith white; rare endemic of North Carolina and South Carolina . . . .*C. asperifolia* (s)
    2. Leaves, twigs, and branchlets pubescent or glabrous; if scabrous, only on upper leaf surface
        4. Twigs, petioles, or lower surface of leaves pubescent, the hairs spreading or often woolly (alternate step 4, p. 148)
            5. Lower leaf surface pubescent with prominent, often rusty patches in the axils of the veins; flowers yellow, blooming before the leaves; branchlets and twigs glabrous; fruit red . . . .*C. officinalis*
            5. Lower leaf surface pubescent or glabrous, lacking conspicuous patches in axils of the veins; flowers green or yellow-green, blooming after or as the leaves emerge; branchlets and twigs pubescent or glabrous; fruit red, white, or blue
                6. Leaves very broadly ovate, with 6–8 pairs of lateral veins, very densely woolly pubescent on lower surface; twigs streaked with dark purple; fruit blue or blue-gray . . . .*C. rugosa*
                6. Leaves lanceolate to broadly ovate, with 3–5 (rarely 6) pairs of veins, woolly pubescent to nearly glabrous; twigs brown, to green or red, not streaked; fruit red, white, or blue

7. Leaves with 5–6 pairs of lateral veins; twigs appressed-pubescent, usually glaucous; lateral branches longer than main axis; bark rough, dark brown, with alligator-skin pattern; trees; flowers in tight clusters subtended by 4 large showy white bracts; fruit in dense heads, red . . . . ***C. florida***

7. Leaves with 3–4 pairs of veins; twigs usually spreading-pubescent in part, not glaucous; lateral branches shorter than the main axis; bark various, lacking alligator-skin pattern; shrubs; flower clusters loose, lacking showy bracts; fruit in loose cymes

  8. Leaves scabrous on upper surface; branchlets and branches gray to light brown; fruit white . . . . ***C. drummondii***

  8. Leaves smooth or nearly so on upper surface; branchlets and branches purple (often becoming olive or gray later); fruit white or blue

    9. Pith brown; pubescence on veins on lower surface of leaves spreading, but between veins mostly appressed; fruit blue . . . . ***C. amomum***

    9. Pith white; pubescence between veins on lower surface of leaves woolly; fruit white

      10. Leaves acuminate . . . . ***C. stolonifera***

      10. Leaves acute . . . . *C. alba*

4. Twigs, petioles, and lower surface of leaves glabrous or appressed-pubescent, the hairs short, not woolly

  11. Bark flaking, mottled with patches of gray, green, or red-brown; fruit connate into a globose head; flowers subtended by large creamy white bracts . . . . *C. kousa*

  11. Bark uniformly brown or dark brown; fruit in cymes, umbels, panicles, or solitary, distinct; flowers with subtending bracts minute or lacking

    12. Branches and branchlets red-purple, occasionally olive; pith white; leaves glaucous on lower surface; fruit in cymes, white or black

      13. Lenticels absent or inconspicuous; fruit purple to blue or black

        14. Shrubs; twigs and branchlets becoming bright red; fruit black . . . . ***C. sanguinea*** (ON)

        14. Small to large trees; twigs remaining brown or deep purple-brown; fruit purple to black

          15. Branchlets often glaucous; small to large trees; fruit blue-black . . . . *C. macrophylla* (CS)

          15. Branchlets glabrous or pubescent, not glaucous; small trees; fruit purple to black . . . . *C. coreana* (OC)

      13. Lenticels sometimes prominent; fruit white

16. Leaves acuminate . . . .*C. stolonifera*
16. Leaves acute . . . .*C. alba*
12. Branches and usually also branchlets gray; pith white or brown;
leaves pale or green on lower surface, not glaucous; fruit solitary,
in cymes, umbels, or panicles, white, purple-blue, or blue-black
17. Large trees; twigs dark brown to black; fruit
blue-black . . . .*C. controversa* (oc)
17. Small trees or shrubs; twigs light brown or olive to brown or
gray; fruit white or red
18. Leaves broadly ovate; lateral buds divaricate; lateral branchlets often
short (2-leaved), nearly at right angles to the branch; fruit solitary or
in umbels, red . . . .*C. mas*
18. Leaves elliptical to ovate-lanceolate; lateral buds appressed; lateral
branchlets elongated, at acute angles; fruit in cymes or panicles,
white, on red pedicels . . . .*C. foemina*

Winter key

1. Leaf scars alternate, mostly crowded near the ends of twigs . . . .*C. alternifolia*
1. Leaf scars opposite
2. Twigs and branchlets scabrous, the pith white . . . .*C. asperifolia* (s)
2. Twigs and branchlets pubescent or glabrous, not scabrous, the
pith white or brown
3. Twigs and branchlets green or red-brown, streaked or speckled with
dark purple . . . .*C. rugosa*
3. Twigs and branchlets uniform in color or with different color
on lower and upper surface, not streaked or speckled, green,
red-purple, red-brown, or gray
4. Lateral buds hidden behind raised leaf scars; enlarged depressed
globose flower buds often present, 5–8 mm in diameter; twigs usually
glaucous, terete or flattened; bark dark brown, rough, with
alligator-skin pattern . . . .*C. florida*
4. Lateral buds distinctly visible; enlarged flower buds present
or not; twigs glabrous or pubescent, usually terete; bark
various, lacking alligator-skin pattern
5. Lateral buds divergent; enlarged globose flower buds often present;
twigs usually green . . . .*C. mas*
5. Lateral buds appressed; flower buds not prominently
enlarged; twigs red-purple, brown, gray, or black
6. Twigs and branchlets bright red-purple; lenticels few, 8–20
per internode, separate and conspicuous on twigs,
branchlets, and branches; pith white; plants may be
stoloniferous
7. Young twigs green, branchlets bright red;
fruit black . . . .*C. sanguinea* (on)

7. Young twigs usually red or brown, occasionally green; branchlets
darker red-purple; fruit white to blue

   8. Shrubs stoloniferous, often forming dense thickets; branches tending to
be decumbent; seeds ovoid, rounded at the base . . . .***C. stolonifera***

   8. Shrubs, lacking stolons; branches erect, spreading; seeds somewhat
flattened, narrowed at the base . . . .*C. alba*

6. Twigs and branchlets dull purple, brown, gray, or black; lenticels
absent, or conspicuous but minute (occasionally few and
conspicuous in *C. controversa*) and numerous on twigs and
branchlets, often confluent on branchlets; pith white, pale brown,
or brown; plants lacking stolons

   9. Lenticels prominent but minute on twigs and branchlets; twigs light
brown; branchlets gray, the pith white or pale brown, the
pubescence appressed . . . .***C. foemina***

   9. Lenticels absent on most twigs, sometimes becoming abundant
on branchlets; twigs and branchlets purple, red-brown,
light brown, gray, or black, the pith white or brown, the
pubescence spreading or appressed

      10. Twigs and branchlets purple; 3–4 year branchlets often streaked by
confluent lenticels, the pith brown, the youngest internodes
densely pubescent . . . .***C. amomum***

      10. Twigs light brown, red-brown, gray, or olive; branchlets and
branches light brown, gray, or black, rarely streaked with
lenticels, the pith white or brown, the internodes glabrous
to pubescent

         11. Twigs dark brown to black; large tree; fruit
blue-black . . . .*C. controversa* (oc)

         11. Twigs red-brown to light brown, gray, or olive; shrubs,
small or large trees; fruit red to purple, white,
blue, or black

            12. Twigs red-brown; branchlets light brown or gray;
fruit white or light blue . . . .***C. drummondii***

            12. Twigs light brown, gray, or olive; branchlets brown to
dark brown or black; fruit red, purple, or blue

               13. Bark tight or close; fruit purple to blue

                  14. Branchlets often glaucous; small to large trees; fruit
blue-black . . . .*C. macrophylla* (cs)

                  14. Branchlets glabrous or pubescent, not glaucous; small trees;
fruit purple to black . . . .*C. coreana* (oc)

               13. Bark flaking or curling; fruit red

                  15. Bark flaking, mottled with patches of gray, green, or brown;
branchlets black; fruit connate into a globose head . . .*C. kousa*

                  15. Bark curling, uniformly brown; branchlets dark brown or
gray; fruit distinct, oblong . . . .*C. officinalis*

Cortaderia selloana　　pampas grass (CS)

Corylopsis　　corylopsis

1. Leaves greater than 2 times as long as wide . . . . *C. sinensis*
1. Leaves less than 2 times as long as wide
　2. Leaves mostly less than 5 cm long . . . . *C. pauciflora*
　2. Leaves mostly greater than 5 cm long
　　3. Leaves toothed, the margins between the teeth scalloped or concave, the
　　　lower surface sparsely pubescent, the petioles of young leaves rusty-
　　　pubescent; twigs rusty-pubescent or glabrous . . . . *C. glabrescens*
　　3. Leaves toothed or undulate, the margins between the teeth or veins
　　　straight or convex, the lower surface moderately to densely pubescent,
　　　the petioles of young leaves yellow-pubescent; twigs
　　　yellow-pubescent . . . . *C. spicata*

Corylus　　hazelnut, filbert

1. Bud scales, except the 2 lower ones, as long as the bud; fruit husk
　(involucre) elongated into a tube, the tube greater than 1½ times
　as long as the nut; twigs and petioles glabrous or
　glandular-pubescent
　2. Petioles usually less than 1 cm long; fruit usually 1–2, occasionally 3 to
　　a cluster . . . . **C. cornuta**
　2. Petioles usually greater than 1 cm long; fruit usually 3 to
　　a cluster . . . . *C. sieboldiana* (OC)
1. Bud scales of varying lengths, several scales less than the bud length;
　twigs often coarsely glandular-pubescent; fruit husk (involucre)
　usually not elongated into a tube, if so less than 1½ times as long
　as the nut; twigs and petioles mostly coarsely glandular-pubescent
　3. Petioles mostly greater than 15 mm long; fruit husks divided into many
　　linear segments . . . . *C. colurna*
　3. Petioles mostly less than 15 mm long; fruit husks merely toothed
　　or lobed
　　4. Leaves often broadly truncate, emarginate or obcordate at the apex, the
　　　apex often shorter than the side lobes; twigs light gray to dark gray; fruit
　　　husks with triangular teeth about 5 mm long . . . . *C. heterophylla* (OC)
　　4. Leaves acuminate; twigs light gray to dark brown or black;
　　　fruit husks irregularly and deeply lobed
　　　5. Leaves greater than 1½ times as long as wide; twigs light gray, light
　　　　brown, or dark gray; buds small with no more than 5 scales exposed,
　　　　the scales mostly lacking hairs at the margins; fruit husk twice as
　　　　long as the nut . . . . **C. americana**
　　　5. Leaves less than 1½ times as long as wide; twigs mostly
　　　　brown, dark brown, or black; buds often with more than 5

scales exposed, the scales ciliate; fruit husk (involucre) less
than twice as long as the nuts

  6. Fruit husk closed into a tube, concealing the nut, much longer than
the nut . . . .*C. maxima*

  6. Fruit husk open, exposing the nut, usually shorter than
the nut . . . .*C. avellana*

Cotinus   smoketree

1. Small trees to 12 m tall; branches few; leaves pubescent at least along the veins
on lower surface, mostly greater than 7 cm long . . . .*C. obovatus* (s)
1. Shrubs usually less than 5 m tall; branches many and spreading; leaves
glabrous, mostly less than 7 cm long . . . .*C. coggygria* (ON)

Cotoneaster horizontalis   rock cotoneaster

There are no native cotoneaster species, but there are many species very similar to
rock cotoneaster in cultivation, mostly as hillside or ground covers.

Crataegus   hawthorn

This taxonomically complex genus is not easily deciphered at the species level.
Hybridization, polyploidy, and asexual reproduction make species identification
difficult if not impossible. Hundreds of native northeastern North American
species have been described with no universal acceptance. From these Cronquist
has pulled twenty-two names for the complexes he describes. His treatment is
adapted for use here. Six native species with strictly southern distribution are
not keyed here: *C. aestivalis*, *C. collina*, *C. flava*, *C. harbisonii*, *C. schuettii*,
and *C. triflora*

1. Leaves with veins meeting the margin at the sinuses and the apices
of the lobes
  2. Nutlets usually 1, rarely 2–3 . . . .*C. monogyna* (N)
  2. Nutlets 3–5
    3. Leaves acute or acuminate, often cordate at the base,
2–5 cm wide . . . .*C. phaenopyrum*
    3. Leaves rounded or almost acute, narrowed or attenuate at the
base, often less than 2 cm wide
      4. Leaves glabrous, those on flowering branches mostly narrowly
obovate, 5–15 mm wide, the lobing barely more than
a tendency . . . .*C. spathulata* (s)
      4. Leaves densely pubescent when young, those on the flowering
branches mostly deltoid or ovate, greater than 15 mm wide,
obviously lobed . . . .*C. marshallii* (s)

1. Leaves with veins extending to the margin only at the apices of the lobes or the larger teeth
    5. Branches with thorns usually less than 2 cm long; fruit purple-black; nutlets pitted . . . .*C. douglasii*
    5. Branches with thorns usually longer than 3 cm; fruit red, yellow, or green; nutlets pitted or not
        6. Nutlets 2–3, with deep pit on inner surface, rounded at the ends
            7. Twigs tomentose or villous; leaves 5–9 cm long, 4–8 cm wide . . . .*C. calpodendron*
            7. Twigs glabrous or subglabrous; leaves 3–6 cm long; 2–5 cm wide . . . .*C. succulenta*
        6. Nutlets 2–5, lacking pits (except slightly so in *C. brainerdii*), acute or rounded
            8. Fruits 1–3; shrubs . . . .*C. uniflora* (s)
            8. Fruits usually at least 3, rarely 2–3; trees or shrubs
                9. Leaves on flowering branches mostly with base broad and rounded, truncate, cordate, or subcordate, usually lobed or incised
                    10. Leaves strongly pubescent, at least along the veins on the lower surface; fruit pubescent, at least at the ends . . . .*C. mollis*
                    10. Leaves glabrous or glabrate; fruit glabrous
                        11. Leaves often cordate
                            12. Leaves glabrous, with crisped margins . . .*C. coccinioides* (s)
                            12. Leaves short-pubescent on upper surface when young, margins flat, not crisped . . . .*C. dilatata*
                        11. Leaves rounded, truncate, or subcordate
                            13. Sepals usually deeply glandular-serrate . . . .*C. coccinea*
                            13. Sepals entire or sometimes a few teeth and glands present
                                14. Leaves glabrous when young . . . .*C. pruinosa*
                                14. Leaves with a few short hairs on upper surface when young . . . .*C. flabellata*
                9. Leaves on flowering branches mostly with narrow, acute or attenuate base, with or without lobes
                    15. Leaves on flowering branches mostly elliptic, rhombic, ovate, or occasionally obovate, usually with lobes
                        16. Petiole and blade base bearing glands; fruit usually yellow or bronze-green, hard and dry
                            17. Leaves on flowering branches usually less than 3 cm long, sometimes lobed above the middle . . . . . .*C. michauxii* (s)
                            17. Leaves on flowering branches usually 3–5 cm long, lobed nearly to the base or occasionally lacking lobes . . . .*C. intricata*

16. Petiole and blade base lacking glands; fruit red or occasionally
    yellow, often becoming succulent
    18. Leaves usually abruptly acuminate; nutlets with a very shallow pit
    on inner surface . . . .***C. brainerdii***
    18. Leaves rounded, obtuse, or subacute; nutlets
    lacking pits . . . . ***C. chrysocarpa***
15. Leaves on flowering branches mostly obovate to oblong-elliptic,
    mostly lacking lobes
    19. Leaves lustrous on upper surface; fruit hard and dry; nutlets
    usually 1–2
        20. Leaves pubescent at least when young; twigs and branchlets
    pubescent or glabrous . . . .***C. berberifolia*** (s)
        20. Leaves glabrous; twigs and branchlets glabrous . . . .***C. crus-galli***
    19. Leaves dull on upper surface; fruit often becoming succulent;
    nutlets 3–5
        21. Leaves thin, the upper surface with veins obscure; fruit
    5–8 mm long . . . .***C. viridis***
        21. Leaves thick, the upper surface with veins impressed; fruit
    8–25 mm long
            22. Leaves on flowering branches mostly 2–6 cm long . . . . .***C. punctata***
            22. Leaves on flowering branches mostly less than
    3 cm long . . . .***C. michauxii*** (s)

Cryptomeria japonica　　　cryptomeria (oc)

Cunninghamia lanceolata　　　cunninghamia, China-fir (oc)

×*Cupressocyparis leylandii*　　　leyland-cypress

Cupressus arizonica　　　cypress (oc)

Cydonia oblonga　　　common quince

Cyrilla racemiflora　　　titi, leatherwood (s)

Cytisus scoparius　　　Scotch-broom (n)

Daphne　　　daphne

1. Leaves less than 3 cm long; shrubs procumbent, to 1 m tall . . . .*D. cneorum*
1. Leaves greater than 3 cm long; shrubs erect, usually greater than 2 m tall
    2. Leaves less than 8 mm wide . . . .*D.* ×*burkwoodii*
    2. Leaves greater than 8 mm wide
        3. Leaves persistent, coriaceous . . . .*D. odora*
        3. Leaves deciduous, thin . . . .***D. mezereum*** (n)

Daphniphyllum macropodum   daphniphyllum (CS)

Davidia involucrata   dovetree (OC)

Decumaria barbara   decumaria (S)

Deutzia   deutzia

There are a number of other species and hybrid species in cultivation that fall somewhere between the two species keyed here.

1. Leaves ovate to oblong, denticulate, nearly glabrous on lower surface, the petioles 1–4 mm long . . . .**D. scabra** (ON)
1. Leaves lanceolate, sharply serrulate, stellate-pubescent on lower surface, the petioles greater than 4 mm long . . . .*D. gracilis*

Diapensia lapponica   diapensia

Diervilla   bush-honeysuckle

1. Leaves with petioles 5–10 mm long; twigs ridged or quadrangular . . . .**D. lonicera**
1. Leaves sessile or with petioles less than 3 mm long; twigs usually terete . . . .**D. sessilifolia**

Diospyros   persimmon

1. Leaves cuneate, the petiole less than 2 cm long; buds oblong, acute . . . .*D. lotus* (OC)
1. Leaves rounded to cuneate at the base, the petiole 2–3 cm long; buds ovoid, blunt . . . .**D. virginiana**

Dirca palustris   leatherwood

Edgeworthia papyrifera   paperbush, mitsumata (OC)

Ehretia acuminata   ehretia (OC)

Elaeagnus   oleaster, Russian-olive, elaeagnus

1. Leaves persistent; flowering in fall; plants usually spiny . . . . .**E. pungens** (ON)
1. Leaves deciduous; flowering in spring; plants unarmed or spiny
   2. Leaves, branchlets, twigs, and buds covered with silver scales only . . . .**E. angustifolia** (N)
   2. Leaves, branchlets, twigs, and buds covered with brown and silver scales

Keys to Species

3. Leaves remaining covered with silver scales on both surfaces, occasionally with a few brown scales on lower surface; fruit remaining silver . . . .*E. commutata*
3. Leaves with silver scales on upper surface only when young, usually with brown scales mixed with silver scales on lower surface; fruit becoming red
    4. Branchlets covered with red-brown scales; leaves usually not crisped at the margin; fruiting pedicels 19–25 mm long; floral tube longer than the limb . . . .***E. multiflora*** (ON)
    4. Branchlets yellow-brown with some silver scales; leaves often crisped at the margin; fruiting pedicels about 6 mm long; floral tube equal to the length of the limb . . . . ***E. umbellata*** (N)

Eleutherococcus sieboldianus   five-leaf-aralia, spiny-panax (OC)

Empetrum   crowberry

1. Twigs tomentose or villous, lacking glands; fruit red or purple . . . .*E. rubrum*
1. Twigs with tiny stipitate glands; fruit black . . . .*E. nigrum*

Enkianthus   enkianthus

1. Leaves glabrous or sparsely pubescent along midrib on upper surface; capsules on straight erect stalks . . . .*E. perulatus*
1. Leaves with bristly hairs on the midrib and veins on upper surface; capsules nodding on curved stalks . . . .*E. campanulatus*

Ephedra distachya   ephedra

Epigaea   trailing arbutus

1. Leaves rounded at the apex; flowers greater than 1 cm wide . . . .***E. repens***
1. Leaves acuminate; flowers less than 1 cm wide . . . .*E. asiatica*

Erica   heath

There are many European species of this genus that are cultivated in rock gardens and other ornamental situations. Some of the larger-flowered South African species are cultivated in the South. ***E. tetralix*** has naturalized as well as a few other more rarely occurring species.

1. Shrubs greater than 1 m high . . . .*E. scoparia*
1. Shrubs less than 1 m high
    2. Leaves ciliate . . . . ***E. tetralix*** (N)
    2. Leaves glabrous
        3. Leaves in whorls of 3; flowers in whorls . . . .***E. cinerea*** (ON)
        3. Leaves in whorls of 4–5; flowers solitary or in pairs in the axils

4. Shrubs low, tufted, less than 20 cm high ....*E. carnea*
4. Shrubs 15–30 cm high ....*E. vagans* (ON)

Eriobotrya japonica   loquat, Japanese-medlar (CS)

Eucommia ulmoides   eucommia (OC)

Euonymus   burning bush, spindle tree, euonymus

1. Leaves persistent, often white or noticeably lighter along the
   principal veins; stems trailing, climbing, or erect; branchlets lacking
   wings, smooth or warty
   2. Stems trailing or climbing; branchlets warty, terete ......*E. fortunei* (ON)
   2. Stems erect; branchlets smooth, slightly 4-sided or ridged,
      often striped
      3. Leaves stiff, persisting through the second summer ....*E. japonica*
      3. Leaves somewhat flexible, persisting only through the
         second spring ....*E. kiautschovica* (CS)
1. Leaves deciduous, veins green; stems erect or trailing, not climbing;
   branchlets smooth or winged
   4. Branches and branchlets with corky wings ....*E. alata* (N)
   4. Branches and branchlets smooth, lacking corky wings
      5. Petioles 1–5 mm long; branchlets square in cross section; low
         bushes or stems trailing, rooting at the nodes; fruit red, warty
         6. Leaves obovate, thick, dull green, apex often obtuse; plants trailing,
            the branches creeping or prostrate ....*E. obovata*
         6. Leaves ovate to lanceolate, thin, bright green, apex acute; plants erect,
            straggling, loosely branched ....*E. americana*
      5. Petioles 10–30 mm long; branchlets terete; tall shrubs
         or small trees, stems erect; fruit smooth
         7. Lower leaf surface usually puberulent; fruit purple;
            aril red ....*E. atropurpurea*
         7. Lower leaf surface glabrous; fruit pink; aril orange ...*E. europaea* (N)

Euptelea polyandra   tasseltree, Japanese euptelea (OC)

Exochorda   pearlbush

1. Leaves pale or blue green on lower surface, the petioles less than
   12 mm long; fruit 8 mm long ....*E. racemosa* (ON)
1. Leaves light green, pale only along veins on lower surface, the petiole greater
   than 12 mm long; fruit 8–13 mm long ....*E. giraldii* (ON)

Fagus   beech

1. Leaves denticulate or undulate, with 5–10 veins on each side of
   the midrib . . . .*F. sylvatica*
1. Leaves serrate, with 10–20 veins on each side of the midrib . . . .*F. grandifolia*

×*Fatshedera lizei*   aralia-ivy, tree-ivy (cs)

Fatsia japonica   Japanese fatsia, Formosa rice tree (cs)

Ficus carica   common fig (on, cs)

Firmiana simplex   Chinese parasol tree (n)

Fontanesia   fontanesia

1. Leaves glossy, entire; plants loosely branched . . . .*F. fortunei* (cs)
1. Leaves gray-green, minutely toothed or at least wavy; plants densely branched,
   the branches spreading . . . .*F. phillyreoides* (cs)

Forestiera   swamp-privet

1. Leaves glabrous on the upper surface, glabrous or pubescent on lower surface,
   acuminate or acute; fruit 10–15 mm long . . . .*F. acuminata* (s)
1. Leaves short-pubescent on the midrib on the upper surface, pubescent
   on the lower surface, obtuse or rounded at the apex; fruit
   7–8 mm long . . . .*F. ligustrina* (s)

Forsythia   forsythia, golden bells

1. Branches and branchlets hollow between the nodes . . . .*F. suspensa* (on)
1. Branches and branchlets usually with chambered pith (hollow in
   rapidly growing twigs) between the nodes
   2. Pith chambered at the nodes; leaves always simple . . . .*F. viridissima* (on)
   2. Pith solid at the nodes; leaves sometimes compound . . . .*F. intermedia*

Fothergilla   fothergilla, witch-alder

1. Leaves 5–12 cm long; shrub 2–3 m tall . . . .*F. major* (s)
1. Leaves mostly less than 5 cm long; shrub less than 1 m tall . . . .*F. gardenii*

Franklinia alatamaha   franklinia

Fraxinus   ash

Summer key

1. Leaves simple . . . .*F. excelsior* "Diversifolia" and "Hessei"
1. Leaves compound

2. Twigs acutely 4-angled at internodes . . . .*F. quadrangulata*
2. Twigs terete or obtusely angled
   3. Twigs and petioles pubescent or tomentose, at least
   when young
      4. Twigs and petioles pubescent; petiolules of lower leaflets winged nearly
      to base . . . .*F. pennsylvanica* **var.** *pennsylvanica*
      4. Twigs, leaf rachis, and petioles tomentose; petiolules of
      lower leaflets partially winged or lacking wings
         5. Upper margin of leaf scar deeply notched; fruit terete, winged to
         the middle of the body . . . .*F. profunda*
         5. Upper margin of leaf scar with shallow notch for the bud; fruit
         flattened, winged to the base . . . .*F. caroliniana* (s)
   3. Twigs, leaf rachis, and petioles glabrous or nearly so
      6. Lateral leaflets sessile, 10; buds black, blue-black, or brown
         7. Leaf rachis tomentose at base of leaflets . . . .*F. nigra*
         7. Leaf rachis glabrous . . . .*F. excelsior*
      6. Lateral leaflets stalked, 6; buds rusty, brown, or dark brown
         8. Leaflets mostly 5–7 cm long, often conspicuously rusty-pubescent
         toward the base of the midrib on the lower surface; fruit usually
         2–3 cm long . . . .*F. ornus* (oc)
         8. Leaflets mostly 7–15 cm long, glabrous or
         inconspicuously white-pubescent toward the base of the
         midrib on the lower surface; fruit usually 3–6 cm long
            9. Leaflets elliptic-oval; petiolules of lower leaflets nearly wingless;
            branchlets glabrous; fruit winged only in upper third of
            the body . . . .*F. americana*
           9. Leaflets elliptic-lanceolate; petiolules of lower leaflets winged
           nearly to base; branchlets pubescent; fruit winged on
           upper half of the body, sometimes nearly to
           the base . . . .*F. pennsylvanica* **var.** *lanceolata*

Winter key

1. Twigs acutely 4-angled; buds gray, woolly . . . .*F. quadrangulata*
1. Twigs terete or obtusely angled; buds brown to black
   2. Upper margin of leaf scars deeply concave, or with a sharply
   V-shaped notch
      3. Twigs tomentose at least when young (sometimes glabrous);
      fruit with wing decurrent to middle or more
         4. Fruit body terete, winged to about the middle . . . .*F. profunda*
         4. Fruit body flattened, winged to the base . . . .*F. caroliniana* (s)
   3. Twigs glabrous
      5. Buds puberulent to glabrous, dark brown to black; fruit body winged
      only in upper half, mostly 3–6 cm long . . . .*F. americana*

Keys to Species

5. Buds pubescent, brown; fruit body winged to below the middle, usually
   2–3 cm long . . . .*F. ornus* (OC)
2. Upper margin of leaf scars convex, straight, or seldom
   slightly concave
   6. Lateral buds nearly globose, blue-black, black, or brown; twigs
      dull or glossy, the lenticels conspicuous or not
      7. Buds blue-black; terminal bud mostly obtuse or rounded, dome-
         shaped; twigs somewhat glossy, mostly green, the lenticels
         inconspicuous or lacking; bark on older branches thin or thick, not
         corky; leaf scars semicircular; cultivated . . . .*F. excelsior*
      7. Buds black or sometimes brown; terminal bud acute; twigs dull gray,
         with conspicuous lenticels; bark on older branches corky; leaf scars
         circular; native growing in wet soils . . . .***F. nigra***
   6. Lateral buds reniform in anterior view and also compressed
      parallel to the surface of the twig, brown; twigs glossy or
      velvety; lenticels often raised and conspicuous
      8. Twigs pubescent . . . .*F.* ***pennsylvanica***
      8. Twigs glabrous or glabrate . . . .*F.* ***pennsylvanica*** var. ***lanceolata***

Gardenia jasminoides   common gardenia, cape-jasmine (CS)

Gaultheria   wintergreen, creeping-snowberry

1. Stems that bear leaves nearly erect; fruit red . . . .***G. procumbens***
1. Stems that bear leaves prostrate; fruit white . . . .***G. hispidula***

Gaylussacia   huckleberry

1. Leaves coriaceous, persistent, serrate, lacking resinous dots;
   twigs angled . . . .***G. brachycera***
1. Leaves thin or thick but not coriaceous, deciduous, usually entire,
   with resinous dots; twigs terete
   2. Leaves yellow-green with resinous dots on both leaf surfaces; fruit lacking
      a bloom . . . .***G. baccata***
   2. Leaves green or yellow-green with resinous dots only on lower
      surface; fruit glaucous or not
      3. Leaves glaucous, glabrous, or sometimes pubescent on lower surface;
         twigs glaucous; fruit glaucous, dark blue . . . .***G. frondosa***
      3. Leaves lacking bloom but usually pubescent on lower surface;
         twigs pubescent, usually lacking bloom; fruit pubescent or
         glandular-pubescent, lacking bloom, black
         4. Leaves greater than 5 cm long; stalks of the fruit mostly greater than
            8 mm long; plants greater than 2 m tall . . . .***G. ursina*** (S)

4. Leaves less than 4 cm long; stalks of the fruit less than 3 mm long; plants less than 1 m tall . . . .*G. dumosa*

Gelsemium   jessamine

1. Beak of capsule greater than 2 mm long; seed lacking wings . . . .*G. rankinii* (s)
1. Beak of capsule less than 2 mm long; seed winged . . . .*G. sempervirens*

Genista tinctoria   Dyer's greenwood, woodwaxen (N)
Ginkgo biloba   ginkgo (ON)
Gleditsia   honey-locust

1. Thorns branched or lacking (in most cultivated varieties); fruit linear, 10 cm long or more, pulpy, more than 3-seeded . . . .*G. triacanthos*
1. Thorns mostly simple; fruit oval, about 3–4 cm long, dry, not pulpy, 1–3 seeded . . . .*G. aquatica* (s)

Gordonia lasianthus   loblolly-bay (s)
Gymnocladus dioica   Kentucky coffeetree
Halesia   silverbell

1. Leaves ovate to obovate, mostly abruptly acuminate; fruit 2-winged . . . .*H. diptera* (s)
1. Leaves elliptic to elliptic-obovate, mostly long-acuminate; fruit 4-winged
    2. Shrub or small tree, the trunks several, arching; leaves less than 12 cm long; fruit often less than 3 cm long . . . .*H. tetraptera*
    2. Tree, the trunk usually single, sometimes 2 or 3, straight; leaves mostly greater than 12 cm long; fruit 3–5 cm long . . . .*H. monticola* (s)

Hamamelis   witch-hazel

1. Leaves with lower surface tomentose; twigs tomentose (glabrate in *H. ×intermedia*), the hairs brown or buff; branchlets glabrate; fruit exposed greater than $2/3$ its length; flowering in winter
    2. Twigs tomentose . . . .*H. mollis*
    2. Twigs glabrate . . . .*H. ×intermedia*
1. Leaves with lower surface glabrous or pubescent only on the veins (tomentose all over in some forms of *H. vernalis*); twigs glabrous or sparsely pubescent, the hairs dark brown; branchlets glabrous; fruit exposed less than $2/3$ its length (greater than $2/3$ in *H. japonica*); flowering in fall or winter

Keys to Species

3. Leaves strongly cuneate, wedge-shaped, or long-tapering at the base, often diamond-shaped, the basal half usually entire, veins usually fewer than 6 pairs; flowering in winter . . . .*H. vernalis* (OC)
3. Leaves rounded or only slightly cuneate toward the base, obovate, usually entire for only the basal third or less, veins usually 6 pairs or more; flowering in winter or fall
    4. Leaves lustrous on both surfaces, usually 5–10 cm long; fruit exposed greater than $^2/_3$ its length; flowering in winter; sepals dark red or purple on inner side . . . .*H. japonica*
    4. Leaves dull on upper surface, usually 8–15 cm long; fruit exposed less than $^2/_3$ its length; flowering in fall; sepals yellow-brown on inner side . . . .***H. virginiana***

Harrimanella hypnoides   moss-heather

Hedera helix   English ivy (ON)

Hemiptelea davidii (OC)

Heptacodium miconioides   seven son flower (OC)

Hibiscus syriacus   rose-of-sharon, althea (N)

Hippophae rhamnoides   sea-buckthorn (ON)

Hovenia dulcis   Japanese raisin tree (ON)

Hudsonia   hudsonia

1. Plant villous-tomentose; leaves appressed, oval or oblong, 2 mm long . . . .***H. tomentosa***
1. Plant finely pubescent, downy; leaves divaricate, subulate, 3–6 mm long
    2. Leaves sparsely pubescent, the hairs in tufts; endemic in mountains of North Carolina . . . .***H. montana*** (S)
    2. Leaves pubescent, the hairs usually not in tufts; Nova Scotia to North Carolina . . . .***H. ericoides***

Hydrangea   hydrangea

1. Leaves 3–7 lobed, red-tomentose . . . .***H. quercifolia*** (S)
1. Leaves serrate, but not lobed; glabrous or brown-pubescent
    2. Leaves broadly ovate or suborbicular, the blade less than $1^1/_2$ times as long as wide, greater than 6 cm wide; stems often climbing and twisting, aerial rootlets present . . . .***H. anomala***

2. Leaves ovate, elliptic, or lanceolate-ovate, the blade greater than
   1½ times as long as wide, less than 6 cm wide, long-acuminate;
   stems erect
   3. Leaves elliptic to lanceolate, 3 times as long as wide, finely,
      often sharply, serrate
      4. Leaves cuneate at the base, scabrous . . . .*H. aspera*
      4. Leaves rounded to slightly cuneate at the base, moderately to densely
         pubescent . . . .*H. heteromala*
   3. Leaves ovate, usually 2 times as long as wide, serrate to
      coarsely dentate
      5. Petioles 2–8 cm long, blades coarsely crenate-dentate, rounded,
         truncate, subcordate, or slightly cuneate at the base; buds conical,
         projecting at right angles to the twig . . . .**H. arborescens**
      5. Petioles mostly less than 2 cm long, blades sharply serrate,
         crenate, or irregular, cuneate to rounded at the base; buds
         appressed, pointing forward along the twig
         6. Leaves regularly serrate or coarsely crenate, strongly cuneate
            or wedge-shaped at the base, glabrous or subglabrous,
            the petiole 2 mm thick; branchlets glabrous or
            slightly pubescent . . . .*H. macrophylla*
         6. Leaves finely or sharply serrate, often irregularly so, rounded to
            slightly cuneate at the base, pubescent on lower surface, the petiole
            1 mm thick; branchlets pubescent . . . .**H. paniculata** (ON)

Hypericum   St.-John's-wort, St.-Peter's-wort, St.-Andrew's-cross

Summer vegetative characters are needed in most cases to key this group of shrubs
and subshrubs.

1. Plants less than 50 cm tall, often decumbent or matted
   2. Leaves greater than 3 cm long . . . .*H. calycinum*
   2. Leaves less than 3 cm long
      3. Leaves rounded, lacking notch at base . . . .**H. buckleyi** (s)
      3. Leaves notched at the base
         4. Bark on older stems shredding . . . .**H. stragalum** (s)
         4. Bark on older stems tight, not shredding
            5. Leaves usually greater than 13 mm long . . . .**H. lloydii** (s)
            5. Leaves usually less than 13 mm long
               6. Leaves less than 1 mm wide . . . .**H. reductum** (s)
               6. Leaves greater than 1 mm wide . . . .**H. suffruticosum** (s)
1. Plants usually greater than 50 cm tall, erect
   7. Branchlets stout, erect; leaves thick, ovate to oblong, somewhat clasping
      at base . . . .**H. stans** (s)

7. Branchlets slender; leaves thin, linear-oblong to
   linear-oblanceolate, narrowed at the base, not clasping
   8. Leaves 3–7 cm long; branchlets terete or 2-angled;
      fruit 3–4 celled
      9. Fruit 5–8 mm long ....*H. nudiflorum* (s)
      9. Fruit 10–15 mm long
         10. Flowering in late summer ....*H. prolificum*
         10. Flowering in spring or early summer ....*H. frondosum* (s)
   8. Leaves mostly 1–4 cm long; branchlets often 4-angled or
      winged; fruit 1–5 celled
      11. Leaves 2–4 cm long; fruit 3–5 celled
         12. Leaves mostly subtending axillary fascicles; fruit lanceolate, less
            than 4 mm long, 3–4 celled ....*H. densiflorum*
         12. Leaves mostly lacking axillary fascicles; fruit
            ovoid, 6–10 mm long, 4–5 celled
            13. Fruit shallowly lobed or not lobed ....*H. kalmianum*
            13. Fruit deeply lobed ....*H. lobocarpum* (s)
      11. Leaves 1–3 cm long; fruit 1–3 celled
         14. Leaves lacking a notch or groove at the base
            15. Leaves oblong, greater than 1 cm wide ....*H. patulum*
            15. Leaves lanceolate, less than 1 cm wide ....*H. cistifolium* (s)
         14. Leaves with a notch or groove at the base
            16. Leaves elliptic, oblong, or linear, usually greater
               than 2 mm wide
               17. Seeds less than 1 mm long ....*H. galioides* (s)
               17. Seeds about 1 mm long ....*H. hypericoides*
            16. Leaves linear-subulate or linear, usually less
               than 2 mm wide
               18. Leaves with 2 deep longitudinal grooves on lower surface,
                  those of axillary branchlets crowded, appearing fasciculate;
                  fruit less than 5 mm long ....*H. fasciculatum* (s)
               18. Leaves with inconspicuous longitudinal grooves, well-spaced,
                  not crowded; fruit usually greater than
                  5 mm long ....*H. nitidum* (s)

Idesia polycarpa   idesia, iigiri tree (OC)

Ilex   holly

Many hybrid hollies are in cultivation. Only some of the most common ones are
included here.

1. Leaves thin or of medium thickness, not coriaceous, deciduous
   2. Nutlets smooth and even on the back; leaves and fruit borne on
      typical branchlets, not on short spurs

3. Leaves mostly ovate or obovate, margin serrate, downy on the lower surface, or with few hairs, chiefly along the veins; fruit 5–7 mm in diameter . . . .*I. verticillata*

3. Leaves mostly lanceolate or oblong-lanceolate, margin serrulate, mostly glabrous on the lower surface; fruit 7–10 mm in diameter . . . .*I. laevigata*

2. Nutlets ribbed on the back; leaves and fruit often borne on spurs

    4. Leaves elliptic or oblong-elliptic; fruit less than 5 mm long . . . .*I. serrata* (OC)

    4. Leaves ovate, obovate, lanceolate, or oblong-lanceolate; fruit greater than 5 mm long

        5. Margin of leaves crenate or bluntly serrate; branchlets roughened by raised lenticels . . . .*I. decidua*

        5. Margin of leaves sharply and finely serrate; branchlets smooth, slender . . . .*I. montana*

1. Leaves coriaceous, persistent

  6. Leaves acuminate to cuspidate, not terminating in a spine, 5–7 cm long, entire or with 2–4 minute sharp teeth on the apical half of each margin (except *I. chinensis*); fruit red

    7. Branchlets pubescent, black . . . .*I. ciliospinosa*

    7. Branchlets glabrous to puberulent, mostly brown or dark brown

        8. Leaves acute to acuminate, mostly with 2–4 minute sharp linear teeth on the apical half of each margin (except *I. chinensis*), occasionally entire

           9. Leaves mostly with 2–4 minute sharp linear teeth on the apical half of each margin, occasionally entire . . . . .*I. coriacea* (S, ON NORTH)

           9. Leaves toothed the length of the margin, never entire . . .*I. chinensis*

        8. Leaves mostly cuspidate, entire

           10. Petioles 1–2 cm long; fruit stalks less than 5 mm long . . . .*I. rotunda* (CS)

           10. Petioles about 1 cm long; fruit stalks greater than 10 mm long . . . .*I. pedunculosa*

  6. Leaves obtuse, rounded, acute, or terminating in a spine, not acuminate or cuspidate, 1–10 cm long, entire or toothed, the teeth spiny and prominent or obtuse and obscure; fruit red or black

    11. Plants low, spreading or creeping . . . .*I. rugosa* (OC)

    11. Plants erect, shrubs or trees

        12. Leaves crenate, serrate, or entire, not spiny, with minute black dots on lower surface (lacking in *I. glabra*), mostly less than 3 cm long (often more in *I. glabra*); fruit black

           13. Leaves entire or serrate only toward the apex

               14. Leaves glabrous, lacking punctate dots, the apex obtuse; fruit black . . . .*I. glabra*

14. Leaves with lower surface punctate-dotted, the apex acute; fruit red, orange-red, or yellow . . . .*I. cassine* (S)

13. Leaves crenate or appressed-crenate from the apex to the base

    15. Branchlets gray; fruit 4–8 mm in diameter; leaves crenate . . . .*I. vomitoria* (S)

    15. Branchlets black or brown; fruit 7–9 mm in diameter; leaves crenate, crenulate, sometimes nearly entire . . . .*I. crenata* (ON)

12. Leaves coarsely spiny-toothed or, if entire, with the apex terminating in a spine, lacking pale dots, mostly greater than 3 cm long (often less in *I. pernyi* and *I.* ×*meserveae*); fruit red (black in *I. cornuta*)

    16. Leaves entire, terminating in a spine; fruit red

        17. Flowers and fruit in axillary clusters on branchlets of the previous year . . . .*I. opaca*

        17. Flowers and fruit 1 to few in axillary cymes on new shoots or twigs . . . .*I. aquifolium* or *I.* ×*attenuata* (OC)

    16. Leaves with coarse spiny teeth; fruit red or black

        18. Leaves with 1–3 teeth on each margin; fruit red or black

            19. Leaves mostly greater than 5 cm long, with marginal teeth or spines restricted to the apical third . . . .*I.* ×*attenuata* (OC)

            19. Leaves mostly less than 5 cm long, with marginal teeth or spines spread over greater than a third of the margin

                20. Branchlets glabrous; leaves about 5 cm long, the 2 spines near the apex and 2 at the base curved down; fruit black . . . .*I. cornuta*

                20. Branchlets pubescent; leaves less than 5 cm long, the spines spaced along each margin, straight or slightly curved; fruit red

                    21. Leaves mostly less than 3 cm long . . . .*I. pernyi*

                    21. Leaves mostly greater than 3 cm long . . . .*I.* ×*aquipernyi* (OC)

        18. Leaves mostly with more than 3 teeth

            22. Branchlets pubescent, black . . . .*I. ciliospinosa*

            22. Branchlets glabrous or pubescent, green to dark brown

                23. Leaves with margins wavy or flat, the petiole flattened or grooved, sometimes winged . . . .*I.* ×*altaclarensis* (OC)

                23. Leaves with margins often crisped, the petiole terete, not winged

                    24. Flowers and fruit 1 to few in axillary cymes on new shoots or twigs . . . .*I. aquifolium*

                    24. Flowers and fruit in axillary clusters on branchlets of the previous year

                      25. Leaves 2–5 cm long . . . .*I.* ×*meserveae*

                      25. Leaves 4–10 cm long . . . .*I. opaca*

Illicium   anise tree

1. Leaves mostly rounded to obtuse at the apex . . . .*I. parviflorum* (CS)
1. Leaves mostly acute to acuminate
    2. Leaves often 15 cm long, the petiole mostly less than
    1 cm long . . . .*I. anisatum* (CS)
    2. Leaves less than 10 cm long, the petiole greater than 1 cm long, often
    nearly 2 cm long . . . .*I. floridanum* (CS)

Indigofera kirilowii   indigo

Itea virginica   Virginia sweetspire

Iva   marsh-elder

1. Leaves greater than 5 cm long . . . .*I. frutescens*
1. Leaves less than 5 cm long . . . .*I imbricata* (S)

Jasminum   jasmine

A number of jasmine species are sometimes cultivated in the southern United
States. The most common ones are keyed below.

1. Leaves alternate; erect shrub . . . .*J. humile* (CS)
1. Leaves opposite; plant climbing or trailing
    2. Leaflets 3, deciduous . . . .*J. nudiflorum*
    2. Leaflets 5–9, persistent or semi-evergreen . . . .*J. officinale* (CS)

Juglans   walnut, butternut

Summer key

1. Leaflets mostly 7–9, margins entire or remotely denticulate; leaves and twigs
    glabrous; pith light brown; bark smooth, gray . . . .*J. regia* (ON)
1. Leaflets mostly 11–23, margins serrate; leaves and usually the twigs
    pubescent; pith light or dark brown; bark rough or with broad
    ridges, light brown, dark brown, or gray
    2. Leaves smooth and glossy above; leaf scar glabrous; pith light brown, bark
    light brown to dark brown or almost black, rough . . . .*J. nigra*
    2. Leaves somewhat pubescent above; leaf scar with a hairy fringe
    along the upper margin; pith light or dark brown; bark of trunk
    gray, with broad ridges
        3. Leaflets 10–15 cm long; pith light brown; fruits in racemes at least
        10 cm long; nuts smooth . . . .*J. ailanthifolia*
        3. Leaflets usually less than 10 cm long; pith dark brown; fruits in short-
        stalked clusters of 3–5, 6–10 cm long; nuts ridged . . . .*J. cinerea*

Winter key

1. Leaf scar with a hairy fringe along the upper margin; pith dark
   brown or light brown; bark of trunk gray, with broad ridges
   2. Pith dark brown, diaphragms thick, nearly as wide as the intervening
      chambers; upper margin of leaf scar straight or at least not
      conspicuously notched . . . .*J. cinerea*
   2. Pith light brown, diaphragms thin, much narrower than the intervening
      chambers; upper margin of leaf scar conspicuously notched . . .*J. ailantifolia*
1. Leaf scar glabrous or slightly short-pubescent along upper margin,
   lacking fringe of hairs; pith light brown; bark of trunk gray and
   smooth or dark brown and rough
   3. Twigs glabrous, not sticky; buds glabrate; bark of trunk light gray,
      remaining smooth . . . .*J. regia* (ON)
   3. Twigs densely gray-downy, somewhat sticky at first; buds pubescent; bark of
      trunk dark brown, ridges narrow and broken crosswise . . . .*J. nigra*

Juniperus   juniper

1. Leaves all needlelike, linear, in whorls of 3, divergent; berrylike
   cones axillary
   2. Leaves joining the twig in 1 place, the upper surface concave or
      grooved, often concealing a white longitudinal band on
      the upper surface
      3. Leaves deeply grooved, the white band usually concealed and narrower
         than the green margins . . . .*J. conferta*
      3. Leaves with upper surface merely concave, the evident white band as
         wide or wider than the green margins . . . .*J. communis*
   2. Leaves long decurrent, the upper surface flat or shallowly concave
      4. Leaves green to green-blue . . . .
         . . . . .*J. procumbens* or *J. chinensis* (or cultivars of other cultivated species)
      4. Leaves usually very blue . . . .*J. squamata*
1. Leaves mostly scalelike, as wide as long, imbricated, appressed;
   berrylike cones terminal
   5. Low shrubs forming dense mats; berrylike cones pendulous on
      curved peduncles; branchlets emitting strong odor when crushed
      6. Leaves blue-green with agreeable odor . . . .*J. horizontalis*
      6. Leaves dark green with disagreeable odor . . . .*J. sabina*
   5. Trees or shrubs, sometimes prostrate in cultivation; berrylike
      cones erect on straight peduncles; branchlets odiferous, but not
      overpoweringly so, when crushed
      7. Shrubs or small trees; apex of scalelike leaves acute . . . .*J. chinensis*
      7. Trees; apex of scalelike leaves tending to be sharp
         8. Branchlets and young branches spreading;
            bark shredding . . . .*J. virginiana*

8. Branchlets and young branches short, close together; bark loose but
   not falling . . . .*J. scopulorum*

Kadsura japonica   kadsura, kadsura vine (cs, oc)

Kalmia   mountain-laurel

1. Leaves deciduous; endemic to North Carolina and
   South Carolina . . . .*K. cuneata* (s)
1. Leaves persistent; widely distributed
   2. Twigs compressed, 2-edged; leaves nearly sessile, white on lower surface,
      margins revolute . . . .*K. polifolia*
   2. Twigs terete; leaves distinctly petioled, green on both surfaces,
      margins flat
      3. Leaves narrowly oblong, mostly obtuse, usually in whorls of
         3 or opposite, less than 2.5 cm wide; fruits in
         lateral corymbs . . . .*K. angustifolia*
      3. Leaves elliptical or oval, acute, alternate or some opposite or
         rarely whorled, greater than 2.5 cm wide; fruits in
         terminal corymbs . . . .*K. latifolia*

Kalopanax pictus   sennoki, kalopanax, castor-aralia (oc)

Kerria japonica   kerria (on)

Koelreuteria paniculata   golden-rain tree (on)

Kolkwitzia amabilis   beautybush

Laburnum   golden chain tree

The three commonly cultivated species cannot be distinguished on the basis of leaf
or branchlet characteristics alone.

1. Flowering racemes 25–40 cm long; main stem or trunk
   usually single   *L. alpinum*
1. Flowering racemes 10–25 cm long; multi-trunked
   2. Flowering racemes less than 20 cm long . . . .*L. anagyroides*
   2. Flowering racemes often to 25 cm long . . . .*L. ×watereri* (oc)

Lagerstroemia   crape-myrtle

1. Leaves mostly 5 cm long, glabrous or pubescent on the veins, lacking tufts of
   hair in the axils of the veins; bark pink, tan, or brown . . . .*L. indica*
1. Leaves mostly 7–8 cm long, with tufts of hair in the axils of the veins of lower
   surface; bark often red . . . .*L. fauriei*

Larix   larch, tamarack

1. Leaves with 2 conspicuous white longitudinal bands on
   lower surface; twigs glaucous
     2. Young shoots brown, glaucous; leaves keeled, the longitudinal
        bands wide . . . .*L. kaempferi*
     2. Young shoots red-brown, less glaucous; leaves only slightly keeled, the
        longitudinal bands narrow . . . .*L. ×eurolepis*
1. Leaves with 2 inconspicuous, pale longitudinal bands on
   lower surface; twigs lacking bloom (*L. laricina* often glaucous)
     3. Branchlets stiff, not drooping; leaves with ridge or angle on upper surface;
        bark of trunk shedding as small irregular scales; cones less than 2 cm long,
        with about 10 scales . . . .*L. laricina*
     3. Branchlets pendulous; leaves flattened on upper surface; bark of trunk
        shedding as large plate-like scales; cones greater than 2 cm long, with more
        than 40 scales . . . .*L. decidua* (ON)

Laurus nobilis   sweet-bay

Lavandula angustifolia   lavender (ON)

Ledum groenlandicum   Labrador-tea

Leiophyllum buxifolium   sand-myrtle

Leitneria floridana   corkwood (S)

Lespedeza   bush clover

1. Leaflet blades broadly elliptical, usually less than twice as long as wide; fruit
   less than 1 cm long . . . .*L. bicolor* (ON, CS)
1. Leaflet blades narrowly elliptical, greater than twice as long as wide; fruit
   greater than 1 cm long . . . .*L. thunbergii* (ON, CS)

Leucothoe   fetterbush

1. Leaves deciduous; racemes terminal, 1-sided, open
     2. Leaves acuminate; branches recurved-spreading; racemes recurved-
        spreading; fruit deeply lobed; seeds winged . . . .*L. recurva* (S)
     2. Leaves acute; branches ascending; racemes mostly erect; fruit unlobed;
        seeds wingless . . . .*L. racemosa*
1. Leaves persistent; racemes axillary, with flowers on all sides, dense
     3. Petioles 10–15 mm long; flowering racemes
        4–10 cm long . . . .*L. fontanesiana* (S)
     3. Petioles 5–10 mm long; flowering racemes mostly 2–4 cm long
          4. Lateral veins on lower leaf surface conspicuous . . . .*L. axillaris* (S)
          4. Lateral veins on lower leaf surface inconspicuous or
             not evident . . . .*L. populifolia* (S)

Leycesteria formosa   Himalayan-honeysuckle (CS)

Licania michauxii   gopher-apple (CS)

Ligustrum   privet

1. Twigs and buds glabrous or glabrate (pubescent when young in
   L. amurense); lateral buds with acuminate scales; leaves thick,
   leathery, sometimes persistent
   2. Leaves mostly less than 6 cm long
      3. Leaves pubescent on midrib of lower surface, the upper surface dull;
         branchlets pubescent when young . . . .*L. amurense*
      3. Leaves glabrous on lower surface, the upper surface lustrous;
         branchlets glabrous . . . .**L. ovalifolium** (ON)
   2. Leaves greater than 7 cm long
      4. Leaves mostly 10–15 cm long . . . .*L. lucidum*
      4. Leaves 7–10 cm long . . . .*L. japonicum*
1. Twigs and buds pubescent or puberulous; lateral buds with obtuse
   scales; leaves thin, deciduous
   5. Lower leaf surface glabrous; twigs glabrous or pubescent . . . .**L. vulgare** (N)
   5. Lower leaf surface pubescent, often only on midrib;
      twigs pubescent
      6. Leaves pubescent over most of the
         lower surface . . . .*L. obtusifolium* var. *regelianum*
      6. Leaves pubescent only along midrib of lower surface
         7. Largest leaves 8 cm long; flowering panicles loose,
            8–10 cm long . . . .*L. sinense*
         7. Largest leaves 5 cm long; flowering panicles less than
            4 cm long . . . .**L. obtusifolium var. obtusifolium** (ON)

Lindera   spicebush

1. Leaves glabrous or glabrate, obovate, the base tapered; twigs and buds
   glabrous or glabrate . . . .**L. benzoin**
1. Leaves pubescent, elliptic or oval to oblong, the base rounded; twigs and
   buds pubescent . . . .**L. melissaefolia** (S)

Linnaea borealis   twinflower

Liquidambar   sweetgum

1. Young twigs glabrous; branchlets becoming corky and winged;
   leaves 5–7 lobed . . . .**L. styraciflua**
1. Young twigs pubescent; branchlets lacking corky wings; leaves
   3–5 lobed

2. Leaves with 3 lobes; large tree to 35 m . . . .*L. formosana* (oc)
2. Leaves with 5 lobes; small tree to 6 m . . . .*L. orientalis* (oc)

Liriodendron tulipifera   tulip tree

Litsea aestivalis   pondspice (s)

Loiseleuria procumbens   alpine-azalea

Lonicera   honeysuckle

1. Twining or trailing vines; upper leaves connate or distinct; fruits
   clustered, sessile in axillary whorls or terminal spikes
   2. Leaves green on both surfaces, often paler on lower surface, not
      glaucous; upper leaves distinct or connate; twigs pubescent;
      berry red or black
      3. Leaves persistent; berry black . . . .*L. japonica* (N)
      3. Leaves deciduous; berry red
         4. Upper leaves connate
            5. Leaves pubescent on upper surface . . . .*L. hirsuta*
            5. Leaves glabrous on upper surface . . . .*L. flava* (s)
         4. Upper leaves distinct . . . .*L. periclymenum* (N)
   2. Leaves glaucous, at least on the lower surface; upper leaves
      connate; twigs glabrous or nearly so; berry red or orange
      6. Leaves glaucous on both upper and lower surface . . . .*L. prolifera*
      6. Leaves green on the upper surface, glaucous on lower surface
         7. Twigs light gray or yellow-gray, rarely glaucous . . . . .*L. sempervirens*
         7. Twigs brown to gray, glaucous
            8. Leaves oblong or elliptic-oblong; berry red . . . .*L. dioica*
            8. Leaves ovate to ovate-elliptic; berry red
               or orange-red . . . .*L. caprifolium* (ON)
1. Erect shrubs; upper leaves distinct; fruit in pairs (appearing solitary
   in *L. villosa*), on axillary peduncles
   9. Pith of branchlets white, solid between the nodes
      10. Leaves persistent, less than 1 cm long . . . .*L. nitida* (CS)
      10. Leaves deciduous, greater than 1 cm long
         11. Leaves pilose-hirsute to densely villous; peduncles less than 1 cm
             long; fruit a single berry formed by 2 flowers; shrubs less than
             1 m high . . . .*L. villosa* (N)
         11. Leaves glabrous, at least on upper surface; peduncles
             1–4 cm long, fruit remaining paired, distinct or connate;
             shrubs over 1 m high, straggling or erect
            12. Berries distinct, purple-black, enveloped by longer involucre,
                formed by bracts longer than the fruit, which later reflexes; leaves
                9–13 cm long, acute to acuminate . . . .*L. involucrata*

12. Berries united at base, red to purple, the involucre or bracts
shorter than the berries; leaves mostly less than 9 cm long,
obtuse, acute, or mucronate

   13. Leaves ovate-lanceolate or oblong, pubescent, at least on
lower surface, ciliate

      14. Leaves lanceolate, ovate-lanceolate, or oblong, acuminate, strongly
ciliate on margins, petiole, and midrib; peduncles 3–15 mm
long, bristly hairy; branchlets and twigs with
scattered hairs . . . .*L. standishii*

      14. Leaves ovate-oblong, often cordate, acute, weakly ciliate on
petiole and margins at the base; peduncles 20–30 mm long,
glabrous; branchlets and twigs glabrous . . . .***L. canadensis***

   13. Leaves ovate, obovate, oblanceolate, or elliptic,
glabrous, eciliate

      15. Leaves oblanceolate, elliptic, or oblong, greater than twice as
long as wide, acute, rarely mucronate; bud
scales acuminate . . . .***L. oblongifolia***

      15. Leaves ovate or obovate, less than twice as long
as wide, usually mucronate; bud
scales acute . . . .***L. fragrantissima*** (ON)

9. Pith of branchlets dark, hollow or solid between the nodes

   16. Leaves acuminate; peduncles 2–3 mm long, shorter than
the petioles . . . .***L. maackii*** (N)

   16. Leaves acute to obtuse; peduncles 5–25 mm long,
longer than the petioles (The *Lonicera* species in key steps
17–20 are very similar, as much hybridization occurs. The
characters used can only be considered tendencies.)

      17. Leaves, buds, and twigs glabrous or subglabrous; buds ovoid; twigs
gray or yellow-brown; branchlets hollow . . . .***L. tatarica*** (N)

      17. Leaves, buds, and twigs pubescent; buds short conical to
fusiform; twigs gray to dark gray-brown; branchlets
hollow or filled with pith

         18. Buds fusiform, with ciliate pubescent scales; twigs puberulent,
dirty gray; leaves mostly broader above the middle;
branchlets hollow . . . .***L. xylosteum*** (N)

         18. Buds short-conical, obtuse, glabrescent; twigs soft-
pubescent; leaves mostly broader below the middle;
branchlets filled with pith

            19. Leaves glabrate; peduncles sparsely pilose
to glabrate . . . .***L. ×bella*** (N)

            19. Leaves pubescent on lower surface, sometimes on
upper surface; peduncles densely pilose

               20. Petioles usually 2–3 mm long; upper surface of leaves
often glabrous . . . .***L. morrowii*** (ON)

20. Petioles usually 4–5 mm long; upper surface of leaves
     often pubescent . . . .*L. ruprechtiana* (oc)

Loropetalum chinense   loropetalum, fringe flower (cs)

Lycium barbarum   matrimony vine (on)

Lyonia   staggerbush

1. Plants evergreen, second-year branchlets with at least some leaves
   2. Leaves with lower surface glabrous . . . .***L. lucida*** (s)
   2. Leaves with shieldlike or peltate scales on lower surface . . .***L. ferruginea*** (s)
1. Plants deciduous, second-year branchlets lacking leaves
   3. Leaves serrate; visible bud scales 2, the inner scales white-silky; buds
      oblong-acute, closely appressed; winter twigs yellow-green, pubescent, at
      least near tip; capsules globose . . . .***L. ligustrina***
   3. Leaves entire; visible bud scales 4–5, the inner scales glabrous; buds
      conical, divergent; winter twigs brown, glabrous; capsules
      urn-shaped . . . .***L. mariana***

Maackia amurensis   maackia (oc)

Maclura pomifera   Osage-orange

Magnolia   magnolia

In the summer key, several cultivated magnolia species and a hybrid are included for which flowers are necessary for identification. These species, *M. kobus* var. *kobus*, *M. denudata*, and *M.* ×*soulangiana*, are not included in the winter key because they lack easily discerned defining characteristics in winter.

Summer key

1. Leaves persistent, coriaceous, upper surface glossy deep green, lower surface
   rusty-tomentose . . . .*M. grandiflora*
1. Leaves deciduous, thin, upper surface dull, lower surface light
   green and glaucous, glabrous, or pubescent, not rusty-tomentose
   2. Leaves crowded near the end of the twig; leaf buds glabrous;
      terminal bud glaucous, purple
      3. Leaf base cuneate . . . .***M. tripetala***
      3. Leaf base auriculate . . . .***M. fraseri*** (s)
   2. Leaves scattered along the twig; leaf buds pubescent; terminal bud
      pubescent or silky white
      4. Lower surface of leaves glaucous; leaves 7–16 cm long,
         often persistent . . . .***M. virginiana***
      4. Lower surface of leaves pubescent, not glaucous
         5. Leaves mostly 20–90 cm long

6. Leaves 30–90 cm long, the lower surface white, strongly pubescent, the
base auriculate . . . . *M. macrophylla* (s)
6. Leaves 10–25 cm long, the lower surface green, slightly pubescent; the
base cuneate . . . . *M. acuminata*
5. Leaves mostly 5–15 cm long
   7. Leaves lanceolate, broadest at or below the middle; leaf
   buds glabrous . . . . *M. salicifolia* (oc)
   7. Leaves elliptic, oblong, or obovate, mostly broadest above the
   middle; leaf buds pubescent
      8. Leaves acuminate; petioles 1–5 cm long . . . . *M. acuminata*
      8. Leaves mucronate, rounded or retuse at apex with abrupt
      sharp point; petioles mostly 1–2 cm long
         9. Tepals 5–7; large, open tree . . . . *M. kobus*
         9. Tepals 9 or 15–33
            10. Tepals 15–33; small, compact tree . . . . *M. kobus* var. *stellata*
            10. Tepals 9
               11. Petioles tomentose . . . . *M. denudata*
               11. Petioles glabrous
                  12. Flowers white to pink . . . . *M.* ×*soulangiana*
                  12. Flowers mostly purple or white and purple . . . . . *M. liliiflora*

Winter key

1. Leaves persistent
   2. Lower surface of leaves rusty-tomentose; twigs
   rusty-tomentose . . . . *M. grandiflora*
   2. Lower surface of leaves glaucous; twigs glabrous, green . . . . *M. virginiana*
1. Leaves deciduous
   3. Leaf buds glabrous; terminal bud glaucous, purple; leaf scars
   crowded near end of annual growth
      4. Terminal bud 2–3 cm long; leaf scars with minute resin glands on
      surface; 2-year-old branchlets brown, very stout, 8 mm thick;
      bark light gray . . . . *M. tripetala*
      4. Terminal bud 3–5 cm long; leaf scars lacking resin glands; 2-year-old
      branchlets gray, 5 mm thick; bark brown . . . . *M. fraseri* (s)
   3. Leaf buds pubescent; terminal bud hairy or silky, white; leaf scars
   scattered along the twigs
      5. Terminal bud 3–5 cm long; twigs brown, very stout, about
      8 mm thick . . . . *M. macrophylla* (s)
      5. Terminal bud 1–2 cm long; twigs brown or
      green, 2–5 mm thick
         6. Terminal bud covered with fine silky hairs or glabrous; twigs green,
         2–4 mm thick . . . . *M. virginiana*
         6. Terminal bud covered with long white hairs; twigs brown,
         4–5 mm thick . . . . *M. acuminata*

Mahonia   Oregon-grape, holly-grape

1. Leaflets 5–9, with greater than 7 teeth per margin ....*M. aquifolium* (N)
1. Leaflets 9–15, with fewer than 7 teeth on each margin
  2. Petioles mostly less than 2 cm long; shrub to 3 m tall ....*M. pinnata*
  2. Petioles mostly greater than 2 cm long; shrub to 1 m tall .... .*M. bealei* (N)

Malus   apple, crabapple, crab

Summer vegetative characters are needed here to key the apple species.

1. Leaves on elongated shoots, lacking lobes or notches, acute to acuminate
  2. Leaves crenate-serrate; fruits large, apples ....*M. pumila* (N)
  2. Leaves sharply serrulate to serrate; fruits small, crabapples
    3. Leaves folded in bud ....*M. floribunda* (N)
    3. Leaves rolled together in bud
      4. Leaves pubescent, even at maturity ....*M. prunifolia* (N)
      4. Leaves glabrous, sometimes pubescent when young or with some hairs
        persisting along the veins ....*M. baccata* (N)
1. Leaves on elongated shoots with lobes or notches or the apex acute
  5. Leaves glabrous at maturity or with some hairs persisting along the veins
    6. Leaves lanceolate-oblong, obtuse, or subacute; fruit less than
      3 cm wide ....*M. angustifolia* (S)
    6. Leaves ovate to broadly lanceolate, sharply acute or acuminate; fruit
      greater than 3 cm wide ....*M. coronaria*
  5. Leaves pubescent, even at maturity
    7. Leaves rarely lobed ....*M. sargentii*
    7. Leaves mostly 3–5 lobed or some lacking lobes
      8. Leaves folded in bud, 3–6 cm long ....*M. sieboldii* (ON)
      8. Leaves rolled together in bud, 6–10 cm long ....*M. ioensis*

Melia azedarach   Chinaberry (N)

Menispermum canadense   moonseed

Menziesia pilosa   menziesia, minniebush

Mespilus germanica   medlar

Metasequoia glyptostroboides   dawn-redwood

Michelia figo   banana shrub (CS)

Microbiota decussata   microbiota

Mimosa pudica   mimosa, sensitive brier (N)

Miscanthus sinensis   eulalia (N)

Mitchella repens   partridgeberry

Morus   mulberry

1. Leaves glabrous and glossy above, lobes, if present, rounded; buds appressed, more or less flattened, mostly less than 4 mm long; bud scales red-brown, with darker margins not evident; bark yellow-brown . . . .*M. alba* (N)
1. Leaves rough to the touch on upper surface, lobes, if present, pointed; buds divergent, not at all or only slightly flattened, mostly greater than 5 mm long; bud scales green-brown, with darker margins; bark dark brown . . .*M. rubra*

Myrica   bayberry

1. Leaves pale green; terminal bud absent; lateral buds conical-ovoid or oblong; nuts 2-winged, in cone-like catkins, ovoid, resin-dotted, not waxy . . . .*M. gale*
1. Leaves bright green; terminal bud present; lateral buds subglobose; nuts scattered, globose, encrusted with white wax
   2. Leaves mostly 8–15 mm wide, the apex acute; resinous dots common on both surfaces; buds yellow-glandular, 1 mm long . . . .*M. cerifera* (S)
   2. Leaves mostly 15–40 mm wide, apex obtuse; resinous dots often scarce or absent on the upper surface; buds brown, greater than 1 mm long
      3. Leaves deciduous, membranous, dull on upper surface; bark of branches white or gray; fruit pubescent, 3.5–4.5 mm in diameter . . . .*M. pensylvanica*
      3. Leaves persistent, coriaceous, lustrous on upper surface; bark of branches black; fruit glabrous, 3.0–3.5 mm in diameter . . . . .*M. heterophylla* (S)

Nandina domestica   nandina, heavenly-bamboo (ON)
Neillia sinensis   neillia (OC)
Nemopanthus   mountain-holly

1. Leaves entire or nearly so, mucronate, acute, or obtuse, mostly 3–5 cm long, widely distributed . . . .*N. mucronatus*
1. Leaves glandular-serrate, acuminate, often greater than 5 cm long; rare, North Carolina, Virginia, and West Virginia only . . . .*N. collinus*

Nerium oleander   common oleander, rosebay (CS)
Nestronia umbellulata   nestronia (S)
Neviusia alabamensis   Alabama snow wreath (S)
Nyssa   tupelo, blackgum

1. Leaves mostly less than 12 cm long, sparsely pubescent, mostly along veins on lower surface; fruit 10–15 mm long . . . .*N. sylvatica*

1. Leaves mostly greater than 12 cm long, pubescent on lower surface, especially when young; fruit 20–30 mm long . . . .***N. aquatica***

Oemleria cerasiformis   Indian-plum, osoberry (cs)

Oplopanax horridus   devil's club

Osmanthus   devilwood

1. Leaves often with a few spiny teeth, usually less than
   5 cm long . . . .*O. heterophyllus*
1. Leaves entire or finely toothed or with at least 20 large spiny teeth;
   usually greater than 5 cm long
   2. Leaves entire, 5–15 cm long . . . .***O. americanus*** (s)
   2. Leaves entire or finely toothed or with at least 20 large spiny
      teeth, usually 5–10 cm long
      3. Leaves entire or finely toothed . . . .*O. fragrans*
      3. Leaves with 10–12 large spiny teeth on each margin . . . .*O. ×fortunei* (cs)

Ostrya   hophornbeam

1. Leaves usually subcordate at the base, the petiole glandular and less than
   5 mm long; bark pale brown to dark brown; nutlets narrowed at each end,
   glabrous at the apex . . . .***O. virginiana***
1. Leaves rounded at the base, the petiole glabrous and greater than 5 mm long;
   bark gray; nutlets ovoid, with tuft of hairs at the apex . . . .*O. carpinifolia* (oc)

Oxydendrum arboreum   sourwood

Paeonia suffruticosa   tree peony

Paliurus spina-christi   Christ thorn

Parrotia persica   Persian-ironwood (oc)

Parrotiopsis jacquinata   parrotiopsis (oc)

Parthenocissus   Virginia creeper

1. Leaves simple or trifoliate . . . .*P. tricuspidata* (n)
1. Leaves palmately 5-foliate
   2. Tendrils 5–12 branched, tipped with expanded,
      adhesive discs . . . .*P. quinquefolia*
   2. Tendrils mostly 3–5 branched, mostly lacking expanded,
      adhesive discs . . . .***P. vitacea***

Paulownia tomentosa   empress tree, princess tree (N)

Paxistima canbyi   cliff-green

Periploca graeca   silk vine (ON, CS)

Persea   swamp-bay

1. Twigs densely rusty shaggy-pubescent; leaves with pubescent midrib on upper
   and lower surface . . . .*P. palustris* (s)
1. Twigs sparsely pale yellow pubescent; leaves glabrous on upper surface,
   inconspicuously pale yellow pubescent on lower surface . . . . . .*P. borbonia* (s)

Phellodendron   cork tree

1. Lower leaflet surface pubescent, at least on the veins
    2. Leaflets rounded at the base; inflorescence compact, higher
       than wide . . . .*P. chinense* (OC)
    2. Leaflets oblique at the base; inflorescence loose, as wide
       as high . . . .*P. japonicum* (ON)
1. Lower leaflet surface glabrous or sparsely pubescent only at base
    3. Branchlets red-brown; upper leaflet surface dull; trunk dark brown, the
       bark separating into thin plates . . . .*P. sachalinense* (OC)
    3. Branchlets yellow-brown; upper leaflet surface shining; trunk light gray, the
       bark thick and corky . . . .*P. amurense*

Philadelphus   mock-orange

This popularly cultivated genus includes many interspecific hybrids. The following key includes the common ones. There are not definitive enough characteristics
for an adequate winter key.

1. Leaves less than 4 cm long
    2. Leaves mostly 1–2 cm long, mostly entire, especially those on flowering
       shoots; flowers usually solitary . . . .*P. microphyllus* (s)
    2. Leaves mostly 2–4 cm long, usually sparsely toothed;
       flowers usually clustered . . . .*P. ×lemoinei*
1. Leaves mostly greater than 4 cm long
    3. Lower surface of leaves strongly pubescent
        4. Vegetative shoots pubescent; axillary buds evident . . . .*P. hirsutus*
        4. Vegetative shoots subglabrous; axillary buds concealed
            5. Bark on second-year branches gray, remaining tight,
               not exfoliating . . . .*P. pubescens* (s)
            5. Bark on second-year branches brown or gray, usually exfoliating
                6. Leaves villous or densely pubescent on
                   lower surface . . . .*P. ×virginalis*

    6. Leaves softly gray-pubescent on lower surface ....*P. tomentosus* (on)
3. Lower surface of leaves sparingly pubescent or glabrous
    7. Leaves mostly entire
        8. Bark of second-year growth exfoliating; flowers or fruits
           1–3 in cymes ....*P. inodorus*
        8. Bark of second-year growth adhering to the branchlet; flowers or fruits
           7–11 in dense racemose clusters ....*P. lewisii*
    7. Leaves rarely entire, mostly dentate or denticulate
        9. Hairs on lower leaf surface restricted to veins or vein axils
           10. Bark of second-year growth adhering to the branchlet; leaves of
               flowering shoots usually 5–7 cm long ....*P. coronarius* (N)
           10. Bark of second-year branchlets exfoliating; leaves of flowering
               shoots usually 2–3 cm long ....*P. lewisii*
        9. Hairs on lower leaf surface evenly distributed
           11. Lower surface of leaves villous ....*P. ×virginalis*
           11. Lower surface of leaves scattered pubescent
               12. Bark of second-year growth adhering to
                  the branchlet ....*P. lewisii*
               12. Bark of second-year branchlets exfoliating ....*P. floridus* (cs)

Phillyrea   phillyrea

1. Leaves serrate, less than 7 cm long; fruit less than
    7 mm long.......................................*P. latifolia* (oc)
1. Leaves mostly entire, greater than 7 cm long; fruit about
    13 mm long ....*P. decora* (oc)

Phoradendron leucarpum   American mistletoe

Photinia   photinia

1. Leaves thin, deciduous
    2. Young twigs glabrous; leaves ovate, 2–5 cm long; flower or fruit stalks
        smooth, lacking lenticels ....*P. parvifolia* (oc)
    2. Young twigs pubescent; leaves obovate to lanceolate-ovate, 3–7 cm long;
        flower or fruit stalks with warty lenticels ....*P. villosa*
1. Leaves leathery, thick, persistent
    3. Leaves 10–20 cm long ....*P. serratifolia*
    3. Leaves 5–9 cm long
        4. Leaves broadly cuneate at the base, 7–9 cm long ....*P. ×fraseri*
        4. Leaves cuneate (occasionally broadly so) at the base,
          5–8 cm long ....*P. glabra*

Picea 181

Phragmites australis   common reed
Phyllodoce caerulea   mountain-heath
Phyllostachys   bamboo

1. Culms and stems becoming black . . . .*P. nigra* (cs)
1. Culms and stems green, yellow, or yellow-green
  2. Plants with culms openly spaced, not
    clump-forming . . . .*P. aureosulcata* (cs)
  2. Plants with culms in compact clumps
    3. Internodes well developed . . . .*P. bambusoides* (cs)
    3. Internodes at the base of the stem very short
      4. Stems shiny, golden yellow . . . .*P. aurea* (cs)
      4. Stems dull green . . . .*P. aureosulcata* (cs)

Physocarpus opulifolius   ninebark
Picea   spruce

1. Leaves 2–4 cm long, glaucous-blue, stiff, pungent; branchlets light
  orange-brown to pale yellow-brown, rigid; branches
  stiff, horizontal . . . .*P. pungens* (ON)
1. Leaves mostly less than 2 cm long, green or, if glaucous, gray-white,
  sharp-pointed but not stiff; branchlets brown to gray-brown or
  dark brown, flexible or stiff but not rigid; branches flexible,
  horizontal or pendulous
  2. Leaves wider than high, white longitudinal bands on upper surfaces only;
    tree narrowly conical to columnar, the branches often so pendulous as to be
    nearly parallel to the trunk . . . .*P. omorika*
  2. Leaves as high as wide, white longitudinal bands on all surfaces;
    tree conical, or if columnar, the branches horizontal
    3. Branches long, sweeping; branchlets pendulous, glabrous; cones
      9–18 cm long; leaves 12–20 mm long . . . .*P. abies* (N)
    3. Branches short, horizontal or ascending; branchlets mostly
      horizontal, glabrous or pubescent; cones 1–5 cm long;
      leaves 6–17 mm long
      4. Branchlets glabrous; buds glabrous; basal bud scales obtuse, acute, or
        short-acuminate; leaves 11–17 mm long; female cones
        4–5 cm long . . . .*P. glauca*
      4. Branchlets glandular-pubescent; buds pubescent; basal bud
        scales long-acuminate or aristate; leaves 6–15 mm long;
        female cones mostly less than 4 cm long
        5. Leaves mostly 6–12 mm long, glaucous at least when young, straight;
          female cones persistent for several years, mostly 2.0–2.5 cm long,
          the scales strongly erose . . . .*P. mariana*

Keys to Species

5. Leaves mostly 12–15 mm long, lacking bloom, curved; female cones
deciduous in first or second year, mostly 3–4 cm long, the scales entire or
slightly erose . . . .*P. rubens*

Pieris andromeda

1. Leaves lacking marginal hairs; twigs glabrous or lightly
short-pubescent . . . .*P. japonica*
1. Leaves ciliate; twigs obviously pubescent . . . .*P. floribunda* (s)

Pinckneya pubens fever tree

Pinus pine

1. Leaves in fascicles of 5, thin, soft, flexible; sheaths at the base of
leaf fascicles (leaf sheaths) deciduous (except in *P. aristata*)
  2. Leaves entire
    3. Leaves mostly greater than 4 cm long, lacking resin spots, leaf sheaths
deciduous; branches flexible . . . .*P. flexilis* (oc)
    3. Leaves mostly less than 4 cm long, often with sticky white resin spots,
leaf sheaths persistent 2–3 years; branches stiff . . . .*P. aristata* (oc)
  2. Leaves minutely denticulate
    4. Branchlets pubescent
      5. Branchlets sparsely pubescent; leaves 2–4 cm long,
usually twisted . . . .*P. parviflora* (oc)
      5. Branchlets moderately to densely pubescent; leaves greater
than 4 cm long, mostly straight
        6. Leaves light green; main trunk often forked; female cones
10–15 cm long . . . .*P. koraiensis* (oc)
        6. Leaves dark green; main trunk solitary, not forked; female cones
5–9 cm long . . . .*P. cembra* (oc)
    4. Branchlets glabrous
      7. Branchlets glaucous; leaves 14–20 cm long . . . .*P. wallichiana* (oc)
      7. Branchlets lacking bloom; leaves 7–12 cm long
        8. Branchlets green to green brown; leaves flexible; cones
8–20 cm long . . . .*P. strobus*
        8. Branchlets yellow-brown or red-brown; leaves stiff; cones
15–30 cm long . . . .*P. monticola* (oc)
1. Leaves in fascicles of 2 or 3, thick, coarse, stiff; sheaths at base of
leaf fascicles (leaf sheaths) persistent (deciduous in *P. cembroides*
and *P. bungeana*)
  9. Membranous sheaths at the base of the leaf fascicles deciduous
    10. Leaves 2–5 cm long, entire; bark brown,
not flaking . . . .*P. cembroides* (oc)
    10. Leaves 5–10 cm long, minutely denticulate; bark flaky, with green and
white patches . . . .*P. bungeana* (oc)

9. Membranous sheaths at base of the leaf fascicles persistent
   11. Leaves mostly in fascicles of 3
      12. Leaf sheaths less than 1 cm long; female cones mostly 5–8 cm
         long, persisting for many years; leaves 5–20 cm long
         13. Leaf sheaths about 5 mm long; leaves mostly 7–8 cm, sometimes
            to 13 cm long . . . .*P. rigida*
         13. Leaf sheaths 7–9 mm long; leaves 6–20 cm long . . . . . .*P. serotina* (s)
      12. Leaf sheaths mostly 1–3 cm long; female cones 5–25 cm
         long, remaining on the tree for 1–4 years; leaves 8–30
         cm long
         14. Female cones 15–25 cm long, falling within the first year;
            terminal buds greater than 15 mm long; leaves 15–30 cm long;
            leaf sheaths 2 cm long; terminal buds at least 15 mm
            long, white . . . .*P. palustris* (s)
         14. Female cones 5–15 cm long, persisting 3–4 years; leaves
            mostly 8–22 cm long; leaf sheaths 1–2 cm long;
            terminal buds less than 12 mm long, brown or red
            15. Young twigs glaucous; leaves blue-green; female
               cones lateral . . . .*P. taeda*
            15. Young twigs lacking bloom; leaves yellow-green;
               female cones subterminal
               16. Cones sessile; native to western
                  North America . . . .*P. ponderosa*
               16. Cones short stalked; native to
                  southeastern U.S. . . . .*P. elliottii* (cs)
   11. Leaves mostly in fascicles of 2
      17. Twigs glaucous
         18. Leaves straight, not twisted, always in fascicles of 2; leaf sheaths
            3–6 mm long; juvenile leaves deciduous, 3–4 mm long;
            buds resinous . . . .*P. virginiana*
         18. Leaves twisted, occasionally in fascicles of 3; leaf sheaths 5–8 mm
            long; juvenile leaves persistent, 1.0–1.5 mm long; buds not
            very resinous . . . .*P. echinata*
      17. Twigs lacking bloom
         19. Leaves mostly greater than 8 cm long; leaf sheaths
            mostly greater than 7 mm long (except in *P. densiflora*)
            20. Leaves snapping or breaking when bent; cones
               lacking prickles . . . .*P. resinosa*
            20. Leaves merely folding when bent, not breaking;
               cones with or without prickles
               21. Twigs often glaucous, slender; bark on younger
                  stems orange-red; cones 5 cm long; leaves green to
                  light green, somewhat flexible; shrubs or
                  small trees . . . .*P. densiflora* (oc)

21. Twigs lacking bloom, stout; bark brown to black; cones 5–10
    cm long; leaves dark green, thick, stiff; small or large trees
    22. Buds often white, dry, not resinous; trunk and branches often
        contorted or forked; small spreading trees; cone scales with or
        without prickles . . . .*P. thunbergiana*
    22. Buds pale brown, resinous; trunk straight, growth uniform; large
        trees; cone scales with prickles . . . .*P. nigra*
19. Leaves mostly less than 8 cm long; leaf sheaths mostly less than 5 mm
    23. Branchlets dull green-yellow or yellow-brown; leaves straight
        or if twisted then blue-green; cones asymmetrical
        24. Leaves blue-green, twisted; rapidly growing trees; bark on upper trunk
            red, flaking; bud scales free or fringed at the tip . . . . .*P. sylvestris* (N)
        24. Leaves green to dark green, slightly or not at all twisted;
            slow-growing shrubs or small trees; bark dark gray, tight;
            bud scales appressed
            25. Shrubs, multistemmed, many-branched; leaves green; bud scale
                tips brown . . . .*P. mugo*
            25. Trees, single-stemmed; leaves very dark green; bud scale
                tips pale . . . .*P. leucodermis* (OC)
    23. Branchlets orange or orange-brown; leaves mostly stiff and
        twisted; cones symmetrical or asymmetrical
        26. Leaves 2–4 cm long; leaf sheaths mostly less than 2 mm long; cones
            incurved, pointing forward along the branchlet; cone scales mostly
            lacking prickles . . . .*P. banksiana*
        26. Leaves 3–9 cm long; leaf sheaths mostly 2–4 mm long;
            cones mostly perpendicular to the branchlets, not pointing
            forward; cone scales with prickles
            27. Leaves in fascicles of 2 only, yellow-green; cones subterminal,
                3–5 cm long, each scale with a slender prickle . . . .*P. contorta* (OC)
            27. Leaves mostly in fascicles of 2 but also in fascicles of 3, dark
                blue-green; cones lateral, 6–9 cm long, each scale with a stout
                hooked prickle . . . .*P. pungens*

Pistacia chinensis   pistachio (CS)

Pittosporum tobira   Japanese pittosporum, Australian-
    laurel (CS)

Planera aquatica   planer tree, water-elm

Platanus   sycamore, plane tree

1. Leaves usually deeply lobed, the lobes longer than wide; fruits 2–6
   on a stalk . . . .*P. orientalis*
1. Leaves usually shallowly lobed, the lobes as long as wide or wider;
   fruits 1, 2, or 3 on a stalk

2. Lobes of leaves wider than long; fruits mostly solitary . . . .*P. occidentalis*
2. Lobes of leaves about as long as wide; fruits mostly 2 . . . .*P. ×hybrida* (N)

Platycarya strobilacea   cluster-walnut, winged-hickory (OC)

Platycladus orientalis   platycladus, Oriental-arborvitae

Podocarpus   podocarpus, southern-yew

1. Leaves less than 2 cm wide, linear-lanceolate . . . .*P. macrophyllus* (CS)
1. Leaves greater than 2 cm wide, oblong-lanceolate to ovate . . . .*P. nagi* (CS)

Polygonum cuspidatum   Japanese knotweed,
Japanese-bamboo (N)

Poncirus trifoliata   trifoliate-orange, hardy-orange

Populus   poplar, aspen
Summer key

1. Leaves white-tomentose on the lower surface, with 3–5 large teeth
or small lobes . . . .*P. alba* (ON)
1. Leaves green, not white-tomentose when mature, with many teeth
  2. Petioles terete, often grooved, easily rolled between fingers
    3. Leaves densely tomentose when young and expanding, the apex blunt or
    rounded; buds slightly resinous . . . .*P. heterophylla*
    3. Leaves glabrous when young and expanding, the apex acute or
    acuminate; buds very resinous
      4. Petioles and twigs pubescent; leaves mostly cordate or
      subcordate . . . .*P. ×jackii* (ON)
      4. Petioles and twigs glabrous or puberulous; leaves mostly rounded or
      cuneate at the base . . . .*P. balsamifera*
  2. Petioles flattened, difficult to roll between fingers
    5. Leaves bright or yellow-green, deltoid to ovate-deltoid,
    long-acuminate
      6. Leaves with 2 glands near the apex of the petiole, the margins usually
      densely ciliolate; lateral buds divergent; branches horizontal or
      ascending but not fastigiate . . . .*P. deltoides*
      6. Leaves lacking glands, the margins lacking hairs; lateral
      buds appressed; branches strongly fastigiate
        7. Leaves with small teeth slightly or not at all incurved, the base
        mostly cuneate . . . .*P. nigra* (ON)

Keys to Species

7. Leaves with prominent teeth with incurved callous tips, the base
usually truncate . . . .*P.* ×*canadensis* (ON)
5. Leaves dull or gray-green, ovate to orbicular, obtuse, acute, or
abruptly short-acuminate
   8. Buds gray-pubescent; leaves mostly greater than 8 cm long
      9. Leaves truncate to cordate at the base with irregular broad, rounded
teeth or lobes, white-tomentose on lower surface
when young . . . .*P.* ×*canescens* (ON)
      9. Leaves truncate at base, coarsely dentate, with acute or narrow teeth,
pale or green, glabrous or slightly pubescent on lower surface
when young . . . .*P. grandidentata*
   8. Buds glossy brown; leaves mostly less than 8 cm long
      10. Leaves finely dentate . . . .*P. tremuloides*
      10. Leaves coarsely dentate . . . .*P. tremula* (OC)

Winter key

1. Buds downy or silky
   2. Buds white-downy; twigs white-downy . . . .*P. alba* (ON)
   2. Buds pale-downy; twigs glabrous or downy
      3. Twigs downy . . . .*P.* ×*canescens* (ON)
      3. Twigs glabrous or slightly pubescent . . . .*P. grandidentata*
1. Buds glabrous or resinous
   4. Twigs yellow, yellow-brown, or olive-brown; buds
lacking fragrance
      5. Lateral buds appressed, mostly less than 8 mm long;
tree fastigiate . . . .*P. nigra* (ON)
      5. Lateral buds divergent, usually at least 10 mm long; tree spreading,
not fastigiate . . . .*P. deltoides*
   4. Twigs dark brown or red-brown; buds fragrant or not
      6. Buds less than 10 mm long, glossy, slightly resinous but
lacking fragrance
         7. Margin of bud scales scarious . . . .*P. tremuloides*
         7. Margin of bud scales solid, brown
            8. Branchlets 4–7 mm thick, gray, red-brown, dark brown, or
nearly black . . . .*P. heterophylla*
            8. Branchlets less than 4 mm thick, usually gray . . . . . .*P. tremula* (OC)
      6. Buds greater than 10 mm long, resinous, fragrant
         9. Lateral buds with 2 visible scales; twigs glabrous . . . . . .*P. balsamifera*
         9. Lateral buds with 3 visible scales; twigs glabrous or hairy,
at least at base
            10. Twigs hairy, at least at the base . . . .*P.* ×*jackii* (ON)
            10. Twigs glabrous . . . .*P.* ×*canadensis* (ON)

Potentilla   cinquefoil

1. Leaflets 3–7, entire, shrub to 100 cm tall . . . .*P. fruticosa*
1. Leaflets 3, entire except for the toothed apex; prostrate shrub to
30 cm tall . . . .*P. tridentata*

Prinsepia   prinsepia

1. Leaves greater than 1 cm wide . . . .*P. sinensis* (OC)
1. Leaves less than 1 cm wide . . . .*P. uniflora* (OC)

Prunus   cherry, plum, peach, apricot
Summer key

1. Leaves evergreen, the teeth obscure, few or none . . . .*P. laurocerasus*
1. Leaves deciduous, the teeth abundant and usually prominent
  2. Lower leaf surfaces tomentose or downy
    3. Leaves finely and evenly serrate . . . .*P. maritima*
    3. Leaves coarsely and unevenly serrate
      4. Fruit sessile or subsessile, tomentose, lacking furrow or bloom; leaves
      conduplicate; terminal bud present . . . .*P. tomentosa* (N)
      4. Fruit stalked, glabrous, furrowed, glaucous; leaves convolute; terminal
      bud lacking . . . .*P. domestica* (ON)
  2. Lower leaf surfaces glabrous or pubescent only along the veins
    5. Leaves with teeth sharply pointed and spreading
      6. Leaf teeth long-pointed or in fine hair-like points
        7. Leaves yellow-green when young, the lower leaf surface pubescent
        along the veins, the petiole pubescent; young
        twigs pubescent . . . .*P. yedoensis* (OC)
        7. Leaves often red or bronze when young, the lower leaf surface
        glabrous, the petiole glabrous . . . .*P. serrulata*
      6. Leaf teeth sharp-pointed but not long and drawn out
        8. Leaves widest below or near the middle
          9. Leaves doubly serrate, mostly 5–10 cm long; fruit
          lacking bloom . . . .*P. subhirtella*
          9. Leaves singly serrate; fruit glaucous
            10. Leaves narrowly elliptic, gradually acuminate; fruit 1 cm
            long, dark purple, the pit turgid . . . .*P. alleghaniensis*
            10. Leaves lanceolate-ovate to oblong-obovate,
            abruptly long-acuminate; fruit 2–3 cm long, red
            or yellow, the pit flattened
              11. Leaves glabrous on upper surface, obtuse to cuneate
              at the base . . . .*P. americana*
              11. Leaves shortly pubescent on upper surface, rounded to
              broadly cuneate at the base . . . .*P. mexicana*

Keys to Species

8. Leaves widest above the middle
    12. Petioles less than 8 mm long; leaves coarsely serrate . . . .*P. triloba*
    12. Petioles greater than 8 mm long; leaves often more
        finely serrate
        13. Leaves shortly pubescent on upper surface . . . .*P. mexicana*
        13. Leaves glabrous on upper surface
            14. Petioles mostly lacking glands, branches often with thorns, fruit
                2–3 cm long, red or yellow . . . .*P. americana*
            14. Petioles with glands near the apex; branches with or
                without thorns; fruit less than 1 or 1–3 cm long
                15. Twigs with a few hairs; leaves hairy along the veins on lower
                    surface; fruit in clusters of 4–6 . . . .*P. yedoensis* (oc)
                15. Twigs glabrous or hairy only when young; leaves
                    glabrous or with axillary tufts on lower surface; fruit
                    numerous in racemes
                    16. Leaves dull green on upper surface; fruit black, with
                        rough pits . . . .*P. padus* (N)
                    16. Leaves lustrous dark green on upper surface; fruit red or
                        red-purple, with smooth pits . . . .*P. virginiana*
5. Leaves with teeth mostly blunt, rounded, or appressed
    17. Leaf blades usually 3 or more times longer than wide
        18. Trees; leaves 10–23 cm long, lacking glands; fruit 3–8 cm
            long, yellow . . . .*P. persica* (N)
        18. Shrubs; leaves 3–8 cm long, the teeth with glands at the tip; fruit
            1–2 cm long, blue-black . . . .*P. fruticosa* (N)
    17. Leaf blades usually less than 3 times longer than wide
        19. Shrubs, often forming thickets (low stocky trees in
            P. umbellata); leaves usually less than 7 cm long
            20. Leaves pale green on lower surface, entire at the base; fruit purple
                to black . . . .*P. pumila*
            20. Leaves green on lower surface, toothed to the base;
                fruit red, blue, or black
                21. Leaves somewhat folded with a trough at the midrib, evenly
                  toothed, the teeth often with a gland at the tip, petiole
                  often red . . . .*P. angustifolia* (s)
                21. Leaves flat, unevenly or doubly toothed, the teeth
                lacking glands, petiole usually green
                  22. Leaves mostly less than twice as long as wide; branchlets with
                    thorns; fruit blue or black . . . .*P. spinosa* (ON)
                  22. Leaves mostly more than twice as long as wide;
                  branchlets lacking thorns; fruit red or black
                      23. Shrubs less than 2 m tall, fruit red . . . .*P. glandulosa*
                  23. Low stocky trees, usually in colonies;
                      fruit black . . . .*P. umbellata* (s)

19. Small or large trees; leaves greater or less than 7 cm long
  24. Leaves mostly less than twice as long as wide
    25. Leaves broadly ovate to elliptic, the apex curved and
      long acuminate
      26. Branchlets often with thorns; fruit ellipsoid, on pedicels
        1–2 cm long . . . .*P. nigra*
      26. Branchlets lacking thorns; fruit globose, nearly sessile, occasionally
        with pedicels less than 1 cm long . . . .*P. armeniaca*
    25. Leaves ovate, obovate, or orbicular, abruptly short-pointed,
      not curved
      27. Leaves mostly less than 5 cm long; fruit mostly
        2–3 cm long . . . .*P. cerasifera*
      27. Leaves mostly greater than 5 cm long; fruit mostly less
        than 2 cm long
        28. Leaves mostly blunt or obtuse; fruit less than
          1 cm long . . . .*P. mahaleb* (ON)
        28. Leaves mostly acute; fruit 1–2 cm long . . . .*P. cerasus* (ON)
  24. Leaves mostly greater than twice as long as wide
    29. Petioles usually 2–5 cm long
      30. Leaves glabrous or sparsely pubescent along the veins on the lower
        surface, crenate-serrate, the glands at the petiole apex large,
        conspicuous; fruit red . . . .*P. avium* (N)
      30. Leaves densely pubescent along the veins on the lower surface,
        crenate, petiole glands small and inconspicuous; fruit
        variously colored . . . .*P. domestica* (ON)
    29. Petioles mostly less than 2 cm long
      31. Petioles pubescent; fruit 2–4 cm long, in clusters
        32. Leaves evenly toothed, with red glands on the teeth,
          with inconspicuous veins on lower surface,
          oblong-lanceolate . . . .*P. munsoniana* (s)
        32. Leaves irregularly toothed, lacking glands, with
          prominent veins on lower surface, oblong-ovate to
          elliptic-ovate . . . .*P. hortulana* (s)
      31. Petioles glabrous or subglabrous; fruit less than 3 cm long,
        solitary or in pairs, clusters, or racemes, red or black
        33. Leaves with dense red-brown tomentum along midrib toward
          the base on the lower surface, flat or at least not wrinkled,
          acuminate or acute; fruit in racemes of 10 or
          more, black . . . .*P. serotina*
        33. Leaves glabrous, lacking red-brown pubescence,
          sometimes somewhat wrinkled, long-acuminate; fruit
          solitary or in clusters of 2–7, red or yellow
          34. Twigs red-brown to gray; petioles less than 1 cm long;
            fruit red . . . .*P. pensylvanica*

34. Twigs green; petiole greater than 2 cm long;
    fruit yellow . . . .*P. mume* (cs)

Winter key

1. Fruit large (to 8 cm in diameter), yellow; buds woolly pubescent; trees; bark
   rough, black . . . .***P. persica*** (N)
1. Fruit smaller (less than 5 cm long), red, yellow, blue, purple, or
   blue-black; buds glabrous or pubescent, not woolly; shrubs or trees;
   bark rough or smooth, brown, red-brown, bronze, or black
   2. Leaves evergreen . . . .*P. laurocerasus*
   2. Leaves deciduous
      3. Terminal bud lacking; fruit furrowed on 1 side, often glaucous
         (alternate step 3, p.)
         4. Fruit pubescent, at least when young, separating from the
            stalk at maturity, the stone furrowed on the margin
            5. Twigs green . . . .*P. mume* (cs)
            5. Twigs dark brown, gray, or black . . . .*P. armeniaca*
         4. Fruit glabrous, usually falling with the pedicel
            6. Fruits mostly 1–2 in a cluster, the pit often rough
               7. Twigs glabrous or glabrate; fruit 2–3 cm long
                  8. Fruit blue, blue-black, or purple, the
                     stalks pubescent . . . .***P. domestica*** (ON)
                  8. Fruit yellow or red, the stalks glabrous . . . .*P. cerasifera*
               7. Twigs pubescent; fruit less than 2 cm long
                  9. Branches and branchlets very thorny; dark brown to black;
                     fruit blue-black, usually greater than
                     1 cm long . . . .***P. spinosa*** (ON)
                  9. Branches and branchlets with a few thorns or
                     thorns lacking, black or red-brown; fruit dark
                     purple or yellow, usually about 1 cm long
                     10. Branchlets red-brown; fruit dark purple,
                         often solitary . . . .***P. alleghaniensis***
                     10. Branchlets black; fruit yellow, 1–2 . . . .***P. umbellata*** (s)
            6. Fruits mostly 3 or more in a cluster, the pit
               usually smooth
               11. Young twigs pubescent; branches and branchlets
                   lacking thorns; fruit purple . . . .*P. maritima*
               11. Young twigs glabrous; branches and branchlets
                   usually thorny; fruit red, orange-red, or yellow
                   12. Fruit ellipsoid, light red, orange-red, or yellow . . . . . .***P. nigra***
                   12. Fruit globose or subglobose, red or yellow
                       13. Fruit on red stalks
                           14. Widespread shrub or small tree to 8 m, spreading from
                               roots, forming thickets . . . .***P. americana***

14. Midwestern tree to 12 m tall, not spreading from roots ....*P. mexicana*

13. Fruit on green stalks

    15. Fruit lacking bloom and dots, 2–3 cm long; roots sprouting, plants forming thickets ....*P. hortulana* (s)

    15. Fruit slightly glaucous, covered with pale dots, mostly 2 cm or less long

       16. Fruit 2 cm long; branches red-brown ....*P. munsoniana* (s)

       16. Fruit less than 2 cm long; branches gray to black ....*P. angustifolia* (s)

3. Terminal bud present; fruit lacking a furrow, lacking bloom, the pit turgid or nearly round

    17. Fruit furrowed on 1 side, pubescent, becoming glabrous; twigs glabrous; shrub ....*P. triloba*

    17. Fruit lacking furrow, pubescent or glabrous, twigs glabrous or pubescent; trees or shrubs

       18. Fruit sessile or subsessile, the pedicels no more than 2 mm long; shrubs ....*P. tomentosa* (N)

       18. Fruit with distinct pedicels greater than 3 mm long; trees or shrubs

          19. Fruits 12 or more, in racemes; trees

             20. Bark foul-smelling; fruit black, the pit rough ......*P. padus* (N)

             20. Bark lacking foul scent; fruit black, red, or red-purple, the pit rough or smooth

                21. Buds about 6 mm long; scales rounded at apex, their margins light gray; twigs stout, light brown, dull; tall shrub or small tree; bark smooth ....*P. virginiana*

                21. Buds about 4 mm long; scales blunt or pointed at apex, margins red-brown; twigs slender or thicker, red-brown, glossy; tree; bark rough and scaly on old trunks ....*P. serotina*

          19. Fruits 1 to few, in clusters; trees or shrubs

             22. Shrubs; buds sometimes in groups of 3

                23. Young twigs glandular-puberulent ....*P. mahaleb* (ON)

                23. Young twigs glabrous or pubescent, not glandular

                   24. Plant mound-shaped; fruit 5–7 mm long, deep red-purple ....*P. fruticosa* (N)

                   24. Plant spreading or erect; fruit usually greater than 7 mm long, red

                      25. Branchlets in a zigzag growth pattern; young twigs pubescent ....*P. glandulosa*

                      25. Branchlets straight; young twigs glabrous ......*P. pumila*

             22. Trees; buds solitary

                26. Young twigs glandular-puberulent ....*P. mahaleb* (ON)

                26. Young twigs glabrous or pubescent, not glandular

Keys to Species

27. Buds clustered at the ends of slender red twigs and short spurs; twigs glossy . . . .*P. pensylvanica*
27. Twigs with buds scattered, not clustered at ends unless spurs are absent, the spurs with buds scattered or clustered at ends; twigs dull or somewhat glossy
    28. Branchlets and twigs black; fruits large; large or small trees
        29. Twigs stout; buds chestnut-brown; tall tree with persistent central stem; bark smooth, often peeling in rolls . . . .*P. avium* (N)
        29. Twigs slender; buds brown; low-branching tree without persistent central stem; bark rough, with small scales . . . .*P. cerasus* (ON)
    28. Branchlets and twigs light brown, gray, gray-brown to dark red-brown; fruits small; small trees; the commonly cultivated flowering cherries that are trees
        30. Buds mostly 3 mm long, the scales spreading slightly and sometimes sparsely hairy; twigs light brown or gray-brown to red brown . . . .*P. subhirtella*
        30. Buds mostly 4–7 mm long, the scales appressed, glabrous, sometimes shining; twigs black, dark red-brown, brown, or gray
            31. Buds mostly 6–7 mm long; twigs dark red-brown . . . .*P. serrulata*
            31. Buds mostly 4–5 mm long; twigs brown or gray . . . .*P. yedoensis* (OC)

Pseudolarix kaempferi   golden-larch (OC)

Pseudotsuga menziesii   Douglas-fir (ON)

Ptelea trifoliata   hop tree, wafer-ash

Pterocarya stenoptera   Chinese wingnut (OC)

Pteroceltis tartinowii   pteroceltis (OC)

Pterostyrax   epaulette tree

1. Leaves with fine, sharp teeth; shrubs or trees; fruit 5-winged, stellate pubescent . . . .*P. corymbosus* (OC)
1. Leaves with shallow, more blunt teeth; trees; fruit 10-ribbed, densely bristly pubescent . . . .*P. hispidus* (OC)

Pueraria lobata   kudzu vine, kudzu (N)

Punica granatum   pomegranate (CS)

Pyracantha   firethorn

1. Leaves tomentose on lower surface . . . .*P. angustifolia* (OC)
1. Leaves glabrous or sparsely pubescent on lower surface

2. Petioles pubescent; fruit stalks pubescent . . . .*P. coccinea* (ON)
2. Petioles glabrous or sparsely pubescent; fruit stalks glabrous
   3. Leaves acute, mucronate, oblong to oblong-lanceolate; twigs rusty-
      pubescent; fruit orange-red . . . .*P. crenulata* (OC)
   3. Leaves obtuse, often emarginate, usually obovate-oblong; twigs
      rusty-pubescent, gray-pubescent, or glabrous; fruit red
      4. Twigs gray-pubescent or glabrous; leaves entire . . . .*P. koidzumii*
      4. Twigs rusty-pubescent; leaves
      crenulate-serrulate . . . .*P. atalantioides* (OC)

Pyrularia pubera   buffalo nut, oil nut (s)

Pyrus   pear

1. Leaves setose-serrate; fruit spherical or apple-shaped, very hard
  when ripe . . . .*P. pyrifolia*
1. Leaves crenate-serrate; fruit pyriform (pear-shaped), soft when ripe
   2. Twigs generally lacking thorns; small tree; leaves usually acuminate or
      cuspidate, the apex sometimes exaggerated into a long, wavy, arching tip;
      fruit globose, small (1–2 cm long) . . . .*P. calleryana*
   2. Twigs often with thorn-like leafy spurs; small to large trees; leaves
      apiculate to acuminate; fruit pear-shaped, sometimes apple-
      shaped, often large (10 cm long)
      3. Leaves crenate-serrulate, nearly entire . . . .*P. communis* (ON)
      3. Leaves serrulate to crenate-serrulate . . . .*P. ×lecontei*

Pyxidanthera barbulata   pyxie, flowering-moss

Quercus   oak

The winter key here does not include seven cultivated species that are included in the summer key. Neither key includes any of the many hybrid species in this genus.

Summer key

1. Leaves regularly or irregularly lobed, usually divided more than $^1/_3$
  the distance from margin to midrib, or with only 3 large lobes
  toward the apex (alternate step 1, p. 196)
   2. Mature leaves with lobes bristle-tipped, at least some
      bristles 2–3 mm long (alternate step 2, p. 195)
      3. Leaves 3-lobed, the base often narrowed
      4. Mature leaves densely and minutely appressed-stellate and like velvet
      on lower surface . . . .*Q. pagoda* (s)
      4. Mature leaves glabrous or with tufts of hairs in the axils
      of the veins on lower surface

5. Twigs less than 2 mm in diameter, glabrous; leaves narrowly cuneate; acorn cup flat or saucer-shaped, enclosing just the base; growing in wet areas . . . . *Q. nigra* (s)

5. Twigs greater than 2 mm in diameter, puberulent; leaves broadly cuneate at the base; acorn cup cup-shaped, enclosing at least $1/3$ the length of the nut; growing in sand or other drier habitats . . . . *Q. marilandica*

3. Leaves with more than 3 lobes, the base rounded to truncate, usually not gradually narrowed

  6. Buds glabrous, occasionally ciliate on the margins

    7. Twigs pubescent; leaves with lower surface minutely and densely pubescent . . . . *Q. ilicifolia*

    7. Twigs glabrous; leaves with lower surface appearing glabrous at maturity or occasionally with tufts of hairs restricted to the axils of the veins

      8. Shrubs or small trees; Georgia and South Carolina . . . *Q. georgiana* (s)

      8. Trees; widespread

        9. Twigs light brown to yellow-green; buds yellow-brown, waxy, usually acute . . . . *Q. shumardii*

        9. Twigs dark brown to red-brown; buds dark brown to red-brown, dull, not waxy

          10. Leaves with upper surface dull, not shiny, shallowly lobed, usually less than half the distance toward the midrib, the lobes acute or tapered, the adjacent lobes separated by open, divergent sinuses . . . . *Q. rubra*

          10. Leaves with upper surface shiny, lustrous, deeply lobed, usually more than half the distance to the midrib, the lobes club-shaped, the adjacent lobes often convergent at the apex, the sinuses closed

            11. Lowest branches horizontal or ascending; acorn cup turbinate, deeply cup-shaped; Ohio, Michigan, Minnesota, Wisconsin, and Ontario . . . . *Q. ellipsoidalis*

            11. Lowest branches descending; acorn cup flat, saucer-shaped; widely distributed . . . . *Q. palustris*

  6. Buds pubescent, even if only toward the apex

    12. Buds rounded at the apex, pubescent only toward the apex . . . . *Q. coccinea*

    12. Buds acute or long-pointed at the apex, pubescent all over

      13. Leaves with lower surface persistently pubescent; twigs persistently pubescent . . . . *Q. falcata*

      13. Leaves with lower surface glabrate, sometimes with persistent hairs in the axils of the veins; twigs glabrate or pubescent

        14. Leaves with irregular, curved, elongate lobes, the lobes at least twice as long as wide; small, open trees in open, sandy habitats near the coast . . . . *Q. laevis* (s)

14. Leaves regularly lobed, the lobes less than twice as long as wide; inner bark orange; widespread . . . .*Q. velutina*

2. Mature leaves with lobes lacking bristles

  15. Mature leaves with lower surface glabrous or appearing so to the naked eye or with isolated tufts of hairs in the axils of the veins; twigs glabrous

    16. Leaves moderately to deeply lobed, the lobes acute; acorn cup bur-like, enclosing the nut or with only the nut apex visible . . . .*Q. lyrata* (s)

    16. Leaves sinuate or shallowly to deeply lobed, the lobes rounded, sometimes retuse; acorn cup with bumps or smooth scales, not bur-like, enclosing less than half the nut

      17. Leaves strongly cordate at the base; acorn stalks usually 3–10 cm long . . . .*Q. robur* (N)

      17. Leaves cuneate or narrowed at the base, acorn nearly sessile or usually less than 3 cm long

        18. Leaves moderately to deeply lobed their entire length; cuneate at the base . . . .*Q. alba*

        18. Leaves sinuately lobed or narrowly obovate, 3-lobed toward the apex, at least $1/3$ of each margin entire or at most sinuate, or narrowed at the base

          19. Leaves with lower surface often silver, minutely appressed-stellate, the shade leaves sometimes glabrate; acorn cup shallow, saucer-shaped, enclosing less than $1/4$ of the nut . . . .*Q. sinuata* (s)

          19. Leaves glabrate, the lower surface sometimes with persistent stalked, multi-radiate hairs near the midrib; acorn cups often enclosing at least $1/3$ of the nut

            20. Leaves about as wide as long at their widest point, the base usually rounded or cordate . . . .*Q. marilandica*

            20. Leaves twice as long as wide, the base cuneate to attenuate . . . .*Q. austrina* (s)

  15. Mature leaves conspicuously pubescent on 1 or both surfaces or strongly glandular or waxy on lower surface; twigs glabrous, stellate, or tomentose

    21. Twigs pubescent, tomentose, or stellate (often visible only with hand lens, especially on shade leaves)

      22. Leaves pubescent but not stellate on upper surface; branchlets with corky wings; buds glabrous, red-brown; acorn cup fringed with long soft awns . . . .*Q. macrocarpa*

      22. Leaves stellate on upper surface; branchlets smooth, lacking corky wings; buds pubescent, chestnut-brown; acorn cup lacking fringe of soft awns . . . .*Q. stellata*

    21. Twigs initially glabrous or becoming glabrous in first season; acorn cup lacking awns or the awns short, stout, and recurved

23. Lower leaf surface sparsely to densely pubescent, the hairs erect
   and like velvet

   24. Leaves with mixed erect and appressed-stellate hairs on lower surface;
      acorns on stalks to 6 cm long, the scales near cup rim often with
      recurved awns; large trees of wet areas . . . .***Q. bicolor***

   24. Leaves with only erect hairs on lower surface; acorns subsessile, the
      scales lacking awns; shrubs and small trees of
      sand areas . . . .***Q. margaretta*** (s)

23. Lower leaf surface silver, minutely appressed-stellate, often
   glabrate in shade leaves

   25. Leaves deeply lobed, often extending more than halfway to the midrib,
      the lobes often blunt-acute, with 2 of them large, blocky, and squared;
      acorn stalk less than 1 cm long . . . .***Q. lyrata*** (s)

   25. Leaves entire to shallowly or deeply lobed, the lobes small,
      rounded, usually extending less than halfway to the midrib,
      lacking 2 large, blocky, squared lobes; acorn stalk less than
      1 cm long or 2–4 cm long

      26. Leaves with erect hairs and appressed-stellate hairs; acorn stalk
         2–4 cm long . . . .***Q. bicolor***

      26. Leaves with scattered or dense appressed-stellate hairs or glabrate or
         glabrous, lacking erect hairs; acorn subsessile or with stalk less than
         1 cm long . . . .***Q. sinuata*** (s)

1. Leaves entire, dentate, serrate, or shallowly or sinuately lobed; if
lobed, divided less than $^1/_3$ the distance from margin to midrib,
not 3-lobed toward the apex

   27. Leaves with 6–20 unbranched, rather straight, parallel,
      secondary veins on each side of the midrib, the margins
      regularly toothed, each vein extending into a single tooth, the
      teeth never spinose, soft awns sometimes present (the chestnut
      oaks) (alternate step 27, p. 198)

      28. Leaf teeth tipped with soft awns

         29. Leaves green, glabrous or glabrate on lower surface . . . .*Q. acutissima*

         29. Leaves pale or gray, soft, downy-pubescent
            on lower surface . . . .*Q. variabilis* (oc)

      28. Leaf teeth rounded, acute, or mucronate

         30. Leaves narrowed and cordate to nearly auriculate at the
            base, subsessile, or the petiole less than 8 mm long

            31. Leaves with lower surface densely hairy, or like velvet;
               twigs densely yellow-pubescent;
               buds pubescent . . . .*Q. dentata* (oc)

            31. Leaves with lower surface glabrous or glabrate; twigs glabrous;
               buds glabrous . . . .*Q. mongolica* (oc)

         30. Leaves rounded to subcordate at the base, the petiole
            greater than 8 mm long

32. Leaves nearly all with stipules persistent and turning black; twigs sooty; acorn cups with long curled, awn-like scales over the entire surface . . . .*Q. cerris*

32. Leaves with stipules early deciduous or persisting only near leaves at the end of the twigs; twigs variously colored, lacking sooty or dingy appearance; acorn cups with long curled awn-like scales only near the rim or lacking

  33. Lower leaf surface like velvet, with evenly distributed 1–4 rayed, erect hairs

    34. Leaves regularly toothed, lacking deep irregular divisions near the base, the lower surface with only erect hairs, lacking flat appressed-stellate hairs; acorn greater than 25 mm long, the cup scales lacking awns, the stalk to 2 cm long . . . .*Q. michauxii* (s)

    34. Leaves sometimes with irregular deep divisions near the base, the lower surface with appressed-stellate and erect hairs; acorn less than 25 mm long, the cup scales sometimes with a few stout spinose awns near the cup rim, the stalk 2–7 cm long . . . .*Q. bicolor*

  33. Lower leaf surface glabrous or with appressed-stellate or unevenly distributed erect hairs, not like velvet

    35. Lower leaf surface with appressed-stellate hairs and appressed unbranched hairs oriented parallel to the secondary veins . . . .*Q. glandulifera* (oc)

    35. Lower leaf surface with appressed-stellate hairs, but appressed unbranched hairs lacking or only on the veins

      36. Leaves sometimes irregularly, deeply divided toward the base; acorn with stalk 2–7 cm long, the cup scales often with stout spinose awns near the rim . . . .*Q. bicolor*

      36. Leaves dentate, serrate, or toothed, lacking deep, irregular divisions toward the base; acorn with stalk less than 2 cm long, the cup scales lacking stout spinose awns

        37. Lower leaf surface usually with tufts of spreading hairs along the midrib and scattered, irregular, 2–4 rayed stellate hairs; bark deeply furrowed

          38. Leaves with divisions irregular, of irregular length (rarely, the margins nearly undulate); cultivated . . . .*Q. petraea* (oc)

          38. Leaves with regular teeth, of uniform length; native . . . .*Q. montana*

        37. Lower leaf surface lacking erect hairs along the midrib, with scattered to dense 6–10 rayed, appressed-stellate hairs; bark scaly

          39. Secondary veins mostly more than 10 on each side of the midrib; trees usually on limestone or calcareous soils . . . .*Q. muehlenbergii*

39. Secondary veins mostly fewer than 8 on each side of the midrib; rhizomatous shrubs on dry sand or occasionally on acidic shale on ridges . . . .***Q. prinoides***

27. Leaves with 2–12 irregular, branched, curved, crooked, obscure or not, obviously parallel secondary veins on each side of the midrib, the margins entire or irregularly toothed or spinose

40. Lower leaf surface glabrous or occasionally with inconspicuous tufts of hairs in the axils of the veins

41. Leaves elliptic to rhombic, widest above the middle

42. Leaves with midrib and secondary veins impressed or flush on upper surface, the base usually long-cuneate . . . .***Q. laurifolia*** (s)

42. Leaves with midrib and secondary veins conspicuously raised on upper surface, the base rounded to short-cuneate . . . .***Q. hemispherica*** (s)

41. Leaves narrowly elliptic to oblong-linear, lanceolate, or oblanceolate, with uniform width except for abruptly tapering ends

43. Rhizomatous shrubs; leaves with revolute margins; acorns maturing the first year . . . .***Q. pumila*** (s)

43. Trees; leaves with flat margins; acorns maturing the second year

44. Twigs puberulent; bark of trunk pebbly or checkered, black; small trees of dry habitat . . . .***Q. incana*** (s)

44. Twigs glabrous; bark of trunk fissured with flat-topped longitudinal ridges, dark gray; large trees of wetter soils . . . .***Q. phellos***

40. Lower leaf surface stellate-pubescent or with tufts of erect hairs distributed rather evenly, the hairs sometimes minute, glandular hairs or wax sometimes present

45. Lower leaf surface yellow-green, glandular or glabrate, lacking stellate hairs or tufts of erect hairs, or with spreading hairs only along the midrib . . . .***Q. chapmanii*** (s)

45. Lower leaf surface variously pubescent, not noticeably glandular yellow, the hairs usually pale, the surface then green or pale

46. Leaves usually convexly cupped or with revolute margins, the lower surface with prominent or obscurely raised veins near the margins, the secondary veins often impressed on upper surface

47. Lower leaf surface with minute appressed-stellate hairs, often with scattered, straight tufts of erect hairs like velvet; trees; acorns stalked, the stalks 1–10 cm long . . . .***Q. geminata*** (s)

47. Lower leaf surface lacking stellate hairs, with scattered to dense tufts of erect, often curly hairs; low rhizomatous shrubs; acorns sessile . . . .***Q. pumila*** (s)

46. Leaves flat (revolute in *Q. geminata*), the lower surface lacking prominent, raised veins, the secondary veins not impressed on upper surface

    48. Lower leaf surface with erect tufts of hairs or erect to semi-erect, stellate hairs, often like velvet to touch

        49. Bark scaly, papery; buds globose . . . . *Q. oglethorpensis*(s)

        49. Bark tight, furrowed; buds acute

            50. Buds dull, brown; leaves large, mostly greater than 4 cm wide . . . . *Q. imbricaria*

            50. Buds shiny, brown to red-brown; leaves narrow, willow-like, less than 4 cm wide . . . . *Q. incana* (s)

    48. Lower leaf surface with obvious or minute appressed-stellate hairs, sometimes appearing glaucous, glabrous, or yellow-glandular, not like velvet to touch

        51. Acorns subsessile or on stalks to 5 mm, the cotyledons distinct . . . . *Q. sinuata* (s)

        51. Acorns on stalks 3–30 mm long, the cotyledons distinct or connate

            52. Low shrubs; twigs mostly glabrous; leaves asymmetric . . . . *Q. minima* (s)

            52. Trees or shrubs; twigs puberulent or stellate-pubescent; leaves symmetric

                53. Leaves convex, the margins strongly revolute, the veins impressed on the upper surface, the lower surface usually with scattered erect tufts of hairs and appressed-stellate hairs . . . . *Q. geminata* (s)

                53. Leaves flat, the margins flat, the veins even, not impressed on upper surface, the lower surface with only appressed-stellate hairs . . . . *Q. virginiana* (s)

Winter key

1. Acorns mature or lacking on the current year's growth, having matured in the first season (the white oaks and *Q. pumila*) (alternate step 1, p. 201)

    2. Leaves persistent or tardily deciduous

        3. Low shrubs; twigs nearly glabrous; leaves asymmetrical or symmetrical

            4. Leaves asymmetrical, the base cuneate to rarely truncate or cordate . . . . *Q. minima* (s)

            4. Leaves symmetrical, the base acute to rounded . . . . . . *Q. pumila* (s)

        3. Trees and shrubs; twigs puberulent or stellate-pubescent; leaves symmetrical

            5. Leaves convex, strongly revolute, the veins impressed on upper surface, the lower surface usually with scattered, erect tufts of hairs and appressed-stellate hairs . . . . *Q. geminata* (s)

Keys to Species

5. Leaves flat, the margins flat, the veins even, not impressed on upper surface, the lower surface with only appressed-stellate hairs . . . .***Q. virginiana*** (s)

2. Leaves deciduous (some brown leaves often persisting)

  6. Twigs or buds pubescent

    7. Buds glabrous

      8. Shrubs less than 1 m tall . . . .***Q. pumila*** (s)

      8. Large trees

        9. Branchlets with corky, rough wings; twigs finely tomentulose; acorn cup with fringed scale . . . .***Q. macrocarpa***

        9. Branchlets smooth, lacking corky wings; twigs with minute spreading hairs, sometimes appearing glabrous; acorn cup with blunt scales, lacking awns or fringe . . . .***Q. michauxii*** (s)

    7. Buds pubescent, at least toward the apex

      10. Twigs pubescent

        11. Trees . . . .***Q. stellata***

        11. Shrubs less than 2 m tall . . . .***Q. chapmanii*** (s)

      10. Twigs glabrous

        12. Plants growing in calcareous soils, often limestone bluffs; acorn cup brown, the scales loose . . . .***Q. muehlenbergii***

        12. Plants growing in sandy, bottomland soils; acorn cup pale, tomentose . . . .***Q. austrina*** (s)

  6. Twigs and buds glabrous (or appearing glabrous in ***Q. michauxii***)

    13. Low shrubs, less than 2 m tall . . . .***Q. prinoides***

    13. Trees greater than 2 m tall

      14. Bark with deep longitudinal furrows with relatively flat or smooth intervening plates . . . .***Q. montana***

      14. Bark scaly, exfoliating, papery, checkered or rough, lacking deep longitudinal furrows

        15. Peduncles greater than 1 cm long

          16. Widely distributed, cultivated and naturalized trees; acorn cup scales lacking awns . . . .***Q. robur*** (N)

          16. Native trees of wet soils; acorn cup often with short, stout, recurved awns irregularly placed at the margin . . . . . .***Q. bicolor***

        15. Peduncles less than 1 cm long

          17. Acorn cups enclosing more than $2/3$ of the acorn; plants growing in wet bottomlands . . . .***Q. lyrata*** (s)

          17. Acorn cups enclosing less than half the acorn; plants growing in a variety of habitats

            18. Acorn cups nearly flat, enclosing only $1/5$ to $1/4$ of the acorn . . . .***Q. sinuata*** (s)

            18. Acorn cups deeper, enclosing more than $1/4$ of the acorn

              19. Twigs with sparse spreading hairs visible only with a hand lens; acorns large, the cup scales loose . . .***Q. michauxii*** (s)

19. Twigs glabrous, lacking sparse spreading hairs; acorns smaller, the cup scales tight
    20. Shrubs or small trees restricted to open sands ......*Q. margaretta* (s)
    20. Small to large trees on rich bottomlands or dry uplands, not on sandy soils
        21. Plants widespread ....*Q. alba*
        21. Plants on rich bottomlands in Georgia, South Carolina, and Mississippi ....*Q. oglethorpensis* (s)
1. Small, undeveloped acorns present on the current year's growth, maturing the second season (the red oaks except *Q. pumila*)
    22. Twigs or buds pubescent (alternate step 22, p. 202)
        23. Buds glabrous
            24. Shrubs or small trees to 6 m tall ....*Q. ilicifolia*
            24. Small or large trees greater than 6 m tall
                25. Twigs brown or red-brown, sparsely pubescent to tomentose; acorns $1/4$ to $1/3$ enclosed by the cup; plants growing in well-drained sandy soils; Virginia, North Carolina, South Carolina, Alabama, and Mississippi ....*Q. incana* (s)
                25. Twigs green-brown to brown, glabrous to sparsely pubescent; acorns $1/3$ to $1/2$ enclosed by the cup; plants growing on dry to mesic slopes or ravines and bottomlands; Deleware, Illinois, Indiana, Kentucky, Maryland, New Jersey, North Carolina, Ohio, Pennsylvania, South Carolina, Tennessee, Virginia, and West Virginia ....*Q. imbricaria*
        23. Buds pubescent
            26. Twigs glabrous
                27. Buds pubescent toward the apex
                    28. Buds usually acute and often densely pubescent in apical half ....*Q. coccinea*
                    28. Buds usually rounded and sparsely pubescent in apical $1/3$ ....*Q. rubra*
                27. Buds evenly pubescent all over
                    29. Twigs thick, 3–5 mm in diameter; inner bark orange; buds angled, large, 6–12 mm long, acute; acorn cups cup-shaped, 7–14 mm high ....*Q. velutina*
                    29. Twigs thin, 2–3 mm in diameter; inner bark pale, not orange; buds lacking angles, 3–6 mm long, ovoid; acorn cups saucer-shaped, 2–6 mm high ....*Q. nigra* (s)
            26. Twigs pubescent
                30. Shrubs or small, irregular, strikingly open trees of sandy soils; Alabama, Georgia, Mississippi, North Carolina, South Caroline, and Virginia ....*Q. laevis* (s)
                30. Larger trees of various habitats and more widely distributed; New York to South Carolina, Illinois to Mississippi

31. Twigs thick; acorn cups 6–10 mm long; plants of dry or poor
soils, widely distributed . . . .*Q. marilandica*
31. Twigs thin; acorn cups usually less than 6 mm long; plants of
dry, sandy uplands or poorly drained bottoms and mesic slopes
  32. Plants of dry or sandy uplands . . . .*Q. falcata*
  32. Plants of poorly drained bottomlands or mesic slopes . . .*Q. pagoda* (s)
22. Twigs and buds glabrous
  33. Leaves persistent or tardily deciduous
    34. Plants of drier sandy soils and sandhills . . . .*Q. hemisphaerica* (s)
    34. Plants of sandy floodplains, bottomlands, and riverbanks,
occasionally on poorly drained uplands . . . .*Q. laurifolia* (s)
  33. Leaves deciduous
    35. Small trees to 15 m tall, restricted to dry slopes and grantic outcrops
in Alabama, South Carolina, and Georgia . . . . *Q. georgiana* (s)
    35. Large trees of widespread or northern distribution
      36. Buds straw-colored, waxy; twigs yellow-brown . . . .*Q. shumardii*
      36. Buds brown to red-brown; twigs gray to red-brown
or brown
        37. Lowest branches descending; acorn cups flat,
saucer-shaped . . . .*Q. palustris*
        37. Lowest branches horizontal or ascending; acorn cups
saucer-shaped to deeply cup-shaped
          38. Trunk often with stubs of dead branches; inner bark orange;
acorn cups usually deeply cup-shaped; distribution restricted
to Ohio, Michigan, Minnesota, Wisconsin,
and Ontario . . . .*Q. ellipsoidalis*
          38. Trunk clean, lacking stubs; inner bark pale or
brown; acorn cups saucer-shaped or cup-shaped;
widely distributed
            39. Acorns 8–12 mm long, the cup thin, shallowly saucer-
shaped; New York south to Georgia, west
to Illinois . . . .*Q. phellos*
            39. Acorns 15–30 mm long, the cup thick, saucer-shaped to cup-
shaped; widely distributed in the entire region . . . .*Q. rubra*

Raphiolepis   Indian-hawthorn

1. Leaves sharply serrate, acute or acuminate . . . .*R. indica* (cs)
1. Leaves sparsely dentate to entire, usually obtuse . . . .*R. umbellata* (cs)

Rhamnus   buckthorn

1. Buds naked
  2. Leaf margins serrate or serrulate; leaves glabrous on
lower surface . . . .*R. caroliniana* (s)

2. Leaf margins entire or undulate; leaves pubescent
on lower surface . . . .*R. frangula* (N)
1. Buds scaly
   3. Leaves and buds opposite or sub-opposite; twigs mostly ending in sharp
     black thorns . . . .*R. cathartica* (N)
   3. Leaves and buds alternate; twigs not ending in thorns
      4. Leaves ovate; twigs glabrous, red or brown in winter; low shrub, usually
        less than 1 m tall . . . .*R. alnifolia*
      4. Leaves oblong-lanceolate; twigs often downy, gray in winter; tall shrub,
        1–2 m tall . . . .*R. lanceolata*

Rhododendron   rhododendron, azalea

Some cultivated non-native species are not included in the following key. Most of the commonly planted species are hybrids.

Summer key

1. Leaves coriaceous, persistent, entire
   2. Leaves and twigs dotted with rusty-brown scales; stems prostrate
     or erect, loose shrubs
      3. Leaves mostly less than 2 cm long; stems prostrate, to
        30 cm tall . . . .*R. lapponicum*
      3. Leaves to 13 cm long; stems erect, to 4 m tall . . . .*R. minus* (s)
   2. Leaves and twigs lacking rusty-brown scales; erect shrubs
     or small trees
      4. Leaves less than 4 cm long . . . .*R. indicum*
      4. Leaves greater than 4 cm long
        5. Leaf base tapering . . . .*R. maximum*
        5. Leaf base rounded . . . .*R. catawbiense*
1. Leaves thin, deciduous, ciliate
   6. Twigs glabrous
      7. Leaves with rusty-brown scales or dots on
        each surface . . . .*R. mucronulatum*
      7. Leaves glabrous or pubescent, lacking rusty-brown scales
        8. Leaves pubescent
          9. Leaves with gray strigose hairs, especially on midrib
            of lower surface . . . .*R. canadense*
          9. Leaves with scattered hairs on upper surface and on the veins
            of lower surface . . . .*R. japonicum*
        8. Leaves mostly glabrous (sometimes a few stiff hairs on
         the midrib)
          10. Branchlets mostly with bark peeling and shredding; fruit and fruit
            stalks pubescent . . . .*R. vaseyi* (s)

10. Branchlets with bark tight, not peeling or shredding; fruit and fruit stalks glabrous

 11. Leaves obovate to elliptic, mostly less than 6 cm long . . .*R. arborescens*

 11. Leaves oblanceolate, mostly greater than 7 cm long . . . .*R. japonicum*

6. Twigs pubescent

 12. Leaves glabrous or the midrib with a few stiff hairs or with bristly brown hairs

  13. Twigs strigose-setose, lacking glands . . . .*R. periclymenoides*

  13. Twigs glandular-pubescent, the glandular hairs often shedding early

   14. Twigs densely glandular-pubescent; shrub 20–300 cm tall . . . .*R. viscosum*

   14. Twigs with or without glandular hairs, the hairs shedding early; stoloniferous, colony-forming shrub, 20–60 cm tall . . . .*R. atlanticum*

 12. Leaves with lower surface softly, often sparsely pubescent, the midrib strigose or densely white-pubescent

  15. Leaves with a few stiff, strigose, often closely appressed hairs along the midrib of lower surface . . . .*R. bakeri*

  15. Leaves softly pubescent over entire lower surface

   16. Lower leaf surface with brown hairs, often with scattered strigose hairs on midrib . . . .*R. serrulatum* (s)

   16. Lower leaf surface with white or pale brown hairs, generally lacking strigose hairs on midrib

    17. Basal $^{1}/_{2}$ to $^{2}/_{3}$ of midrib of lower leaf surface densely hairy, the apical $^{1}/_{3}$ nearly glabrous . . . .*R. calendulaceum*

    17. Midrib on lower leaf surface hairy its entire length

     18. Fruit pubescent . . . .*R. canescens*

     18. Fruit glandular-pubescent . . . .*R. prinophyllum*

Winter key

1. Leaves coriaceous, persistent, entire

 2. Leaves and twigs dotted with rusty-brown scales; stems prostrate or erect, loose shrubs

  3. Leaves mostly less than 2 cm long; stems prostrate, to 30 cm tall . . . .*R. lapponicum*

  3. Leaves to 13 cm long; stems erect, to 4 m tall . . . .*R. minus*

 2. Leaves and twigs lacking rusty-brown scales; erect shrubs or small trees

  4. Leaves less than 4 cm long . . . .*R. indicum*

  4. Leaves greater than 4 cm long

5. Leaf base tapering . . . .*R. maximum*
5. Leaf base rounded . . . .*R. catawbiense*
1. Leaves deciduous
  6. Twigs glabrous
    7. Branchlets with bark peeling and shredding . . . .*R. vaseyi* (s)
    7. Branchlets with bark tight, not peeling or shredding
      8. Fruit glaucous; bud scales glabrous to puberulent, not ciliate; leaf scars
        raised; fruit glaucous . . . .*R. canadense*
      8. Bud scales ciliate, occasionally glandular-pubescent; leafs
        scars not significantly raised; fruit lacking bloom
        9. Buds glabrous . . . .*R. mucronulatum*
        9. Buds puberulous to glandular-pubescent . . . .*R. arborescens*
  6. Twigs pubescent
    10. Twigs tomentulose, at least at tip . . . .*R. canescens*
    10. Twigs glabrate, strigose, strigose-setose, canescent-pilose, or
      with stalked glands or bristly brown hairs
      11. Twigs glandular-pubescent, the glandular hairs often
        shedding early
        12. Twigs densely glandular-pubescent; shrub
          20–300 cm tall . . . .*R. viscosum*
        12. Twigs with glandular hairs sparse or lacking; stoloniferous, colony-
          forming shrub, 20–60 cm tall . . . .*R. atlanticum*
      11. Twigs strigose-setose, lacking glands
        13. Bud scales glabrous; fruit sparsely strigose . . . .*R. periclymenoides*
        13. Bud scales pubescent; fruit glandular-setose or finely
          pubescent or strigose
          14. Fruit glandular-setose; twigs canescent-pilose and often
            somewhat strigose . . . .*R. prinophyllum*
          14. Fruit finely pubescent or strigose, lacking glands;
            twigs strongly or weakly strigose
            15. Young shoots glabrous . . . .*R. bakeri*
            15. Young shoots strigose or mildly strigose
              16. Shrub to 5 m tall; floral bud scales uniformly colored, not
                mucronate . . . .*R. calendulaceum* and *R. japonicum*
              16. Shrub 2–7 m tall; flower bud scales with darker
                borders, mucronate . . . .*R. serrulatum* (s)

Rhodotypos scandens   jetbead (ON)

Rhus   sumac

Summer key

1. Leaves with 3 leaflets, each with 6–12 teeth, aromatic
  when bruised . . . .*R. aromatica*

1. Leaves with 6 or more leaflets, each with 12–30 teeth or entire, not aromatic
   2. Leaflets entire, the rachis decurrent or winged . . . .*R. copallina*
   2. Leaflets serrate, rachis not decurrent, rarely narrowly winged
      3. Leaves and twigs glabrous . . . .*R. glabra*
      3. Twigs pubescent
         4. Leaflets 9–13, pale but not glaucous on lower surface . . . .*R. michauxii* (s)
         4. Leaflets 15–31, pale, glaucous on lower surface . . . .*R. typhina*

Winter key

1. Leaf scars round; catkins present
   2. Twigs glabrous or puberulent . . . .*R. aromatica*
   2. Twigs densely pubescent . . . .*R. typhina*
1. Leaf scars shaped as a shield, heart, crescent, C, or U; catkins absent
   3. Leaf scars broadly crescent- or shield-shaped; fruit in axillary clusters . . . .*R. copallina*
   3. Leaf scars C- or U-shaped, almost surrounding the buds; fruit in terminal clusters
      4. Twigs glabrous, angled or three-sided . . . .*R. glabra*
      4. Twigs pubescent, terete or nearly so
         5. Twigs puberulent or downy; lenticels prominent, corky, raised . . . .*R. copallina*
         5. Twigs densely pubescent; lenticels inconspicuous
            6. Twigs densely hirsute, the hairs long and straight . . . . . .*R. typhina*
            6. Twigs densely pubescent, the hairs shorter and somewhat tangled . . . .*R. michauxii* (s)

Ribes   gooseberry, currant

1. Stems and twigs lacking spines or prickles; fruit in racemes (**R. hirtellum**, which has black fruit, solitary or in clusters of 2–4, may lack spines)
   2. Leaves, twigs, and buds with scattered, sessile, yellow resin glands; berry black
      3. Leaves with glands or dots on lower surface . . . .*R. hudsonianum*
      3. Leaves pubescent or glabrous, lacking glands or dots on lower surface
         4. Resin glands present on upper surface of leaf; twigs glabrate; inner bark lacking fetid odor; leaf scar with decurrent ridge . . . .*R. americanum*
         4. Resin glands absent on upper surface of leaf; twigs puberulent; inner bark with fetid odor; leaf scar lacking decurrent ridge . . . .*R. nigrum* (ON)

2. Leaves, twigs, and buds lacking yellow resin glands
  5. Erect bushy shrubs; berry red or black
    6. Branchlets light brown, puberulent; leaves convolute in bud;
      berry black . . . .*R. odoratum*
    6. Branchlets olive-brown to gray, glabrate; leaves plicate in bud;
      berry red . . . .*R. rubrum* (ON)
  5. Low, prostrate or creeping shrubs; berry red
    7. Leaves with 5–7 prominent lobes; inner bark with fetid odor;
      buds green or red-purple, fusiform; berry
      glandular-bristly . . . .*R. glandulosum*
    7. Leaves with 3–5 prominent lobes; inner bark lacking fetid odor; buds
      gray-brown, ovoid; berry glabrous . . . .*R. triste*
1. Stems and twigs with spines, or prickles, or both; fruit solitary or
2–5 in a cluster
  8. Branches and branchlets usually densely covered with
  internodal prickles
    9. Branchlets red-brown; fruit hairy, black, in racemes . . . .*R. lacustre*
    9. Branchlets gray or dark brown; fruit spiny, red, in clusters
      of 1–4 . . . .*R. cynosbati*
  8. Branches and branchlets with few or no internodal prickles
    10. Berry spiny or hairy
      11. Spines at branchlet nodes slender, 5–10 mm long; berry
        spiny, red . . . .*R. cynosbati*
      11. Spines at branchlet nodes stout, 10–15 mm long; berry pubescent or
        glandular-bristly; red or green . . . .*R. uva-crispi* (ON)
    10. Berry smooth, lacking spines or hairs; dark purple
      or black
      12. Petiole hairs branched; lobes of leaves acute . . . .*R. hirtellum*
      12. Petiole hairs simple, glandular, or absent; lobes of
        leaves obtuse
        13. Branchlets usually with a few internodal spines of varying lengths;
          Michigan, Minnesota, and north . . . .*R. oxyacanthoides*
        13. Branchlets with spines and nodes only, lacking
          internodal spines; south of Minnesota, south of
          Massachusetts
          14. Spines at nodes usually solitary; Minnesota
            to Tennessee . . . .*R. rotundifolium*
          14. Spines at nodes usually 2 or 3 together; Massachusetts to
            North Carolina . . . .*R. missouriense*

Robinia   locust

1. Twigs densely hispid; axillary buds usually visible in winter; low shrub; fruit
  glandular-hispid . . . .*R. hispida*

1. Twigs glabrous, slightly pubescent or sticky-glandular, not hispid; buds usually imbedded beneath surface of leaf scars in winter; trees or shrubs or low shrubs; fruit glabrous or glandular-hispid
    2. Trees; twigs glabrous or puberulent; spines with dilated base; fruit glabrous . . . .*R. pseudoacacia*
    2. Shrubs or small trees; twigs glabrous, puberulent, canescent, or covered with sticky glandular hairs; spines weak, base not dilated; fruit hispid or sparsely glandular-hispid
        3. Twigs covered with sessile and short-stalked glands; small trees or shrubs usually greater than 3 m tall
            4. Twigs covered with sessile and short-stalked sticky glands all over; leaflets 13–25, glabrous to puberulent on lower surface; fruit sparsely glandular-hispid . . . .*R. viscosa*
            4. Twigs densely puberulent with short-stalked glands; leaflets usually 7–13, densely villous on lower surface; fruit densely glandular-hispid . . . .*R. hartwigii* (s)
        3. Twigs glabrous, puberulent, canescent, or glabrate; shrubs less than 3 m tall
            5. Branchlets canescent, becoming glabrous . . . .*R. elliottii* (s)
            5. Branchlets glabrous or glabrate, minutely puberulent at first
                6. Shrub less than 1 m tall . . . .*R. nana* (s)
                6. Shrub 1–3 m tall
                    7. Branchlets puberulent, becoming glabrous . . . . .*R. boyntonii* (s)
                    7. Branchlets glabrous . . . .*R. kelseyi* (s)

Rosa   rose

The hybridization and polyploidy included in this genus cause much taxonomic confusion. Many species from Europe and Asia have naturalized in North America and actively participate in this taxonomic nightmare. A meaningful key that relies only on vegetative material has eluded me. The following key uses vegetative material, but certain steps require knowledge of the flower parts. If a flower or fruit is not present, the specimen must be keyed through each of the choices at step 3 to reach a determination. As for a winter key, I suggest that you mark the plant and come back in summer, because this key, in many cases, only narrows your choice to a few species rather than to the single correct species. The key is adapted from Voss's treatment in *Michigan Flora*.

1. Stipules free from the petiole for at least half their length
    2. Stipules free from the petiole nearly their entire length; leaflets 5–9 . . . .*R. bracteata* (N)
    2. Stipules free from the petiole for half their length; leaflets 3–5 . . . .*R. banksiae*
1. Stipules fused to the petiole more than half their length

3. Outer 3 sepals pinnatifid or with narrow lateral lobes (alternate
    step 3, p. 210)
    4. Stipules deeply pinnatifid; fruit several to many . . . .***R. multiflora*** (N)
    4. Stipules entire, dentate, or with glandular margins; fruit 1–4
       5. Stipules dentate
          6. Leaflets nearly always 5; prickles of different sizes, often curved
             or hooked . . . .***R. damascena*** (S)
          6. Leaflets usually 5–11; prickles equal or of different sizes,
             usually straight
             7. Stipules dentate, the teeth of different sizes; plants climbing
                or trailing, semievergreen; styles united into
                a column . . . .*R. wichuraiana*
             7. Stipules glandular-dentate to subentire; plants erect, deciduous;
                styles free from each other . . . .***R. carolina***
       5. Stipules entire
          8. Leaflets mostly 9–11 . . . .***R. arkansana***
          8. Leaflets mostly 5–7
             9. Flowers double, usually pendulous . . . .*R. centifolia*
             9. Flowers single, erect
                10. Largest leaflets less than 3 cm long
                   11. Leaves with lower surface glabrous or nearly glabrous; fruit
                       with spreading, deciduous sepals . . . .***R. canina*** (N)
                   11. Leaves with lower surface stipitate glandular; fruit
                       with erect, persistent sepals or spreading,
                       deciduous sepals
                       12. Styles villous; fruit with erect,
                           persistent sepals . . . .***R. eglanteria*** (N)
                       12. Styles glabrous; fruit with spreading,
                           deciduous sepals
                           13. Leaves lacking resinous odor, pubescent to tomentose
                               on lower surface; prickles hooked . . . .***R. micrantha*** (N)
                           13. Leaves with resinous odor, densely tomentose on lower
                               surface; prickles straight or curved . . . .***R. tomentosa*** (ON)
                10. Largest leaflets usually greater than 3 cm long
                   14. Stipules with conspicuous stipitate glands at the margins,
                       widening at the apex . . . .***R. canina*** (N)
                   14. Stipules with stipitate glands lacking or no more
                       conspicuous than pubescence on lower leaflet
                       surface, widening at the apex or not
                       15. Leaflets doubly serrate; spines or thorns at the
                           nodes lacking or, if present, of equal
                           prominence to other spines and thorns
                           16. Leaflets densely pubescent on lower surface, glandular;
                               fruit greater than 12 mm long . . . .*R. villosa*

16. Leaflets densely pubescent on midrib, otherwise sparsely pubescent on lower surface; fruit less than 12 mm long . . . .*R. gallica* (ON)

15. Leaflets usually singly serrate; spines and thorns at the nodes often more prominent than other spines and thorns

17. Thorns at the nodes stout, reflexed; leaflets mostly with at least 40 teeth . . . .*R. palustris*

17. Thorns at the nodes slender, straight; leaflets with about 30 teeth, some of them sometimes doubly serrate . . . .*R. carolina*

3. Outer 3 sepals entire

18. Leaflets mostly 3

19. Leaflets pubescent on the veins of the lower surface; prickles few . . . .*R. setigera*

19. Leaflets glabrous on lower surface; prickles many . . . . .*R. laevigata* (s)

18. Leaflets mostly 5–11

20. Leaflets, twigs, and thorns densely pubescent . . . .*R. rugosa* (N)

20. Leaflets, twigs, and thorns glabrous, slightly pubescent, or stipitate glandular

21. Thorns at the nodes reflexed, stout; internodal spines lacking

22. Sepals erect on mature fruit; pedicels glabrous . . . .*R. cinnamomea* (N)

22. Sepals spreading on mature fruit; pedicels glandular-hispid

23. Leaves coarsely serrate; the teeth in the middle of the leaf projecting about 1 mm . . . .*R. virginiana*

23. Leaves finely serrate, the teeth in the middle of the leaf projecting about 0.5 mm . . . .*R. palustris*

21. Thorns at the nodes straight, slender, or lacking; internodal spines similar to nodal thorns

24. Leaflets less than 15 cm long, glabrous or subglabrous; stems bearing thorns and prickles

25. Stems bearing slender broad-based thorns and needlelike prickles . . . .*R. spinosissima*

25. Stems bearing thorns and prickles of similar dimensions and shapes . . . .*R. nitida*

24. Leaflets at least 15 mm long; usually pubescent, if only on midrib of lower surface; stems smooth or with thorns and prickles

26. Thorns and prickles straight, narrowly tapered; fruit with spreading, usually deciduous sepals . . . .*R. carolina*

26. Thorns and prickles of various sizes and shapes; fruit with erect, persistent sepals

27. Prickles present, of different size and shape than thorns

28. Prickles slender, straight; flowers single . . . .*R. woodsii*

28. Prickles broad-based, arching; flowers double . . . .*R. majalis* (N)
27. Prickles lacking or similar to thorns
    29. Leaflets mostly 9–11, sometimes 7 . . . .*R. arkansana*
    29. Leaflets mostly 5–7
        30. Lateral branches smooth; leaflets singly serrate, the teeth sharp, often straight or concave on inner side; stipules with glands lacking, sparse, or dense . . . .*R. blanda* (N)
        30. Lateral branches bristly; leaflets usually doubly serrate, especially toward apex, the teeth ovate; stipules usually densely glandular . . . .*R. acicularis*

Rosmarinus officinalis   rosemary (ON)

Rubus   raspberry, blackberry, dewberry

The following key is largely a summer key, although some winter characters are used.

1. Leaves simple, palmately lobed
    2. Leaves rounded at the apex, the largest less than 4 cm wide; plants less than 25 cm tall . . . .*R. chamemorus*
    2. Leaves acute or acuminate, the largest to 20 cm wide; plants much taller than 25 cm
        3. Leaves acuminate to long acuminate, the petiole lightly to densely pubescent; bark exfoliating in thin layers; widespread . . . . . .*R. odoratus*
        3. Leaves acute or abruptly acuminate, the petiole mostly glabrous; bark mostly tight, not exfoliating; Ontario, Michigan, Minnesota, and west . . . .*R. parviflorus*
1. Leaves compound
    4. Stems lacking thorns or prickles
        5. Leaflets rounded or obtuse at the apex; petals rose-pink, usually greater than 10 mm long . . . .*R. acaulis*
        5. Leaflets acute or acuminate at the apex; petals white, usually less than 8 mm long . . . .*R. pubescens*
    4. Stems with thorns or prickles, or at least bristles (sometimes glabrous in *R. canadensis*)
        6. Leaves glaucous-white on the lower surface; stems terete, glaucous (if not glaucous, then erect and bristly hairy)
            7. Stems erect, somewhat glaucous when young, covered with stiff, straight, bristly hairs or glabrous; fruit red
                8. Inflorescence and stems with glands or small bristles . . .*R. strigosus*
                8. Inflorescence and stems mostly lacking glands and bristles . . . .*R. idaeus* (N)
            7. Stems recurved, very glaucous, with recurved prickles; fruit black or red

Keys to Species

9. Stems and petioles armed with weak or stout hooked prickles, lacking bristles; fruit black
  10. Stems low-arching or trailing, with weak bristles; lower surface of leaves lightly glaucous-pubescent; fruit tight to the receptacle . . . .*R. caesius* (N)
  10. Stems long and high-arching but not trailing, with strong hooked bristles; lower leaf surface strongly glaucous; fruit separating easily from receptacle when ripe . . . .*R. occidentalis*
9. Stems, petioles, and midribs armed with stout prickles interspersed with slender bristles and gland-tipped, red hairs; fruit red . . . .*R. phoenicolasius* (N)
6. Leaves green or with white or pale pubescence on lower surface, lacking white bloom; stems angular (if terete, then trailing or leaves densely white- or gray-tomentose on lower surface), lacking bloom
  11. Leaves densely gray- or white-tomentose on lower surface
    12. Stems often canescent toward the tip; prickles strongly curved . . . .*R. discolor* (ON)
    12. Stems mostly glabrous; prickles nearly straight . . . .*R. bifrons* (ON)
  11. Leaves glabrous, softly, lightly, or densely pubescent or velvety on lower surface
    13. Stems trailing or prostrate, terete; pedicels with prickles
      14. Prickles on stems slender, bristly, not enlarged at base; leaves subcoriaceous, glossy on upper surface; fruit red-purple
        15. Bristles of the stem lacking glands . . . .*R. hispidus*
        15. Bristles of the stem glandular . . . .*R. trivialis* (s)
      14. Prickles on stems stout, enlarged at base; leaves membranous, dull on upper surface; fruit black
        16. Young stems glaucous; prickles few . . . .*R. caesius* (N)
        16. Young stems lacking bloom; prickles numerous
          17. Inflorescence mostly lacking full-sized leaves . . . .*R. recurvicaulis*
          17. Inflorescence leafy
            18. Terminal leaflet of the trifoliolate leaves on the flowering branches ovate, the base rounded to subcordate, the apex acute to long-acuminate . . . .*R. flagellaris*
            18. Terminal leaflet of the trifoliolate leaves on the flowering branches oblanceolate or ovate, the base cuneate, the apex rounded or obtuse to abruptly acuminate
              19. Leaflets greater than 40 cm long, mostly acute at the apex, mostly toothed more than $2/3$ the length . . . . . .*R. enslenii*
              19. Leaflets less than 30 cm long, mostly rounded at the apex, toothed only in apical $1/2$ to $2/3$ . . . .*R. cuneifolius*

13. Stems erect or recurved, not trailing, angular; pedicels with or
without prickles
  20. Leaflets mostly pinnately lobed, cleft nearly to
the midrib . . . .*R. laciniatus* (N)
  20. Leaflets toothed or shallowly lobed
    21. Leaflets lanceolate, greater than 3 times as long as wide
      22. Leaflets sessile . . . .*R. illecebrosus* (ON)
      22. Leaflets stalked . . . .*R. betulifolius* (S)
    21. Leaflets ovate or obovate, usually 2 times as long as wide
or less
      23. Stems with bristles or prickles not broadened at the base; plants
mostly less than 1 m tall . . . .*R. setosus*
      23. Stems armed with stout prickles broadened at the base
(few and small in *R. canadensis*)
        24. Petioles, young shoots, and inflorescence stipitate-glandular;
lower leaf surface pubescent . . . .*R. allegheniensis*
        24. Petioles, young shoots, and inflorescence lacking
glands; lower leaf surface pubescent or glabrous
          25. Leaflets glabrous or nearly so . . . .*R. canadensis*
          25. Leaflets pubescent on lower surface
            26. Leaflets densely short-pubescent or velvety on the lower
surface; mostly rounded at the base . . . .*R. pensilvanicus*
            26. Leaflets softly pubescent along the veins on the
lower surface; rounded or cuneate at the base
              27. Leaflets mostly cuneate at the base, the terminal 1 less
than 2 times as long as wide . . . .*R. pensilvanicus*
              27. Leaflets mostly rounded at the base, the terminal 1
often greater than 2 times as long as wide . . . . .*R. argutus*

Ruscus aculeatus   butcher's broom

Ruta graveolens   common rue (N)

Sabal   palmetto

1. Plants lacking trunk or trunk only to 3 m tall; blade lacking midrib; leaf
segments rarely filamentose . . . .*S. minor* (S)
1. Plants with trunks to 15 m tall; blade with midrib; leaf segments
filamentose . . . .*S. palmetto* (S)

Salix   willow

The following key, adapted from Voss's *Michigan Flora*, includes some winter characteristics, but the current state of the taxonomy and extensive hybridization in the
genus make these plants difficult to key even in summer.

1. Low, matted, or prostrate shrubs; northern or alpine; leaves mostly less than 3 cm long
   2. Leaves elliptic to obovate; fruit glabrous . . . .*S. uva-ursi*
   2. Leaves obovate to suborbicular; fruit glabrous or pubescent
      3. Leaves obovate, less than 2 cm long, greater than 2 times as long as wide, glaucous, glabrous on lower surface, crenulate to subentire; fruit pubescent . . . .*S. arctophila*
      3. Leaves obovate to suborbicular, greater than 2 cm long, greater than 2 times as long as wide, glabrous or pubescent, not glaucous on lower surface, serrulate; fruit glabrous . . . .*S. herbacea*
1. Trees or erect shrubs; distribution not limited to extreme north and high altitudes; leaves mostly greater than 3 cm long
   4. Leaves entire, sometimes obscurely toothed toward the apex; fruits mostly pubescent (not in *S. pedicellaris* and *S. myricoides*) (alternate step 4, p. 215)
      5. Leaves glabrous
         6. Axillary buds on shoots mostly less than 5 mm long; leaves with or without a few obscure teeth near the apex; fruits often glabrous; inflorescences with green bracts or leaves
            7. Branchlets glabrous, stipules lacking; fruits red tinged or bright green; leaves lacking any obscure teeth, slightly revolute . . . .*S. pedicellaris*
            7. Branchlets pubescent, often densely so; stipules present or lacking; fruits yellow-brown to yellow-green; leaves with a few obscure teeth at the apex, flat . . . .*S. myricoides*
         6. Axillary buds on shoots mostly greater than 5 mm long; leaves often obscurely toothed toward apex; fruits pubescent; inflorescences lacking bracts or with minute green or brown bracts at the base
            8. Leaves mostly subopposite, thick, green to purple-green; oblanceolate to narrowly oblong; stipules lacking . . . .*S. purpurea* (ON)
            8. Leaves alternate, thin, green; obovate or oblanceolate to elliptic; stipules lacking or present
               9. Leaves glossy dark green on upper surface with veins tending to be parallel; fruits mostly less than 6 mm long, sessile . . . .*S. planifolia*
               9. Leaves duller green, the lateral veins irregular; fruits mostly greater than 6 mm long, stalked . . . .*S. discolor*
      5. Leaves pubescent on 1 or both surfaces
         10. Branchlets quite glaucous, glabrous; stipules lacking . . . . . .*S. pellita*
         10. Branchlets lacking bloom, pubescent or glabrous; stipules present or lacking
            11. Leaves with straight, silky, appressed hairs
               12. Margin of leaves revolute; fruit nearly sessile . . .*S. viminalis* (ON)

12. Margin of leaves flat or slightly thickened, not revolute; fruit with 1–3 mm stalk

    13. Leaves mostly less than 4 cm long; plants less than 2 m tall . . . .*S. argyrocarpa*

    13. Leaves greater than 4 cm long; plants to 7 m tall

        14. Mature leaves with lateral veins prominent on lower surface; young leaves silky, lacking red-brown hairs; twigs puberulent; fruit ovoid . . . .*S. sericea*

        14. Mature leaves with lateral veins faint on lower surface; young leaves nearly glabrous or with few to many red-brown hairs mixed with white hairs; twigs puberulent or glabrous; fruit lanceolate . . . .*S. petiolaris*

11. Leaves with curled, tomentose or woolly hairs

    15. Young branchlets mostly flocculent-tomentose; leaves mostly 5–12 times as long as wide; capsule densely white-tomentose, the stalk less than 1 mm long . . . .*S. candida*

    15. Young branchlets glabrous or puberulent to villous; leaves mostly less than 5 times as long as wide; capsule silky-pubescent, the stalk 1–6 mm long

        16. Leaves rugose, the veins impressed on upper surface and prominent on lower surface, red-brown hairs lacking . . . .*S. bebbiana*

        16. Leaves flat to slightly rugose, the veins easily visible but not impressed or prominent, red-brown hairs evident, especially on young leaves

        17. Twigs glabrous

            18. Leaves glossy dark green on upper surface with veins tending to be parallel; fruits mostly less than 6 mm long, sessile . . . .*S. planifolia*

            18. Leaves duller green, the lateral veins irregular; fruits mostly greater than 6 mm long, stalked . . . .*S. discolor*

        17. Twigs pubescent, at least in small patches above the nodes

            19. Leaves often slightly rugose, the veins impressed on upper surface, the margins mostly slightly revolute . . . .*S. humilis*

            19. Leaves flat or the veins slightly elevated on upper surface; margins flat

                20. Twigs smooth beneath the bark; common . . . .*S. discolor*

                20. Twigs with distinct ridges beneath the bark; rare escape from cultivation . . . .*S. cinerea* (ON)

4. Leaves finely serrate or crenate nearly the entire length of the margin; fruit pubescent or glabrous

    21. Leaves green (sometimes white early in season) on lower surface

        22. Leaves linear-oblong to narrowly lanceolate, less than 1 cm wide, often with fewer than 4 teeth per cm of margin

23. Leaves less than 5 mm wide, 20 times as long as wide . . . .*S. exigua*
23. Leaves greater than 5 mm wide, 10 times as long
    as wide . . . .*S. babylonica* (ON)
22. Leaves narrowly lanceolate to ovate or elliptic, greater than
    1 cm wide, usually with at least 4 teeth per cm of margin
    24. Petioles lacking glands at or near the base of the blade
        25. Leaves mostly narrowly lanceolate or very narrowly elliptic, the base
            acute or rounded . . . .*S. nigra*
        25. Leaves lanceolate to elliptic, the base rounded to truncate
            or cordate
            26. Leaves densely pubescent to glabrate, sharply serrate (sometimes
                doubly), the teeth tipped with prominent glands, the base
                truncate to cordate . . . .*S. cordata*
            26. Leaves mostly glabrous, crenulate-serrate, the teeth with or
                without small, sunken glands, the base rounded
                to truncate . . . .*S. eriocephala*
    24. Petioles bearing 2 or more glands at or near the base of
        the blade
        27. Petioles usually greater than 5 mm long . . . .*S. ×blanda*
        27. Petioles usually less than 5 mm long
            28. Leaves glabrous on lower surface at maturity, the largest to 18 cm
                long; twigs brittle at the base . . . .*S. fragilis* (N)
            28. Leaves remaining pubescent on lower surface, the
                largest less than 10 cm long; twigs tough or somewhat
                flexible, not brittle at the base
                29. Twigs brown to dark brown . . . .*S. alba* (ON)
                29. Twigs yellow-brown to gray-brown . . . .*S. matsudana* (ON)
21. Leaves white on lower surface
    30. Petioles bearing prominent glands or bumps at or near the base
        of the blade
        31. Leaves linear-oblong, lanceolate, or linear-lanceolate, apex
            attenuate; fruit less than 4 mm long
            32. Petioles usually greater than 5 mm long . . . .*S. ×blanda*
            32. Petioles usually less than 5 mm long
                33. Leaves silky, especially on lower surface, at least 6 teeth per cm of
                    margin; twigs somewhat flexible, at least not brittle at the base;
                    fruit sessile or subsessile . . . .*S. alba* (ON)
                33. Leaves glabrous (silky when young), mostly 4–6 teeth per cm of
                    margin; twigs brittle at the base; fruit borne on stalks nearly
                    1 mm long . . . .*S. fragilis* (N)
        31. Leaves broadly lanceolate to ovate-elliptic, the apex acute or
            acuminate; fruit often greater than 6 mm long
            34. Leaves with blades 1½ to 4 times as long as wide, petioles and
                young leaves glabrous; stipules usually present . . . .*S. pentandra*

34. Leaves with blades 2 to 6 times as long as wide, petioles
 often pubescent, at least when young; stipules present
 or lacking
    35. Leaves, especially on sucker shoots, caudate-acuminate, lustrous;
    petioles and young leaves mostly pubescent; stipules often present on
    shoots; fruit mostly less than 7 mm long . . . .*S. lucida*
    35. Leaves merely acuminate, duller green; petioles and young leaves
    glabrous; stipules lacking; fruit mostly greater than
    7 mm long . . . .*S. serissima*
30. Petioles lacking glands
    36. Inflorescence scales yellow, pilose only at the base and margins,
    falling before ripening of the fruit; fruit glabrous
        37. Leaves long-tapering at apex, drooping; twigs slender; fruit sessile
        or subsessile . . . .*S. amygdaloides*
        37. Leaves acuminate, not long-tapering, not drooping; fruit
        mostly with stalks at least 1 mm long
            38. Leaves glabrous, the teeth coarse, 3–6 per cm of margin, projecting
            mostly greater than 0.5 mm from the margin . . . .*S. fragilis* (N)
            38. Leaves silky, especially on lower surface, the teeth finer, 7–12 per cm
            of margin, projecting less than 0.5 mm from
            the margin . . . .*S. alba* (ON)
    36. Inflorescence scales brown or black, pilose, persistent; fruit
    glabrous or pubescent
        39. Leaves crenulate-undulate, occasionally sharply
        undulate-serrate . . . .*S. caprea* (ON)
        39. Leaves finely or sharply serrate
            40. Lower leaf surface green but covered with dense white
            pubescence, not glaucous
                41. Leaves elliptic to ovate, obscurely or prominently
                undulate-crenate . . . .*S. caprea* (ON)
                41. Leaves linear to lanceolate or narrowly elliptic, sharply,
                finely serrate
                    42. Leaves closely serrate, at least 5 teeth per cm of margin, the
                    teeth tipped with large glands; fruits glabrous . . . . . .*S. cordata*
                    42. Leaves distantly serrate, 2–4 teeth per cm of margin, lacking
                    large glands; fruits mostly thinly silky . . . .*S. exigua*
            40. Lower leaf surface white, glaucous, glabrous, or pubescent
               43. Branchlets yellow-brown to gray-brown . . . . .*S. matsudana* (ON)
               43. Branchlets gray, red-brown, or black
                    44. Branchlets gray; twigs gray or red-brown . . . . . .*S. caroliniana*
                  44. Branchlets and young twigs red-brown to black
                      45. Leaves mostly less than 15 mm wide, usually acute at the
                      base, glabrous or silky; stipules mostly lacking; fruit at least
                      sparsely pubescent

46. Leaves narrow, 4–10 mm wide, the young leaves densely silky, the pubescence including few to many red-brown hairs; branchlets mostly glabrous, often fascicled toward ends of branches; fruit lanceolate . . . .*S. petiolaris*

46. Leaves greater than 10 mm wide, the young leaves densely silky but lacking red-brown hairs; branchlets puberulent, rarely conspicuously fascicled; fruit ovoid . . . .*S. sericea*

45. Leaves usually greater than 15 mm wide, mostly rounded to cordate at the base, glabrous except for pubescent midrib; stipules usually present and conspicuous; fruit glabrous

    47. Leaves glabrous, with balsamic or spicy odor; twigs red, shiny, glabrous . . . .*S. pyrifolia*

    47. Leaves often pubescent on petioles and midribs, lacking strong odor; twigs brown, rarely red, not shiny, puberulent above the nodes

        48. Young leaves with white hairs, lacking red-brown hairs; mature leaves thin, green to white on lower surface, not strongly glaucous . . . .*S. eriocephala*

        48. Young leaves glabrous or with red-brown hairs mixed with white hairs; mature leaves thick, usually very glaucous on lower surface . . . .*S. myricoides*

Sambucus   elderberry, elder

1. Pith orange to red-brown; buds ovate, purple, large; mature fruit red . . . .*S. racemosa*

1. Pith white; buds conical, green or brown, small; mature fruit purple-black or black

    2. Leaflets 5–11; fruit purple-black, 4–5 mm long . . . .*S. canadensis*

    2. Leaflets mostly 5, sometimes 7; fruit black, shining, 6–8 mm long . . . .*S. nigra*

Sapium sebiferum   Chinese tallow tree (N, CS)

Sarcococca   sweet-box

1. Leaves ovate; fruit red . . . .*S. ruscifolia*

1. Leaves lanceolate to elliptic; fruit black or dark purple

    2. Branchlets glabrous . . . .*S. saligna*

    2. Branchlets pubescent . . . .*S. hookeriana*

Sasa palmata   sasa-bamboo (N)

Sassafras albidum   sassafras

Schisandra   schisandra

1. Leaves entire, sometimes remotely denticulate-undulate in the apical half, ovate to elliptic . . . .*S. glabra* (S)

1. Leaves denticulate or serrate, oblong-elliptic
   to obovate . . . .*S. chinensis*

Schizophragma   vine-hydrangea

1. Leaves entire, sometimes denticulate, pubescent on
   lower surface . . . .*S. integrifolium* (cs)
1. Leaves coarsely dentate, glabrous or sparsely pubescent on
   lower surface . . . .*S. hydrangeoides* (oc)

Sciadopitys verticillata   umbrella-pine (oc)

Sebastiania ligustrina   sebastiana (s)

Securinega suffruticosa   securinega (cs)

Sequoia sempervirens   redwood, coast redwood, California
redwood (oc)

Sequoiadendron giganteum   California big tree,
giant-redwood (oc)

Serenoa repens   saw-palmetto (s)

Serissa japonica   yellow rim (cs)

Sesbania punicea   purple sesban, rattle-box (n, cs)

Shepherdia   buffalo berry

1. Leaves ovate, green on upper surface, silver on lower surface; twigs brown,
   scurfy; stems thornless . . . .***S. canadensis***
1. Leaves oblong, silver on both surfaces; twigs silver to brown, with some scales;
   stems mostly bearing thorns . . . .***S. argentea***

Skimmia japonica   skimmia

Smilax   greenbrier

1. Leaves persistent, oblong-lanceolate to linear, tapered at base
     2. Twigs light red-brown; fruit black . . . .*S. laurifolia*
     2. Twigs pale green; fruit dull red or brown . . . .*S. smallii*
1. Leaves deciduous, ovate or rounded, mostly rounded or cordate
   at base
     3. Leaves glaucous on the lower surface . . . .*S. glauca*
     3. Leaves green on both surfaces
        4. Prickles and lower stems of leader shoots scurfy; leaves contracted above
           the base or 3-lobed . . . .***S. bona-nox***

4. Prickles and lower stems glabrous; leaves ovate or rounded, heart-shaped
    5. Leaves thin; branchlets nearly terete . . . .*S. tamnoides*
    5. Leaves thick; branchlets angled
        6. Stem unarmed or somewhat prickly near the base; leaf margins usually
            revolute; fruit coral-red . . . .*S. walteri*
        6. Stem and branchlets armed with stout spines; leaf margins
            mostly flat; fruit blue-black . . . .*S. rotundifolia*

Solanum dulcamara   bitter-nightshade (N)

Solidago pauciflosculosa   woody goldenrod (S)

Sophora japonica   Japanese pagoda tree, Chinese scholar
  tree

Sorbaria sorbifolia   Ural false-spiraea, sorbaria (ON)

×*Sorbaronia* ×*hybrida*   sorbaronia (OC)

Sorbus   mountain-ash

1. Twigs and lower surface of leaves white-villous; buds white-villous, lacking a
  film, not sticky or glutinous . . . .*S. aucuparia* (N)
1. Twigs and leaves glabrous; buds sticky or covered by a film
    2. Leaflets mostly 2–3 times as long as wide; short-acuminate to
      acute; fruit 7–10 mm long . . . .*S. decora*
    2. Leaflets mostly 3–5 times as long as wide, mostly long-acuminate; fruit
      4–7 mm long . . . .*S. americana*

Spiraea   spirea, meadowsweet, steeplebush, bridal wreath

The following key includes cultivated species and hybrids of this popularly planted
genus, many of which have escaped. Other than a few woody characters used in
the summer key, the woody characters are not sufficiently distinctive for a useful
winter key.

1. Twigs and lower surface of leaves woolly
    2. Twigs and leaves retaining woolly pubescence at maturity;
      fruit pubescent . . . .*S. tomentosa*
    2. Twigs and leaves with less conspicuous pubescence at maturity;
      fruit glabrous . . . .*S.* ×*billiardii*
1. Twigs and leaves pubescent or glabrous, not woolly
    3. Leaves, at least the larger ones, lobed or incised

4. Leaves rounded apically and basally, the margins revolute
   or thick ....*S. trilobata*
4. Leaves acute apically and basally, the margins thin
   5. Leaves mostly wider above the middle,
      2–5 cm long ....*S.* ×*vanhouttei* (ON)
   5. Leaves mostly widest at the middle, up to 7 cm long ..... .*S. cantoniensis*
3. Leaves toothed, scalloped, or entire
   6. Leaves mostly blunt or rounded
      7. Leaves serrate at least above the middle ....*S. betulifolia* (OC)
      7. Leaves entire or with a few teeth toward the apex ..... .*S. nipponica* (OC)
   6. Leaves mostly acute
      8. Leaves less than 12 mm wide ....*S. thunbergii* (ON)
      8. Leaves usually greater than 12 mm wide
         9. Leaves shining on upper surface ....*S. prunifolia* (N)
         9. Leaves dull on upper surface
            10. Leaves entire at least below the middle
               11. Leaves entire or shallowly toothed above the middle, elliptic,
                  mostly less than 2 cm wide ....*S. virginiana* (S)
               11. Leaves coarsely toothed above the middle, ovate,
                  mostly greater than 2 cm wide (sometimes less in
                  S. septentrionalis)
                  12. Leaves usually less than twice as long
                     as wide ....*S. betulifolia* (OC)
                  12. Leaves usually greater than twice as long
                     as wide ....*S. septentrionalis* (OC)
            10. Leaves toothed, often doubly toothed the entire length
               of the margin
               13. Leaves rounded or broadly wedge-shaped at the base,
                  occasionally acute; fruit in flat or round-topped
                  corymbs
                  14. Leaves glabrous or nearly so; twigs
                     somewhat angled ....*S.* ×*bumalda*
                  14. Leaves pubescent along the veins on lower surface;
                     twigs terete
                     15. Leaves mostly 3 times as long as wide ..... .*S. japonica* (N)
                     15. Leaves mostly twice as long
                        as wide ....*S. chamaedryfolia* (ON)
               13. Leaves acute or narrowly wedge-shaped at the base;
                  fruit in umbels or elongated panicles
                  16. Leaves about twice as long as wide
                     or less ....*S. chamedryfolia* (ON)
                  16. Leaves 3 times as long as wide or greater
                     17. Leaves often irregularly toothed, often 3 times as long as
                        wide; twigs terete; fruit in umbels ....*S. cantoniensis*

17. Leaves usually consistently, evenly singly or doubly, toothed,
    much greater than 3 times as long as wide; twigs angled; fruit in
    elongated panicles
    18. Leaves singly toothed; branchlets red-brown . . . .*S. alba*
    18. Leaves doubly toothed; branchlets yellow-brown . . . . . .*S. salicifolia* (ON)

Stachyurus praecox   stachyurus (CS)

Staphylea   bladdernut

1. Leaflets 3, pubescent at least on the veins on the lower surface
   2. Leaflets shortly scabrous along the midrib or most of the lower surface and
      sometimes also on the upper surface; fruit 3-lobed . . . .*S. trifolia*
   2. Leaflets softly pubescent along midrib on lower surface, glabrous on upper
      surface; fruit 2-lobed . . . .*S. bumalda* (OC)
1. Leaflets 5–7, glabrous on the lower surface
   3. Leaflets often with short stalks, 1–2 mm long; fruit
      2–4 cm long . . . .*S. pinnata*
   3. Leaflets sessile; fruit 4–7 cm long . . . .*S. colchica* (OC)

Stephanandra incisa   cutleaf stephanandra

Stewartia   stewartia

1. Petioles with margins upturned or winglike; buds concealed or nearly so; seeds
   dull, winged or with thin edges . . . .*S. ovata* (S)
1. Petioles with margins flat, not winged; buds visible; seeds
   shining, with angled margins
   2. Petioles and buds mostly less than 5 mm long; leaves usually less than twice
      as long as wide; shrubs . . . .*S. malachodendron* (S)
   2. Petioles and buds mostly greater than 5 mm long; leaves
      usually greater than twice as long as wide; trees
      3. Branchlets finely pubescent, especially
         when young . . . .*S. monadelpha* (OC)
      3. Branchlets glabrous or sparsely pubescent at the nodes
         4. Buds with more than 5 visible scales; leaves
            smooth, crenulate . . . .*S. pseudocamellia*
         4. Buds with 4–5 visible scales; leaves rugose or
            wrinkled, crenate . . . .*S. koreana* (OC)

Stranvaesia davidiana   stranvaesia (OC)

Styrax   storax

1. Petioles 1–4 mm long . . . .*S. americanus*
1. Petioles mostly greater than 5 mm long

2. Leaves broadly ovate to nearly orbicular, the petioles basally enlarged, covering the buds . . . .*S. obassia* (oc)
2. Leaves elliptic to broadly obovate, the petiole bases not enlarged, the buds exposed
   3. Leaves elliptic, remaining pubescent only in the axils of the veins on lower surface; fruit 1–6 in clusters or solitary . . . .*S. japonicus*
   3. Leaves broadly obovate to broadly elliptic, densely tomentose on lower surface; fruit numerous in racemes or panicles . . . .**S. grandifolius**

Symphoricarpos   snowberry, coralberry

1. Pith continuous; fruits red, in dense clusters . . . .**S. orbiculatus**
1. Pith hollow; fruits white
   2. Upper internodes glabrous to minutely puberulent; leaves often entire . . . .**S. albus**
   2. Upper internodes puberulent; leaves often crenate to rounded-dentate . . . .**S. occidentalis**

Symplocos   sweetleaf, sapphire berry

1. Leaves entire or serrulate; fruit orange-brown . . . .**S. tinctoria** (s)
1. Leaves finely serrate; fruit blue . . . .*S. paniculata*

Syringa   lilac

Because lilac is one of our oldest and most popular ornamental shrubs, many species, hybrids, and cultivars can be found in the region. Some of the common species are keyed here.

1. Trees to 10 m tall or large upright or spreading shrubs to 5 m tall; the bark often curling and peeling, with prominent lenticels
   2. Leaves pubescent, sometimes sparsely so, on lower surface . . . .*S. reticulata*
   2. Leaves glabrous
      3. Leaves ovate to ovate-lanceolate, cuneate or narrowed at the base, veins inconspicuous or not prominent . . . .**S. pekinensis** (on)
      3. Leaves ovate to ovate-orbicular, rounded to subcordate at the base, the veins prominent . . . .*S. reticulata*
1. Shrubs to 4 m tall; bark usually not peeling, the lenticels mostly inconspicuous
   4. Terminal bud prominent, terminating the twig, the main axis continuing . . . .*S. villosa*
   4. Terminal bud suppressed, 2 lateral buds terminating the twig, the main axis not continuing
      5. Leaves pubescent on lower surface, at least along the veins, sometimes sparsely pubescent on upper surface
         6. Leaves mostly greater than 4 cm long, glabrous on upper surface . . . .*S. pubescens* (oc)

6. Leaves mostly less than 4 cm long, usually sparsely pubescent
   on upper surface . . . .*S. microphylla* (OC)
5. Leaves glabrous
   7. Leaves mostly less than 4 cm long . . . .*S. meyeri*
   7. Leaves mostly greater than 4 cm long
      8. Leaves truncate to cordate at the base
         9. Leaves ovate-orbicular to orbicular, the length equal to or exceeded by
            the width . . . .*S. oblata* (OC)
         9. Leaves ovate, the length exceeding the width . . . .**S. vulgaris** (ON)
      8. Leaves cuneate or narrowed at the base
         10. Leaves ovate-lanceolate . . . .*S. ×chinensis*
         10. Leaves lanceolate or compound
            11. Leaves simple, entire . . . .*S. ×persica*
            11. Leaves lobed or compound
               12. Leaves usually lobed, often divided nearly to
                  the midrib . . . .*S. laciniata* (OC)
               12. Leaves compound, leaflets 7–11 . . . .*S. pinnatifolia* (OC)

Tamarix   tamarisk

1. Branches dark brown to black . . . .**T. parviflora** (ON)
1. Branches red-brown to brown or purple
   2. Leaves rhombic-ovate, crowded, overlapping, sometimes covering half
      the next leaf . . . .*T. gallica*
   2. Leaves lanceolate to rhombic-ovate, usually spaced well apart on
      the twig, not overlapping
      3. Leaf base narrowed, clasping ⅓ of the
         twig diameter . . . .**T. ramosissima** (ON)
      3. Leaf base wide, clasping half of the twig diameter
         4. Leaves and young shoots light green . . . .**T. parviflora** (ON)
         4. Leaves and young shoots blue-green
            5. Leaves rhombic-ovate, sometimes with scarious or
               brown margins . . . .*T. gallica*
            5. Leaves mostly lanceolate, the margins
               blue-green . . . .**T. chinensis** (ON)

Taxodium   bald-cypress

1. Leaves linear, greater than 12 mm long, spreading; branches and branchlets
   spreading, horizontal or pendulous . . . .**T. distichum**
1. Leaves subulate, less than 12 mm long; appressed;
   branchlets upright . . . .**T. ascendens** (s)

Taxus   yew

1. Bud scales mostly blunt, slightly keeled; leaves acute
   to acuminate . . . .**T. baccata** (ON)

1. Bud scales often sharp-pointed, keeled; leaves abruptly acuminate
    2. Plants low, creeping or ascending; leaves less than
       2 mm wide . . . .***T. canadensis***
    2. Plants erect; leaves 2–3 mm wide
        3. Leaves with midrib prominently raised . . . .***T. cuspidata*** (ON)
        3. Leaves with midrib slightly raised . . . .*T.* ×*media*

Ternstroemia gymnathera   ternstroemia

Tetradium daniellii   evodia (OC)

Thuja   arborvitae

1. Leaves yellow-green, with a prominent gland . . . .***T. occidentalis***
1. Leaves with white markings on lower surface of branchlet, with
   or without a gland
    2. Leaves on older branchlets long-acuminate . . . .*T. plicata* (OC)
    2. Leaves acute or short-acuminate, the apex short, triangular
       and spreading . . . .*T. standishii* (OC)

Thujopsis dulobrata   hiba-arborvitae, false-arborvitae (OC)

Tilia   linden, basswood

1. Leaves pubescent only in the axils of veins on the lower surface
    2. Leaves mostly greater than 10 cm long, the axillary hairs on the lower leaf
       surface absent at or near the base . . . .***T. americana***
    2. Leaves mostly less than 5 cm long, the axillary hairs on the lower leaf
       surface evident over entire leaf, including the base . . . .***T. cordata*** (ON)
1. Leaves pubescent over entire lower surface
    3. Branches erect or horizontal; lower leaf surface lightly pubescent, with
       axillary tufts of hairs . . . .***T. platyphyllos*** (ON)
    3. Branches erect, spreading, or pendulous; lower leaf surface
       white-tomentose, lacking axillary tufts of hairs
        4. Branches erect, spreading; leaves serrate or slightly lobed, the petiole less
           than half the length of the blade . . . .***T. tomentosa*** (ON)
        4. Branches pendulous; leaves finely serrate or serrulate, the petiole greater
           than half the length of the blade . . . .***T. petiolaris*** (ON)

Toona sinensis   toona, cedrela

Torreya nucifera   Japanese torreya (OC)

Toxicodendron   poison-ivy, poison-sumac

Toxicodendron is poisonous and can produce a dermatitis in susceptible persons.

1. Leaflets 7–13; twigs stout; buds sessile, scaly; leaf scars
   U- or V-shaped . . . .***T. vernix***

1. Leaflets 3; twigs slender; buds stalked, naked; leaf scars heart- or
   shield-shaped to round
   2. Scrambling or climbing vine with aerial roots . . . .*T. radicans*
   2. Shrubs 1–3 m tall, lacking aerial roots
      3. Petioles glabrous; Virginia north and west . . . .*T. rydbergii*
      3. Petioles pubescent; New York south . . . .*T. pubescens*

Tracheleospermum   star-jasmine

1. Leaves deciduous, 4–9 cm long . . . .*T. difforme* (s)
1. Leaves persistent, 2–5 cm long . . . .*T. asiaticum*

Trochodendron aralioides   wheel tree

Tsuga   hemlock

1. Leaves with white longitudinal bands on both surfaces . . . .*T. mertensiana* (oc)
1. Leaves with white longitudinal bands on lower surface only
   2. Leaves minutely serrulate
      3. Leaves on twigs tapering, wider at the base than at the apex, mostly in
         several planes . . . .*T. canadensis*
      3. Leaves on twigs with apex about as wide as the base, mostly in
         1 plane . . . .*T. heterophylla* (oc)
   2. Leaves entire
      4. Twigs and branchlets glabrous . . . .*T. sieboldii* (oc)
      4. Twigs and branchlets pubescent
         5. Twigs pubescent throughout . . . .*T. diversifolia* (oc)
         5. Twigs pubescent only in the grooves
            6. Leaves notched at the apex, the white longitudinal bands on lower
               surface inconspicuous . . . .*T. chinensis* (oc)
            6. Leaves rounded or obscurely notched at the apex, the white
               longitudinal bands on lower surface prominent . . . .*T. caroliniana*

Ulex europaeus   gorse (n)

Ulmus   elm

Summer key

1. Leaves very scabrous on the upper surface, sometimes 3-lobed at apex
   2. Leaves less than 4 cm long, the apex rounded or obtuse; flowering and
      fruiting in autumn . . . .*U. crassifolia* (s)
   2. Leaves greater than 6 cm long, the apex acute or acuminate;
      flowering and fruiting in late winter or early spring
      3. Leaves ovate-oblong . . . .*U. rubra*
      3. Leaves obovate . . . .*U. glabra* (on)
1. Leaves smooth or nearly so on the upper surface (sometimes
   somewhat scabrous in *U. alata* and *U.* ×*hollandica*)

4. Twigs and branchlets mostly with corky, winglike ridges
  5. Leaves obtuse, rounded at base, not subcordate; buds 2 mm long (flower buds larger), the scales lacking hairs on the margins; fruit less than 1 cm long
    6. Young twigs glabrous or slightly pubescent; leaves narrowly elliptic or oblong, smooth, lustrous, slightly wavy . . . .*U. alata* (s)
    6. Young twigs pubescent; leaves broadly elliptic to ovate, slightly scabrous or rough on upper surface . . . .*U. minor* (ON)
  5. Leaves often subcordate at the base; buds 3–6 mm long, the scales ciliate; fruit greater than 1 cm long
    7. Leaves 5–9 cm long; flowering and fruiting in autumn . . . .*U. serotina* (s)
    7. Leaves 8–14 cm long; flowering and fruiting in late winter and early spring . . . .*U. thomasii*
4. Twigs and branchlets smooth, lacking corky, winglike ridges
  8. Leaves usually 10–15 cm long
    9. Leaves hairy at the margins . . . .*U. americana*
    9. Leaves glabrous at the margins . . . .*U. ×hollandica*
  8. Leaves mostly 2–6 cm long
    10. Leaves doubly toothed, the margins of the larger teeth with 2–3 teeth
      11. Branchlets and twigs red-brown to dark brown, nearly black . . . .*U. minor* (ON)
      11. Branchlets and twigs gray to brown . . . .*U. carpinifolia* (OC)
    10. Leaves singly toothed or occasionally doubly toothed
      12. Young shoots or branchlets pubescent; flowering and fruiting in late summer and fall . . . .*U. parvifolia*
      12. Young shoots or branchlets glabrous or sparsely pubescent; flowering and fruiting in spring . . . .*U. pumila* (ON)

Winter key

1. Corky winglike ridges usually present on twigs or branchlets
  2. Buds 6–8 mm long . . . .*U. thomasii*
  2. Buds usually less than 6 mm long
    3. Buds obtuse, the margins of the scales lacking hairs; flowering and fruiting in late winter and early spring, the fruit symmetric, 5–10 mm long
      4. Young twigs glabrous or slightly pubescent . . . .*U. alata* (s)
      4. Young twigs pubescent . . . .*U. minor* (ON)
    3. Buds acute to obtuse, the margins usually hairy; flowering and fruiting in autumn, the fruit oval, asymmetric, 10–15 mm long
      5. Buds less than 3 mm long, rounded to obtuse; fruit 10 mm long . . . .*U. crassifolia* (s)

Keys to Species

5. Buds 3–6 mm long, acute; fruit greater than
   10–15 mm long . . . .*U. serotina* (s)
1. Corky winglike ridges lacking; branchlets mostly smooth
   6. Buds globose, branchlets gray to gray-brown . . . .*U. pumila* (on)
   6. Buds ovoid; branchlets gray to dark brown or red-brown
      7. Buds densely rusty-pubescent; twigs gray to buff, scabrous; inner bark
         mucilaginous; fruit lacking stalks or with stalks 1 mm long; inner bark of
         uniform color . . . .*U. rubra*
      7. Buds glabrous or pubescent, but not densely rusty-pubescent;
         twigs red-brown, gray-brown, or dark brown, glabrous or
         pubescent; inner bark dry, not mucilaginous; fruit with stalks
         greater than 1 mm long; inner bark in alternate white and
         brown layers or of uniform color
         8. Buds smoky-brown or almost black, the scales uniformly
            dark throughout
            9. Twigs hispid; bark of trunk light, smooth or with broad flat-topped
               longitudinal ridges; inner bark somewhat
               mucilaginous . . . .*U. glabra* (ON)
            9. Twigs glabrescent; bark of trunk dark, rough, the ridges breaking
               into oblong blocks, inner bark dry,
               not mucilaginous . . . .*U. minor* (ON)
         8. Buds chestnut-brown or red-brown, the scales mostly with
            darker margins or dark band near the apex
            10. Crown open and spreading, not cylindrical or vase-
                shaped; buds divergent, the scales dark red-brown, with
                slightly darker, almost black, margins
                11. Bud scales dark red-brown, with darker band near the
                    apex, ciliate . . . .*U. carpinifolia* (OC)
                11. Bud scales dark brown with inconspicuous darker margins,
                    lacking hairs at the margins . . . .*U. parvifolia*
            10. Crown cylindrical or vase-shaped; buds divergent or
                somewhat appressed, the scales chestnut-brown, with
                dark margins
                12. Crown cylindrical; buds usually divergent, pubescent; inner
                    bark of uniform color or alternating brown and
                    white layers . . . .*U. thomasii*
                12. Crown broadly vase-shaped; buds somewhat appressed,
                    glabrous; inner bark of alternating brown and
                    white layers . . . .*U. americana*

Vaccinium   blueberry, cranberry

1. Leaves persistent; stems creeping or trailing; berry red
   (purple-black in *V. crassifolium*)

2. Leaves with dark bristly points on the lower surface; stems somewhat tufted; growing in alpine habitats . . . .*V. vitis-idaea*
2. Leaves glaucous-white, pale green, or brown on the lower surface, lacking bristly points; stems slender, not tufted; growing in bogs or wet or sandy areas
  3. Leaves flat or slightly revolute, elliptical, blunt or rounded at apex, mostly greater than 8 mm long . . . .*V. macrocarpon*
  3. Leaves strongly revolute, narrowly ovate, pointed at apex, mostly less than 8 mm long
    4. Leaves glaucous-white on lower surface; fruit red; New Jersey and north . . . .*V. oxycoccos*
    4. Leaves brown or pale green on lower surface; fruit purple-black; Virginia and south . . . .*V. crassifolium*
1. Leaves deciduous; stems usually erect; berry green, yellow-green, blue, or black (sometimes red in **V. erythrocarpum**)
  5. Plants 2–10 m tall
    6. Leaves obovate to oblong or orbicular, denticulate or entire, glossy; branchlets red-brown; small trees . . . .*V. arboreum* (s)
    6. Leaves ovate to elliptic-lanceolate, entire or serrulate, dull; branchlets green to red, speckled; tall shrubs
      7. Leaves greater than 4 cm long; twigs glabrous . . . .*V. corymbosum*
      7. Leaves less than 4 cm long; twigs puberulent . . . .*V. elliottii* (s)
  5. Stems less than 2 m tall
    8. Plants greater than 1 m tall
      9. Leaves usually greater than 5 cm long, entire, elliptic, oblong, or obovate, petiole 2–3 mm long; fruit yellow-green to purple-black . . . .*V. stamineum*
      9. Leaves usually less than 5 cm long, entire or serrulate, lanceolate, elliptic, or oval, petiole less than 2 mm long; fruit red to purple-black
        10. Leaves lanceolate to lanceolate-ovate, acuminate, serrulate; fruit sometimes red . . . .*V. erythrocarpum* (s)
        10. Leaves elliptic to oval, obtuse or rounded at the apex, entire at least in the apical half; fruit purple-black . . . .*V. ovalifolium*
    8. Plants less than 1 m tall
      11. Plants dwarf, densely covering the ground, less than 10 cm tall; alpine . . . .*V. boreale*
      11. Plants greater than 10 cm tall; not restricted to alpine habitat
        12. Leaves glaucous or very pale on the lower surface, the margins usually entire or only minutely serrulate; berry yellow-green to blue or purple-black
          13. Leaves serrulate, with incurved teeth; fruit purple-black; Michigan, Ontario, north and west . . . .*V. membranaceum*

13. Leaves entire; fruit yellow-green or blue; widespread
    14. Branchlets ascending, glabrous, covered with speckles; berry blue,
        dry, glaucous . . . .***V. pallidum***
    14. Branchlets recurved-spreading, pubescent, speckles mostly lacking;
        berry yellow-green, juicy, glaucous or not . . . .***V. stamineum***
12. Leaves bright green or brown on the lower surface, or the
  margins toothed; berry blue
    15. Leaves serrate or serrulate
      16. Leaves lanceolate, serrulate; branchlets grooved . . . . . .***V. angustifolium***
      16. Leaves obovate or spatulate, crenulate-serrulate; branchlets even,
        lacking grooves . . . .***V. caespitosum***
    15. Leaves entire
      17. Leaves greater than 4 times as long as wide, the lower surface with
        stalked glands . . . .***V. tenellum*** (s)
      17. Leaves less than 4 times as long as wide, the lower surface
        glabrous or pubescent, not glandular
        18. Leaves, twigs, branchlets, and branches strongly hirsute; leaves
            mostly greater than 4 cm long . . . .***V. hirsutum*** (s)
        18. Leaves, twigs, branchlets, and branches glabrous to
            strongly pubescent, not hirsute; leaves less
            than 4 cm long
            19. Leaves mostly 20–40 mm long; leaves and twigs strongly
                pubescent; bark of branchlets tight, covered
                with speckles . . . .***V. myrtilloides***
            19. Leaves mostly 5–20 mm long; leaves and twigs glabrous or
                nearly so; bark of branchlets shredding, speckles lacking
                or few . . . .***V. uliginosum***

Viburnum   viburnum, arrowwood

Summer key

1. Leaves lobed, palmately veined
  2. Petioles bearing glands; stipules present; fruit red
    3. Leaves pubescent on the lower surface . . . .***V. opulus* var. *opulus*** (N)
    3. Leaves glabrous except along the veins . . . . .***V. opulus* var. *americanum***
  2. Petioles lacking glands; stipules present or absent; fruit
    purple-black or red
    4. Stipules present; leaves densely pubescent on lower surface, the teeth
      lacking glands; fruit purple-black . . . .***V. acerifolium***
    4. Stipules absent; leaves glabrate, the teeth near the base bearing
      glands; fruit red . . . .***V. edule***
1. Leaves entire or toothed, lacking lobes, pinnately veined
  5. Leaves with 3 main veins from the base . . . .*V. davidii* (cs)
  5. Leaves with a single main vein, the midrib

6. Buds naked
   7. Leaves with lower surface stellate-pubescent or glabrous; twigs stellate-pubescent; fruit red, changing to almost black
      8. Leaves deciduous, thin, the upper surface slightly stellate-pubescent . . . .***V. lantana*** (ON)
      8. Leaves evergreen, thick, the upper surface lustrous or dull, glabrous
         9. Leaves wrinkled, the veins deeply impressed on upper surface, the lower surface densely gray- to yellow-brown–tomentose; branchlets densely stellate-pubescent . . . .*V. rhytidophyllum*
         9. Leaves flat, smooth on upper surface, the lower surface sparsely pubescent to glabrous; branchlets glabrous
            10. Leaves ciliate, the petioles pubescent . . . .*V. tinus*
            10. Leaves lacking hairs on the margins, the petioles glabrous
               11. Leaves entire, undulate, or minutely toothed in the apical $^1/_3$, the teeth remote from each other . . . .*V. odoratissimum* (including the variety *awabuki*)
               11. Leaves toothed regularly or irregularly $^1/_2$ to $^3/_4$ of the way toward the base . . . .*V. suspensum*
   7. Leaves with lower surface very rusty-scurfy or stellate-pubescent; twigs rusty-scurfy or rusty-scurfy and stellate-pubescent; fruit red, even at maturity or changing to black
      12. Leaves mostly greater than 10 cm long, rusty-scurfy on lower surface; fruit red, even at maturity . . . .***V. alnifolium***
      12. Leaves mostly less than 10 cm long, stellate-pubescent or rusty-scurfy and stellate-pubescent on lower surface; fruit black or purple-black at maturity
         13. Leaves remotely and minutely toothed, nearly entire
            14. Leaves with teeth rounded or crenulate . . . .*V. macrocephalum* (OC)
            14. Leaves serrulate, the teeth sharp but remote . . . . . .*V.* ×*burkwoodii*
         13. Leaves sharply and finely toothed or, if entire, orbicular or suborbicular
            15. Leaves ovate or elliptic, serrate . . . .*V.* ×*carlcephalum*
            15. Leaves suborbicular or orbicular, sometimes entire or subentire . . . .*V. carlesii*
6. Buds with scales
   16. Leaves coarsely toothed, the secondary veins extending to the teeth; outer pair of bud scales as long as or shorter than the bud
      17. Outer pair of bud scales as long as the bud, valvate; leaves obovate, oblong-ovate to orbicular, acute to acuminate
         18. Leaves subcordate, rounded, or subtruncate at the base, pubescent on both surfaces . . . .***V. dilatatum*** (N)
         18. Leaves mostly cuneate at the base, nearly glabrous on upper surface

19. Leaves mostly even at the base, usually with at least 10 pairs of secondary veins ....***V. plicatum*** (ON)
19. Leaves often oblique at the base, with 8 or fewer pairs of secondary veins ....***V. sieboldii*** (ON)
17. Outer pair of bud scales shorter than the bud; leaves ovate to orbicular, acute
   20. Petioles mostly less than 1 cm long; linear stipules present ....***V. rafinesquianum***
   20. Petioles mostly greater than 1 cm long; linear stipules present or not
      21. Leaves mostly cordate, the lowest 2 veins on each side of the midrib meeting opposite each other at or near the midrib; petioles often greater than 3 cm long; older bark exfoliating in thin flakes ....***V. molle***
      21. Leaves rounded or truncate, rarely subcordate, lowest 2 veins on each side of the midrib meeting the midrib at different positions; petioles mostly 1–2 cm long; older bark tight, not exfoliating
         22. Petioles glabrous; lower surface of leaves glabrous or pubescent only in axils of the veins ....***V. recognitum***
         22. Petioles pubescent; lower leaf surface pubescent, especially along major veins ....***V. dentatum***
16. Leaves entire or finely and closely toothed, the secondary veins terminating or branching before reaching the leaf margin; outer pair of bud scales as long as the bud (1 shorter pair in *V. farreri*)
   23. Leaves entire or somewhat wavy-toothed; cymes peduncled
      24. Leaves glossy above, usually entire; peduncle as long as or longer than the cyme ....***V. nudum***
      24. Leaves dull above, usually wavy-toothed; peduncle shorter than the cyme ....***V. cassinoides***
   23. Leaves finely or sharply toothed; cymes sessile or shortly peduncled
      25. Leaves denticulate; inflorescence shortly stalked; fruit remaining red or turning purple-black at maturity
         26. Leaves elliptical, hirsute along the veins on lower surface, the petioles mostly less than 1 cm long; fruit remaining red at maturity ....***V. setigerum*** (ON)
         26. Leaves ovate to obovate, shortly pubescent in the axils of the veins on lower surface, the petioles greater than 1 cm long; fruit red, turning purple-black at maturity ....***V. farreri***
      25. Leaves finely and sharply serrate; inflorescence sessile; fruit changing to blue-black or black at maturity

27. Petioles with very narrow margins, lacking wings; leaves mostly
acute or obtuse . . . . ***V. prunifolium***
27. Petioles narrowly or broadly winged; leaves mostly rounded
or acuminate
   28. Petioles very narrowly and evenly winged; leaves mostly retuse,
rounded, or obtuse . . . . ***V. rufidulum*** (s)
   28. Petioles broadly and unevenly winged; leaves
mostly acuminate . . . . ***V. lentago***

Winter key

1. Buds naked, densely tomentose (alternate step 1, p. 234)
   2. Leaves with 3 main veins from the base . . . . *V. davidii* (cs)
   2. Leaves with a single main vein, the midrib
      3. Leaves persistent
         4. Leaves wrinkled, the veins deeply impressed on upper surface, the
lower surface densely gray to yellow-brown–tomentose; branchlets
densely stellate-pubescent . . . . *V. rhytidophyllum*
         4. Leaves flat, smooth on upper surface, the lower surface
sparsely pubescent (white-tomentose in *V. ×burkwoodii*) to
glabrous; branchlets glabrous
           5. Leaves thin, white-tomentose on lower surface . . . . . *V. ×burkwoodii*
           5. Leaves thick, coriaceous, glabrous to slightly brown-
pubescent on lower surface
              6. Leaves ciliate, the petioles pubescent, with stellate and
simple hairs . . . . *V. tinus*
              6. Leaves lacking hairs on the margins, the petioles
glabrous or, if pubescent, with stellate hairs only
                 7. Leaves entire, undulate or minutely toothed in the apical $^1/_3$,
the teeth remote from each other . . . .
. . . . *V. odoratissimum* (including the variety *awabuki*)
                 7. Leaves toothed regularly or irregularly $^1/_2$ to $^3/_4$ of the way
toward the base . . . . *V. suspensum*
      3. Leaves deciduous; branchlets scurfy-pubescent; cymes
5–12 cm wide
         8. Buds light cinnamon-brown; leaf scars mostly 2–5 mm wide;
twigs glossy purple-brown to dark gray, speckled
with lenticels . . . . ***V. alnifolium***
         8. Buds brown-gray; leaf scars mostly 1 mm wide; twigs dull,
usually light gray to light brown
           9. Twigs light brown to brown . . . . *V. lantana* (on)
           9. Twigs mostly light gray . . . .
. . . . *V. carlesii, V. macrocephalum* (oc), and the hybrid *V. ×carlcephalum*

1. Buds scaly, glabrous, or pubescent
   10. Outer pair of bud scales shorter than the bud
      11. Bark of stem and branches exfoliating . . . .*V. molle*
      11. Bark of stem and branches close, not exfoliating
         12. Twigs pubescent, longitudinal ridges lacking or at least
            less prominent . . . .*V. acerifolium*
         12. Twigs glabrous or sometimes densely puberulent, with or
            without longitudinal ridges
            13. Twigs smooth and even, lacking longitudinal ridges
               14. Buds with 2 short and 2 long scales visible; twigs light brown to
                  yellow-brown or red-brown . . . .*V. rafinesquianum*
               14. Buds with more than 4 scales visible; twigs dark brown
                  to gray . . . .*V. farreri*
            13. Twigs with longitudinal ridges
               15. Lateral buds plump and spreading; twigs and buds
                  sometimes pubescent . . . .**V. dentatum**
               15. Lateral buds slender and appressed; twigs and
                  buds glabrous . . . .**V. recognitum**
   10. Outer pair of bud scales as long as the bud, valvate
      16. Buds glabrous or glutinous, oblong or ovoid
         17. Buds acute or acuminate . . . .*V. setigerum* (ON)
         17. Buds blunt or with rounded apex
            18. Plants tall, erect . . . .*V. opulus* (**var. opulus** and **var. americanum**)
           18. Plants low, straggling . . . .*V. edule*
      16. Buds scurfy, tomentose, pubescent, or stellate-pubescent,
         linear-lanceolate
         19. Buds tomentose, red . . . .*V. rufidulum* (s)
         19. Buds scurfy, pubescent, or stellate-pubescent, gray
           or red-brown
           20. Buds stellate-pubescent
              21. Inflorescences with primary branches greater than 2 cm long,
                distant from each other . . . .**V. sieboldii** (ON)
              21. Inflorescences with primary branches about 1 cm long, often
                quite close to each other, sometimes appearing
                as umbels . . . .**V. plicatum** (ON)
           20. Buds scurfy or pubescent
              22. Fruit ovoid, red, long persistent . . . .**V. dilatatum** (N)
              22. Fruit globose, blue-black
                23. Twigs mostly short and stiff, nearly at right angles to the
                  stem; buds mostly short-pointed, often rusty-pubescent
                  or scurfy . . . .**V. prunifolium**
               23. Twigs mostly long and flexuous, ascending;
                  buds long-pointed, scurfy, red-brown or
                  lead-colored

24. Bud scales thin, broadened at the base; flower buds, before swelling, completely covered by the outer pair of bud scales . . . .*V. lentago*
24. Bud scales thick, not broadened at the base; flower buds, before swelling, only partially covered by outer pair of bud scales
    25. Stalk of the inflorescence shorter than the inflorescence . . . . . .*V. cassinoides*
    25. Stalk of the inflorescence as long as or longer than the inflorescence . . . .*V. nudum*

Vinca   periwinkle

1. Leaves truncate to subcordate at the base, the margins ciliate . . . .*V. major* (N)
1. Leaves narrowed or cuneate at the base, the margins lacking hairs . . . .*V. minor* (N)

Vitex   chaste tree

1. Branchlets terete; leaflets 5–7, lanceolate, 7–10 cm long, entire or with a few coarse teeth . . . .*V. agnus-castus* (ON)
1. Branchlets 4-sided; leaflets 3–5, mostly 5, 3–10 cm long, entire or serrate . . . .*V. negundo* (ON)

Vitis   grape

1. Pith continuous, lacking a partition or diaphragm at the node; bark close, not exfoliating . . . .*V. rotundifolia*
1. Pith interrupted by a diaphragm at the nodes; bark usually exfoliating
    2. Tendrils lacking or only opposite the uppermost leaves . . . .*V. rupestris*
    2. Tendrils present consistently or intermittently but throughout the vine
        3. Tendrils (or inflorescence) opposite the leaf at 3 or more successive nodes; lower leaf surface densely tomentose, often becoming glabrous with age
            4. Leaves with lower surface rusty-woolly, completely hiding the lower surface; twigs rusty-pubescent . . . .*V. labrusca*
            4. Leaves with lower surface pubescent or with silky hairs on the veins; twigs glabrous to pubescent . . . .*V.* ×*novae-angeliae*
        3. Tendrils opposite the leaves at 2 or fewer successive nodes; lower leaf surface glabrous or pubescent
            5. Leaves mostly deeply lobed, the sinuses greater than halfway to the apex of the petiole . . . .*V. vinifera* (ON)
            5. Leaves mostly shallowly lobed (except sometimes on young or vigorous shoots)

6. Lower leaf surface pubescent and floccose, the long silky hairs
often flattened
   7. Branchlets angular; lower surface of leaves at first pubescent, later
pale gray-green . . . .*V. cinerea*
   7. Branchlets terete; lower surface of leaves rust-brown with
floccose hairs or glaucous
      8. Petioles pubescent; lower leaf surface rust-brown, with
floccose hairs . . . .*V. aestivalis* **var.** *aestivalis*
      8. Petioles nearly glabrous, often glaucous, often red; lower leaf
surface glaucous . . . .*V. aestivalis* **var.** *argentifolia*
6. Lower leaf surface mostly glossy, glabrous except for axils of the
veins or with short, straight or spreading hairs on the veins
   9. Twigs purple-red . . . .*V. palmata* (s)
   9. Twigs green, gray, or brown
      10. Diaphragm or partition of lower nodes about 2 mm thick;
leaves mostly with obvious lobes that point forward, the acute teeth
with sides flat or slightly concave, ciliolate;
fruit glaucous . . . .*V. riparia*
      10. Diaphragm or partition at nodes about 1 mm thick; leaves
mostly with lobes barely evident and spreading, the teeth
broadly acute or obtuse, the sides of the teeth convex,
lacking hairs; fruit shining
         11. Branchlets terete; leaves with petioles and veins on lower leaf
surface glabrous . . . .*V. vulpina*
         11. Branchlets angled; leaves with petioles and veins on lower leaf
surface pubescent . . . .*V. baileyana*

Weigela   weigela

1. Leaves glabrous (sometimes sparsely pubescent on petioles and lower surface),
petiole greater than 6 mm long; branchlets glabrous . . . .*W. coraeensis* (oc)
1. Leaves pubescent on 1 or both surfaces, petiole less than 5 mm
long or lacking; branchlets pubescent (sometimes glabrous in
W. praecox and *W. japonica*)
   2. Leaves pubescent on both surfaces
      3. Leaves mostly less than 6 cm long; branchlets pubescent or glabrous;
fruit glabrous . . . .*W. praecox*
      3. Leaves mostly greater than 6 cm long; branchlets pubescent;
fruit pubescent . . . .*W. floribunda*
   2. Leaves glabrous or sparingly pubescent on upper surface, often
only at the midrib, pubescent on lower surface
      4. Leaves densely pubescent all over lower surface . . . .*W. hortensis* (oc)
      4. Leaves mostly pubescent only along veins on lower surface

5. Branchlets with 2 longitudinal lines of pubescence; petioles mostly less than 2 mm long or lacking ....*W. florida*
5. Branchlets sparingly pubescent, nearly glabrous; petioles greater than 2 mm long ....*W. japonica* (OC)

Wisteria   wisteria

1. Twigs densely pubescent; fruit densely pubescent, like velvet
  2. Leaflets mostly greater than 13, not or slightly curved at the apex, long-acuminate; vines twining clockwise ....**W. floribunda** (N)
  2. Leaflets mostly fewer than 13, abruptly curved and obtuse at the apex; vines apparently twining counterclockwise ....**W. sinensis** (N)
1. Twigs glabrous or sparsely pubescent; fruit glabrous
  3. Leaflets less than 2 cm wide; racemes less than 10 cm long ....**W. frutescens** (S)
  3. Leaflets greater than 2 cm wide; racemes greater than 15 cm long ....**W. macrostachya** (S)

Xanthorhiza simplicissima   yellowroot

Yucca   yucca, Spanish dagger, Spanish bayonet

1. Plants with woody trunk obvious, the leaves scattered along it
  2. Leaves with rough edges, lacking threads ....*Y. aloifolia*
  2. Leaves with smooth edges or with threads dangling ....**Y. gloriosa** (S)
1. Plants with woody stem or trunk not visible; leaves in dense basal rosette
  3. Leaves less than 13 mm wide; inflorescence simple ....**Y. glauca** (ON)
  3. Leaves greater than 13 mm wide; inflorescence branched
    4. Leaves firm or stiff ....**Y. filamentosa** (S)
    4. Leaves flexible ....**Y. flaccida** (S)

Zanthoxylum   prickly-ash, toothache tree

1. Leaflets ovate to ovate-lanceolate, glabrous, the base often oblique; buds dark brown or nearly black; fruit in a terminal cyme ....**Z. clava-herculis**
1. Leaflets ovate-oblong, downy when young, the base rounded to cuneate, rarely oblique; buds red; fruit in axillary clusters ....**Z. americanum**

Zelkova   zelkova

1. Leaves 5–13 cm long, sharply serrate ....*Z. serrata*
1. Leaves less than 5 cm long, undulate-serrate

2. Leaves ovate-oblong, with 7–10 pairs of veins . . . .*Z. sinica* (OC)

2. Leaves elliptic to oblong, with 6–8 pairs of veins, pubescent on the veins of lower surface . . . .*Z. carpinifolia* (N)

Zenobia pulverulenta   zenobia (S)

Ziziphus jujuba   common jujube, Chinese-date (OC)

SYSTEMATIC LIST OF SPECIES INCLUDED IN THE KEYS

Note: Native and naturalized species appear in bold type; cultivated species are not bold. The following abbreviations are used: N = naturalized; ON = occasionally or infrequently naturalized; S = rarely native north of North Carolina and Tennessee; CS = rarely cultivated north of North Carolina and Tennessee; OC = occasionally or rarely cultivated; A = native of Asia; E = native of Europe; F = native of Africa; W = native of western North America. Synonyms are indented in parentheses.

(GYMNOSPERMS)

GINKGOPHYTA

Ginkgoaceae

Ginkgo L.
  G. biloba L.   ginkgo, maidenhair tree (ON, A)

PINOPHYTA

Taxaceae

Taxus L.
  T. baccata L.   English yew (ON, E)
  T. canadensis Marshall   American yew, ground-hemlock
  T. cuspidata Siebold & Zucc.   Japanese-yew (ON, A)
  T. ×*media* Rehder (**T. cuspidata Siebold & Zucc.** × **T. baccata L.**)
     hybrid yew, Anglo-Japanese yew
Torreya Arn.
  T. nucifera (L.) Siebold & Zucc.   torreya, Japanese torreya (OC, A)

Araucariaceae

Araucaria Juss.

    A. araucana (Molina) K. Koch   monkey-puzzle tree (CS)

Cephalotaxaceae

Cephalotaxus Siebold & Zucc.

    C. fortunei Hooker   plum-yew (OC, A)

    C. harringtonia (D. Don) K. Koch   plum-yew (OC, A)

Podocarpaceae

Podocarpus L'Her. ex Pers.

    P. macrophyllus (Thunb.) D. Don   Japanese-yew, southern-yew
        (CS, A)

    P. nagi (Thunb.) Zoll. & Moritzi ex Makino   broadleaf
        podocarpus (CS, A)

Pinaceae

Abies Mill.

    A. balsamea (L.) Mill. balsam fir, balsam

    A. concolor (Gordon & Glend.) Lindl. ex Hildebr.   concolor fir,
        white fir (W)

    A. fraseri (Pursh) Poiret   fraser fir, she-balsam (S, CULTIVATED
        NORTH)

Cedrus Trew.

    C. atlantica (Endl.) G. Manetti ex Carriere   atlas-cedar, blue
        atlas-cedar (F)

    C. deodara (Roxb. ex Lamb.) G. Don in Loudon   deodar-cedar
        (A)

    C. libani A. Rich.   cedar-of-Lebanon (A)

Larix Mill.

    L. decidua Mill.   European larch (ON, E)
    (L. europaea DC)

    L. ×*eurolepis* A. Henry (*L. kaempferi* [Lamb.] Carriere × **L. decidua**
        Mill.)   dunkeld larch

    L. kaemferi (Lamb.) Carriere   Japanese larch (A)
        (*L. leptolepis* Gordon)

    L. laricina (Du Roi) Koch   American larch

Picea A. Dietr.

    P. abies (L.) H. Karst.   Norway spruce (N, E)

P. glauca (**Moench**) **Voss**   white spruce, skunk spruce, cat spruce

P. mariana (**Mill.**) **Britton, Stearns & Poggenb.**   black spruce, swamp spruce

P. omorika (Pancic) Purkyne   omorika spruce, Serbian spruce (E)

P. pungens **Engelm.**   Colorado blue spruce, blue spruce (ON, W)

P. rubens **Sarg.**   red spruce

Pinus L.

P. aristata Engelm.   bristlecone pine (OC, W)

P. banksiana **Lamb.**   jack pine, gray pine

P. bungeana Zucc. ex Endl.   lacebark pine (A)

P. cembra L.   Swiss stone pine (OC, E)

P. cembroides Zucc.   pinyon pine (OC, W)
  (*P. cembroides* var. *edulis* Engelm.)

P. contorta Douglas ex Loudon   lodgepole pine (OC, W)

P. densiflora Siebold & Zucc.   Japanese red pine (OC, A)

P. echinata **Mill.**   yellow pine, shortleaf pine

P. elliottii Engelm.   slash pine (CS)

P. flexis E. James   limber pine (OC, W)
  (*P. caribaea* Morelet)

P. koraiensis Siebold & Zucc.   Korean pine (OC, A)

P. leucodermis Antoine   Bosnian pine (OC, E)

P. mugo Turra   mugo pine (E)

P. monticola Lamb.   western white pine (OC, W)

P. nigra J. Arnold   Austrian pine (E)

P. palustris **Mill.**   longleaf pine (S)

P. parviflora Siebold & Zucc.   Japanese white pine (OC, A)

P. ponderosa Douglas ex C. Lawson   western yellow pine, bull pine (W)

P. pungens **Lamb.**   table mountain pine, hickory pine

P. resinosa **Aiton**   red pine, Norway pine

P. rigida **Mill.**   pitch pine

P. serotina **Michx.**   pond pine (S)

P. strobus **L.**   white pine

P. sylvestris **L.**   Scotch pine (N, E)

P. taeda **L.**   loblolly pine, old field pine

P. thunbergiana Franco   Japanese black pine (A)

P. virginiana **Mill.**   Virginia pine, Jersey pine, scrub pine

P. wallichiana A. B. Jacks.　　Himalayan pine (OC, A)
　　(*P. griffithii* McClelland)
Pseudolarix Gordon
　　P. kaempferi Gordon　　golden-larch (OC, A)
Pseudotsuga Carriere
　　P. menziesii (Mirb.) Franco.　　Douglas-fir (ON, W)
　　(P. douglasii [Sabine ex D. Don] Carriere)
　　(P. taxifolia [Lamb.] Britton)
Tsuga Carriere
　　T. canadensis (L.) Carriere　　hemlock, hemlock-spruce
　　T. caroliniana Engelm.　　Carolina hemlock
　　T. chinensis (Franch.) Pritz.　　Chinese hemlock (OC, A)
　　T. diversifolia (Maxim.) Mast.　　northern Japanese hemlock
　　　　(OC, A)
　　T. heterophylla (Raf.) Sarg.　　western hemlock (OC, W)
　　T. mertensiana (Bong.) Carriere　　mountain hemlock (OC, W)
　　T. sieboldii Carriere　　southern Japanese hemlock (OC, A)

Cupressaceae

Calocedrus Kurz
　　C. decurrens (Torr.) Florin　　incense-cedar (OC, W)
　　(*Libocedrus decurrens* Torr.)
Chamaecyparis Spach
　　C. lawsoniana (A. Murr.) Parl.　　Port Orford–cedar, Lawson-
　　　　cypress (W)
　　C. nootkatensis (Lamb.) Spach　　Alaska-cedar, yellow-cypress,
　　　　Nootka-cypress (OC, W)
　　C. obtusa Siebold & Zucc.　　Hinoki-cypress (A)
　　C. pisifera Siebold & Zucc.　　Sawara-cypress (OC, A)
　　C. thyoides (L.) Britton, Stearns, & Poggenb.　　white-cedar
Cryptomeria D. Don
　　C. japonica D. Don　　cryptomeria, Japanese-cedar (A)
Cunninghamia R. Br.
　　C. lanceolata (Lamb.) Hooker　　China-fir, cunninghamia (OC, A)
×*Cupressocyparis* Dallim.
　　×*C. leylandii* (A. B. Jacks. & Dallim.) Dallim.　　(*Cupressus*
　　　　macrocarpa Hartweg × *Chamaecyparis nootkatensis* [D. Don]
　　　　Spach)　　Leyland-cypress
Cupressus L.

C. arizonica Greene   Arizona cypress (OC, W)

Juniperus L.

    J. chinensis L.   Chinese juniper (A)

    J. communis L. var. *communis*   common juniper

    J. communis L. var. *depressa* **Pursh**   spreading juniper, prostrate juniper

    J. conferta Parl.   shore juniper (A)

    J. horizontalis **Moench**   creeping juniper

    J. procumbens (Endl.) Miq. in Siebold & Zucc.   creeping juniper (A)

    J. scopulorum Sarg.   Rocky Mountain juniper (W)

    J. sabina L.   savin (E)

    J. squamata Lamb.   Himalayan juniper (A)

    J. virginiana **L.**   eastern redcedar, savin

Metasequoia Miki

    M. glyptostroboides Hu & Cheng   dawn-redwood (A)

Microbiota Kom.

    M. decussata Kom.   microbiota, Siberian-juniper (A)

Platycladus Spach

    P. orientalis (L.) Franco   Oriental-arborvitae (A) (*Thuja orientalis* L.)

Sciadopitys Siebold & Zucc.

    S. verticillata (Thunb.) Siebold & Zucc.   umbrella-pine (OC, A)

Sequoia Endl.

    S. sempervirens (D. Don) Endl.   redwood, coast redwood (OC, W)

Sequoiadendron Buchholz

    S. giganteum (Lindl.) Buchholz   giant-sequoia, California big tree (OC, W)

Taxodium (L.) Rich.

    T. ascendens **Brongn.**   pond-cypress (S)

    (*T. distichum* **[L.] Rich. var.** *imbricarium* **[Nutt.] Croom)**

    T. distichum **(L.) Rich.**   bald-cypress

Thuja L.

    T. occidentalis **L.**   arborvitae

    T. plicata D. Don   western redcedar, giant arborvitae (OC, W)

    T. standishii (Gordon) Carriere   Standish arborvitae, Japanese arborvitae (OC, A)

Thujopsis (L.f.) Siebold & Zucc.

T. dolobrata Siebold & Zucc.   hiba-arborvitae, false-arborvitae
(OC, A)

GNETOPHYTA
Ephedraceae
Ephedra L.
  E. distachya L.   ephedra (A, E)

MAGNOLIOPHYTA (ANGIOSPERMS)

MAGNOLIOPSIDA (DICOTYLEDONS)
MAGNOLIIDAE
Magnoliaceae
Liriodendron L.
  L. tulipifera L.   tulip tree, yellow-poplar, tulip-poplar
Magnolia L.
  M. acuminata L.   cucumber tree, mountain magnolia
  M. denudata Desr.   yula, yulan magnolia (A)
  M. fraseri Walter   Fraser magnolia, ear-leaved magnolia,
    long-leaved cucumber tree (S)
    (M. pyramidata Bartram ex Pursh)
  M. grandiflora L.   southern magnolia (A)
  M. kobus DC var. *kobus*   kobus magnolia (A)
  M. kobus DC var. *stellata* (Siebold & Zucc.) Blackburn   star
    magnolia (A)
  M. liliiflora Desr.   purple magnolia (A)
  M. macrophylla Michx.   bigleaf magnolia, large-leaved
    cucumber tree (S)
    (M. dealbata Zucc.)
  M. salicifolia (Siebold & Zucc.) Maxim. (OC, A)
  M. ×soulangeana Soul.-Bod. (*M. denudata* Desr. × *M. liliiflora*
    Desr.)   saucer magnolia (A)
  M. tripetala L.   umbrella tree, umbrella magnolia, elkwood
  M. virginiana L.   sweet-bay, swamp-bay
Michelia L.
  M. figo (Lour.) Spreng.   banana shrub (CS, A)
    (*M. fuscata* [Andr.] Blume)
    (*Magnolia fuscata* Andr.)

Annonaceae

Asimina Adans.

A. parviflora (Michx.) Dunal   dwarf pawpaw (s)

A. triloba (L.) Dunal   pawpaw, custard-apple

Calycanthaceae

Calycanthus L.

C. floridus L. var. floridus   Carolina allspice (s)
(*C. fertilis* Walter)

C. floridus L. var. glaucus (Willd.) Torr. & A. Gray   Carolina
allspice

Chimonanthus Lindl.

C. praecox (L.) Link   wintersweet (OC, A)

Lauraceae

Laurus L.

L. nobilis L.   laurel, sweet-bay (E)

Lindera Thunb.

L. benzoin (L.) Blume   spicebush

L. melissaefolia (Walter) Blume   hairy spicebush, Jove's fruit,
pondberry (s)

Litsea Lam.

L. aestivalis (L.) Fernald   pondspice (s)

Persea Mill.

P. borbonia (L.) Spreng.   redbay (s)

P. palustris (Raf.) Sarg.   swamp redbay (s)

Sassafras Nees.

S. albidum (Nutt.) Nees.   sassafras

Aristolochiaceae

Aristolochia L.

A. macrophylla Lam.   Dutchman's pipe
(*A. durior* **Hill**)

A. tomentosa Sims   woolly pipe vine

Illiciaceae

Illicium Thunb.

I. anisatum L.   Chinese star anise (CS, A)

I. floridanum Ellis   purple anise (cs)
I. parviflorum Michx.   star anise (cs)

Schisandraceae

Kadsura Juss.
   K. japonica (L.) Dunal (cs, oc, a)
Schisandra Michx.
   S. chinensis (Turcz.) Baill. (a)
   S. *glabra* (Brickell) Rehder   bay star vine, wild-sarsaparilla
      (s)
      (**S. *coccinea* Michx.**)

Ranunculaceae

Clematis L.
   C. *addisonii* Britton
   C. apiifolia DC. (a)
   C. armandii Franch.   Armand clematis (a)
   C. *catesbyana* Pursh (s)
   C. *crispa* L.   blue-jasmine, marsh clematis, early clematis
   C. *flammula* L. (on, a)
   C. *florida* Thunb. (on, a)
   C. ×*jackmanii* T. Moore (*C. lanuginosa* Lindl. × **C. *viticella* L.**)
      Jackman clematis (n)
   C. ×jouiniana C.K. Schneid. (*C. heracleifolia* DC. × *C. vitalba* L.)
   C. lanuginosa Lindl. (a)
   C. ×lawsoniana T. Moore & A. B. Jacks. (*C. lanuginosa* Lindl. × *C. patens* C. Moore & Decne.)
   C. montana Buch.-Ham. (a)
   C. *occidentalis* DC.   purple clematis, rock clematis
      (**C. *verticillaris* DC.**)
   C. *orientalis* L. (on, a)
   C. patens C. Moore & Decne. (a)
   C. serratifolia Rehder (a)
   C. *tangutica* (Maxim.) Korsh. (on, a)
   C. *terniflora* DC. (on, a)
      (*C. dioscoreifolia* H. Lev. & Vaniot)
      (*C. paniculata* Thunb., not J. F. Gmel.)
   C. texensis Buckley (cs)
   C. *viorna* L.

C. virginiana L.   virgin's bower, woodbine
C. vitalba L. (A, E, F)
C. viticella L. (ON, A, E)
Xanthorhiza Marshall
  X. simplicissima Marshall   yellowroot, shrub yellowroot
    (*X. apiifolia* L'Her.)

Berberidaceae

Berberis L.
  B. canadensis Mill.   American barberry, Alleghany barberry
  B. julianae C. K. Schneider   wintergreen barberry (A)
  B. thunbergii DC.   Japanese barberry (N, E)
  B. vulgaris L.   common barberry (N, E)
Mahonia Nutt.
  M. aquifolium (Pursh) Nutt.   Oregon-grape (N, W)
    (**Berberis aquifolium Pursh**)
  M. bealei (Fortune) Carriere (N, A)
  M. pinnata (Lag.) Fedde (W)
Nandina Thunb.
  N. domestica Thunb.   nandina, heavenly-bamboo, sacred-
    bamboo (ON, A)

Lardizabalaceae

Akebia Decne.
  A. quinata (Houtt.) Decne.   five-leaf akebia (ON, E)

Menispermaceae

Calycocarpum (Nutt.) Spach.
  C. lyonii (Pursh) Nutt.   cupseed (S)
Cocculus DC.
  C. carolinus (L.) DC.   Carolina-moonseed (S)
Menispermum L.
  M. canadense L.   moonseed, yellow-parilla

HAMAMELIDAE

Trochodendraceae

Trochodendron Siebold & Zucc.
  T. aralioides Siebold & Zucc.   wheel tree (CS, A)

Cercidiphyllaceae

Cercidiphyllum Siebold & Zucc.

C. japonicum Siebold & Zucc. ex Hoffm. & Schultes
  katsura, cotton candy tree (ON, A)

Eupteleaceae

Euptelea Siebold & Zucc.

  E. polyandra Siebold & Zucc.   tasseltree, Japanese euptelea
    (OC, A)

Platanaceae

Platanus L.

  P. ×hybrida Brot. (P. occidentalis L. × *P. orientalis L.*)
      London plane (N)
      (*P. ×acerifolia* [Aiton] Willd.)
  P. occidentalis L.   sycamore, buttonball
  P. orientalis L.   Oriental plane (A)

Hamamelidaceae

Corylopsis Siebold & Zucc.

  C. glabrescens Franch. & Sav.   corylopsis (A)
  C. pauciflora Siebold & Zucc. (A)
  C. sinensis Hemsl. (A)
      (*C. platyphylla* Rehder & E. H. Wilson)
      (*C. veitchiana* Bean)
      (*C. willmottiae* Rehder & E. H. Wilson)
  C. spicata Siebold & Zucc. (A)

Fothergilla L.

  F. gardenii J. Murr.   witch-alder, fothergilla
  F. major (Sims) Lodd. (S)

Hamamelis L.

  H. ×intermedia Rehder. (*H. japonica* Siebold & Zucc. × *H. mollis*
      Oliv.)
  H. japonica Siebold & Zucc.   Japanese witch-hazel (A)
  H. mollis Oliv.   Chinese witch-hazel (A)
  H. vernalis Sarg. (OC, W)
  H. virginiana L.   witch-hazel

Liquidambar L.

  L. formosana L.   Formosan sweetgum (OC, A)

L. orientalis Mill.   Oriental sweetgum (OC, A)
L. styraciflua L.   sweetgum
Loropetalum R. Br.
   L. chinense (R. Br.) Oliv.   loropetalum, fringe flower (CS, A)
Parrotia C. A. Mey.
   P. persica C. A. Mey. (OC, A)   Persian-ironwood
Parrotiopsis (Niedenzu) C. K. Schneid.
   P. jacquemontiana (Decne.) Rehder (OC, A)

<h3 style="text-align:center">Daphniphyllaceae</h3>

Daphniphyllum Blume
   D. macropodum Miq. (CS, A)

<h3 style="text-align:center">Eucommiaceae</h3>

Eucommia Oliv.
   E. ulmoides Oliv.   eucommia (OC, A)

<h3 style="text-align:center">Ulmaceae</h3>

Aphananthe Planch.
   A. aspera (Thunb.) Planch.   muku tree (OC, A)
Celtis L.
   C. bungeana Blume (OC, A)
   C. laevigata Willd.   sugarberry (S)
   C. occidentalis L.   hackberry
   C. tenuifolia Nutt.
Hemiptelea Planch.
   H. davidii (Hance) Planch. (OC, A)
Planera J. F. Gmel.
   P. aquatica (Walter) J. F. Gmel.   water-elm
Pteroceltis Maxim.
   P. tartinowii Maxim. (OC)
Ulmus L.
   U. alata Michx.   winged elm (S)
   U. americana L.   American elm
   U. carpinifolia Ruppius ex Suckow   smoothleaf elm (OC, E)
     (*U. campestris* L. in part)
   U. crassifolia Nutt.   cedar elm (S)
   U. glabra Huds.   wych elm, Scotch elm (ON, E)
   U. ×*hollandica* Mill. (*U. carpinifolia* [Ruppius ex Sukow] ×
     U. glabra Huds. × *U. plotii* Druce)   Holland elm,
    Dutch elm

U. minor **Mill.**   English elm (ON, E)
   (*U. campestris* **L.** in part)
   (*U. procera* **Salisb.**)
U. pumila **L.**   Siberian elm, dwarf elm (ON, A)
U. parvifolia Jacq.   Chinese elm (A)
U. rubra **Muhl.**   slippery elm
U. serotina **Sarg.**   September elm, red elm (s)
U. thomasii **Sarg.**   rock elm, cork elm
Zelkova Spach
   Z. carpinifolia **(Pall.) K. Koch** (N, A, E)
   Z. serrata (Thunb.) Makino   Japanese zelkova, saw-leaf zelkova
      (A)
   Z. sinica C. K. Schneid. (OC, A)

Moraceae

Broussonetia L'Her.
   B. papyrifera **(L.) Vent.**   paper-mulberry (ON, A)
Ficus L.
   F. carica **L.**   common fig (ON, CS, A)
Maclura Nutt.
   M. pomifera **(Raf.) C. K. Schneid.**   Osage-orange,
      hedge-apple
Morus L.
   M. alba **L.**   white mulberry (N, A, E)
   M. rubra **L.**   red mulberry

Leitneriaceae

Leitneria Chapm.
   L. floridana **Chapm.**   corkwood (s)

Juglandaceae

Carya Nutt.
   C. aquatica **(F. Michx.) Nutt.**   water hickory (s)
   C. carolinae-septentrionalis **Ashe**   southern shagbark hickory
      (s)
   C. cordiformis **(Wangenh.) Koch**   bitternut, swamp hickory
   C. glabra **(Mill.) Sweet**   pignut
   C. illinoinensis **(Wangenh.) Koch**   pecan
      (*C. pecan* **(Marshall) Engl. & Graebn.**)
   C. laciniosa **(F. Michx.) Loudon**   shellbark hickory

C. myristicaeformis **F. Michx.**   nutmeg hickory (s)
C. ovalis **(Wangenh.) Sarg.**   small pignut
C. ovata **(Mill.) Koch**   shagbark hickory
C. pallida **Ashe**   pale hickory
C. tomentosa **(Lam.) Nutt.**   mockernut
Juglans L.
    J. ailanthifolia Carr.   Japanese walnut (A)
      (*J. sieboldiana* Maxim.)
    J. cinerea **L.**   butternut, white walnut
    J. nigra **L.**   black walnut
    J. regia **L.**   English walnut, Persian walnut (ON, A)
Platycarya Siebold & Zucc.
    P. strobilacea Siebold & Zucc.   cluster-walnut (OC, A)
Pterocarya Kunth
    P. stenoptera C. DC.   Chinese wingnut (OC, A)

Myricaceae

Comptonia L'Her.
    C. peregrina **(L.) J. M. Coult.**   sweetfern
      (*Myrica peregrina* [L.] Kuntze)
Myrica L.
    M. cerifera **L.**   wax-myrtle (s)
    M. gale **L.**   sweet gale
    M. heterophylla **Raf.**   southern bayberry (s)
    M. pensylvanica **Mirbel**   bayberry

Fagaceae

Castanea Mill.
    C. alnifolia **Nutt.**   Florida chinkapin, downy chestnut (s)
    C. crenata Siebold & Zucc.   Japanese chestnut (A)
      (*C. floridana* [Sarg.] Ashe)
    C. dentata **(Marshall) Borkh.**   American chestnut
    C. mollissima **Blume**   Chinese chestnut (ON, A)
    C. pumila **(L.) Mill.**   chinkapin
    C. sativa Mill.   Spanish chestnut (E)
Fagus L.
    F. grandifolia **Ehrh.**   American beech
    F. sylvatica L.   European beech (E)
Quercus L.

Q. acutissima Carruth.   sawtooth oak (A)

Q. alba L.   white oak

Q. austrina Small   bluff oak, bastard white oak (S)

Q. bicolor Willd.   swamp white oak

Q. cerris L.   turkey oak (A, E)

Q. chapmanii Sarg.   Chapman oak (S)

Q. coccinea Muenchh.   scarlet oak

Q. dentata Thunb.   daimyo oak (OC, A)

Q. ellipsoidalis E. J. Hill   northern pin oak

Q. falcata Michx.   Spanish red oak
   (**Q. pagodifolia [Elliott)] Ashe**)

Q. geminata Small   sand live oak (S)

Q. georgiana M. A. Curtis   Georgia oak (S)

Q. glandulifera Blume   konara oak (OC, A)
   (*Q. serrata* Thunb. of authors)

Q. hemisphaerica Bartram ex Willd.   Darlington oak, laurel
   oak (S)

Q. ilicifolia Wangenh.   scrub oak, bear oak

Q. imbricaria Michx.   shingle oak

Q. incana Bartram   bluejack oak (S)
   (**Q. cinerea Michx.**)

Q. laevis Walter   American turkey oak (S)

Q. laurifolia Michx.   laurel oak (S)

Q. lyrata Walter   overcup oak (S)

Q. macrocarpa Michx.   bur oak, mossycup oak

Q. margaretta Ashe   scrubby post oak (S)

Q. marilandica Muenchh.   blackjack oak

Q. michauxii Nutt.   swamp chestnut oak (S)

Q. minima (Sarg.) Small   minimal oak (S)

Q. mongolica Fisch. ex Turcz.   Mongolian oak (OC, A)

Q. montana Willd.   chestnut oak
   (**Q. prinus L.**)

Q. muehlenbergii Engelm.   yellow chestnut oak, yellow oak

Q. nigra L.   water oak (S)

Q. oglethorpensis (Witt.) Duncan (S)

Q. pagoda Raf.   cherrybark oak (S)

Q. palustris Muenchh.   pin oak

Q. petraea L. ex Liebl.   durmast oak (OC, A, E)

Q. phellos L.   willow oak

Q. prinoides Willd.   chinquapin oak

Q. pumila **Walter**   runner oak (s)

Q. robur L.   English oak (N, E)

Q. rubra L.   red oak
  (*Q. borealis* **F. Michx.**)

Q. shumardii **Buckley**   shumard oak

Q. sinuata **Walter**   Durand's white oak (s)
  (*Q. durandii* **Buckley**)

Q. stellata **Wangenh.**   post oak

Q. variabilis Blume (OC, A)

Q. velutina **Lam.**   black oak

Q. virginiana **Mill.**   live oak, southern oak (s)

Betulaceae

Alnus Mill.

  A. glutinosa (**L.**) **Gaertn.**   European black alder (N, A, E, F)

  A. incana (**L.**) **Moench ssp.** *rugosa* (**Du Roi**) **Clausen**
    speckled alder, hoary alder
    (*A. rugosa* (**Du Roi**) **Spreng.**)

  A. maritima (**Marshall**) **Nutt.**   seaside alder

  A. serrulata (**Aiton**) **Willd.**   smooth alder, hazel alder

  A. viridis (**Vill.**) **Lam. & DC.**   green alder, mountain alder
    (*A. crispa* [**Aiton**] **Pursh var.** *crispa*)
    (*A. crispa* [**Aiton**] **Pursh var.** *mollis* [**Fernald**] **Fernald**)

Betula L.

  B. albosinensis Burkill (OC, A)

  B. alleghaniensis **Britton**   yellow birch
    (*B. lutea* **Michx.**)

  B. cordifolia **Regel**   heartleaf birch, mountain white birch
    (*B. papyrifera* **Marshall var.** *cordifolia* [**Regel**] **Dippel**)

  B. davurica Pallas (OC, A)

  B. glandulosa **Michx.**   dwarf birch

  B. grossa Siebold & Zucc.   Japanese cherry birch (A)
    (*B. carpinifolia* Siebold & Zucc.)
    (*B. ulmifolia* Siebold & Zucc.)

  B. lenta **L.**   black birch, cherry birch

  B. maximowicziana Regel   monarch birch (OC, A)

  B. michauxii **Spach**   Newfoundland dwarf birch (s)

  B. minor (**Tuck.**) **Fernald**   dwarf white birch
    (*B. papyrifera* **Marsh. var.** *minor* [**Tuck.**] **S. Watson &**
      J. M. Coult.)

B. nigra L.   river birch
B. papyrifera Marshall   paper birch, canoe birch
B. pendula Roth   European white birch (N, E)
B. platyphylla Sukaczev   Asian white birch (ON, A)
  (**B. mandsurica [Regel] Nakai**)
B. populifolia Marshall   gray birch
B. pubescens Ehrh.   European birch, downy birch (E, A)
  (*B. alba* L.)
B. pumila L.   bog birch, swamp birch
Carpinus L.
  C. betulus L.   European hornbeam (OC, E)
  C. caroliniana Walt.   American hornbeam, blue-beech, ironwood
    (**C. americana** Michx.)
    (**C. caroliniana** ssp. **caroliniana**)
    (**C. caroliniana** ssp. **virginiana [Marshall] Furlow**)
Corylus L.
  C. americana Walter   American hazelnut
  C. avellana L.   European hazelnut (E)
  C. colurna L.   Turkish hazelnut (A, E)
  C. cornuta Marshall   beaked hazelnut
  C. heterophylla Fisch. & Trautv.   Siberian hazelnut (OC, A)
  C. sieboldiana Blume.   Japanese hazelnut (OC, A)
  C. maxima Mill.   giant filbert (A)
Ostrya Scop.
  O. carpinifolia Scop.   European hophornbeam (OC, A, E)
  O. virginiana (Mill.) K. Koch   American hophornbeam

CARYOPHYLLIDAE

Polygonaceae

Polygonum L.
  P. cuspidatum Siebold & Zucc.   Japanese knotweed, Japanese-bamboo (N, A)

DILLENIIDAE

Paeoniaceae

Paeonia L.
  P. suffruticosa (Andr.) Donn.   tree peony (A)

Theaceae

Camellia L.

    C. japonica L.   common camellia (A)

    C. sasanqua Thunb.   sasanqua camellia (A)

Cleyera Thunb.

    C. japonica Thunb.   Japanese cleyera (CS, A)

Franklinia Marshall

    F. alatamaha (Marshall.) Sarg.   franklinia, Franklin tree
        (extirpated; native of Georgia)

Gordonia Ellis

    G. lasianthus (L.) Ellis   loblolly-bay (s)

Stewartia L.

    S. koreana Nakai ex Rehder (OC, A)

    S. ovata (Cav.) Weatherby   mountain stewartia (s)

    S. malachodendron L.   silky stewartia (s)

    S. monadelpha Siebold & Zucc. (OC, A)

    S. pseudocamellia Maxim.   Japanese-camellia

Ternstroemia Mutis ex L. f.

    T. gymnanthera (Wight & Arn.) T. Sprague.   ternstroemia (A)

Actinidiaceae

Actinidia Lindl.

    A. arguta (Siebold & Zucc.) Planch. ex Miq.   bower actinidia,
        tara vine (A)

    A. chinensis Planch.   kiwi, Chinese gooseberry, yang-tao (A)

Clusiaceae (Guttiferae, Hypericaceae)

Hypericum L.

    H. buckleyi M. A. Curtis (s)

    H. calycinum L.   creeping St.-John's-wort, goldflower (A, E)

    H. cistifolium Lam. (s)

    H. densiflorum Pursh   bushy St.-John's-wort

    H. fasciculatum Lam.   sandweed (s)

    H. frondosum Michx. (s)

    H. galioides Lam. (s)

    H. hypericoides (L.) Crantz   St.-Andrew's-cross
        (***Ascyrum hypericoides* L.**)

    H. kalmianum L.   kalm St.-John's-wort

    H. lloydii (Svenson) P. Adans. (s)

H. lobocarpum **Gatt.** (s)
H. nitidum **Lam.** (s)
H. nudiflorum **Michx.** (s)
H. patulum Thunb. (A)
H. prolificum **L.**   shrubby St.-John's-wort
H. reductum **P. Adams** (s)
H. stans **(Michx.) P. Adams & N. Robson**   St.-Peter's-wort
  (s)
  (*Ascyrum stans* **Michx.**)
  (*H. crux-andrae* [**L.**] **Crantz**)
H. stragalum **P. Adams. & N. Robson** (s)
H. suffruticosum **P. Adams. & N. Robson** (s)

Tiliaceae

Tilia L.
  T. americana **L.**   American basswood, American linden
  T. cordata **Mill.**   small-leaved European linden (ON, E)
  T. petiolaris **DC.**   pendent silver linden (ON, A)
  T. platyphyllos **Scop.**   large-leaved linden (ON, A)
  T. tomentosa **Moench**   silver linden (ON, A)

Sterculiaceae

Firmiana Marsili
  F. simplex **(L.) Wight**   Chinese parasol tree (N, A)

Malvaceae

Hibiscus L.
  H. syriacus **L.**   rose-of-sharon (N, A)

Flacourtiaceae

Idesia Maxim.
  I. polycarpum Maxim.   iigiri tree (OC, A)

Cistaceae

Hudsonia L.
  H. ericoides **L.**   heathlike-hudsonia
  H. montana **Nutt.** (s)
  H. tomentosa **Nutt.**   false-heather

Stachyuraceae

Stachyurus Siebold & Zucc.
  S. praecox Siebold & Zucc. (CS, A)

Tamaricaceae

Tamarix L.
  T. *chinensis* Lour.   tamarisk (ON, A)
  T. gallica L.   French tamarisk, manna plant (E)
  T. *parviflora* DC.   tamarisk (ON, E)
  T. *ramosissima* Ledeb.   tamarisk (ON, A, E)
    (**T. *odessana* Steven ex Bunge**)
    (**T. *pentandra* Pallas**)

Salicaceae

Populus L.
  P. *alba* L.   white poplar (ON, A, E)
  P. *balsamifera* L.   balsam poplar, tacamahac, hackmatack
    (**P. *tacamahaca* Mill.**)
    (**P. *candicans* Aiton**)
  **P. ✕*canadensis* Moench (P. *deltoides* Marshall × P. *nigra* L.
    'Italica')**   Carolina poplar (ON)
  P. ✕*canescens* (Aiton) Sm. (P. *alba* L. × P. *tremula* L.)   gray
    poplar (ON)
  P. *deltoides* Bartram ex Marshall   cottonwood
  P. *grandidentata* Michx.   bigtooth aspen, large-toothed aspen
  P. *heterophylla* L.   swamp cottonwood, downy poplar
  P. *nigra* L.   black poplar (ON, E, A)
  P. *nigra* L. 'Italica'   Lombardy poplar (ON, E)
  **P. ✕*jackii* Sarg. (P. *balsamifera* L. × P. *deltoides* Bartram ex
    Marshall)**   balm of Gilead (ON)
  P. tremula L.   European aspen (OC, E)
  P. *tremuloides* Michx.   quaking aspen, aspen
Salix L.
  S. *alba* L.   white willow (ON, E)
  S. *amygdaloides* Andersson   peached-leaved willow
  S. *arctophila* Cockerell
  S. *argyrocarpa* Andersson   silvery willow
  S. *babylonica* L.   weeping willow (ON, A, E)

S. bebbiana **Sarg.**   beaked willow, Bebb's willow

S. ×*blanda* **Andersson** (*S. babylonica* L. × *S. fragilis* L.)
Wisconsin weeping willow

S. candida **Willd.**   hoary willow, sage willow

S. caprea **L.**   goat willow (ON, E)

S. caroliniana **Michx.**   Carolina willow

S. cinerea **L.**   gray willow (ON, A, E)

S. cordata **Michx.**   sand dune willow, furry willow

S. discolor **Muhl.**   pussy willow

S. eriocephala **Michx.**   stiff willow, heartleaf willow, diamond
willow
(*S. rigida* **Muhl.**)

S. exigua **Nutt.**   sandbar willow
(*S. interior* **Rowlee**)

S. fragilis **L.**   crack willow, brittle willow (N, E)

S. herbacea **L.**   herb willow

S. humilis **Marshall**   prairie willow
(*S. tristis* **Aiton**)

S. lucida **Muhl.**   shining willow

S. matsudana **Koidz.**   corkscrew willow, dragon claw willow
(ON, A)

S. myricoides **Muhl.**   blueleaf willow

S. nigra **Marshall**   black willow

S. pedicellaris **Pursh**   bog willow

S. pellita **C. K. Schneid.**   satiny willow

S. pentandra **L.**   bay-leaved willow, laurel willow

S. petiolaris **Sm.**   slender willow, meadow willow

S. planifolia **Pursh**

S. purpurea **L.**   purple willow, basket willow (ON, E)

S. pyrifolia **Andersson**   balsam willow

S. sericea **Marshall**   silky willow

S. serissima (**Bailey**) **Fernald**   autumn willow

S. uva-ursi **Pursh**   bearberry willow

S. viminalis **L.**   silky osier, basket willow (ON, A, E)

Cyrillaceae

Cyrilla **Garden**
C. racemiflora L.   titi, leatherwood

Clethraceae

Clethra L.
    C. acuminata **Michx.**   pepper bush, mountain white-alder
    C. alnifolia **L.**   sweet pepperbush, coast white-alder

Empetraceae

Ceratiola Michx.
    C. ericoides **Michx.**   rosemary (s)
Corema D. Don
    C. conradii **(Torr.) Torr.**   broom-crowberry
Empetrum L.
    E. nigrum **L.**   black crowberry
    E. rubrum **L.**   red crowberry

Ericaceae

Andromeda L.
    A. glaucophylla **Link.**   bog-rosemary
    A. polifolia **L.**   bog-rosemary
Arctostaphylos L.
    A. alpina **(L.) Spreng.**   alpine bearberry
    A. uva-ursi **(L.) Spreng.**   bearberry
Calluna Salisb.
    C. vulgaris **(L.) Hull**   heather (N, E)
Chamaedaphne Moench
    C. calyculata **(L.) Moench**   leatherleaf, cassandra
Enkianthus Lour.
    E. campanulatus (Miq.) Nichols   redvein enkianthus (A)
    E. perulatus (Miq.) C. K. Schneid.   white enkianthus (A)
Epigaea L.
    E. asiatica Maxim.   Japanese ground laurel (A)
    E. repens **L.**   trailing arbutus
Erica L.
    E. carnea L.   spring heath, snow-heather (E)
    E. cinerea **L.**   twisted heath, bell-heather (ON, E)
    E. scoparia L.   broom heath (E)
    E. tetralix **L.**   cross-leaved heath, bog-heather (N, E)
    E. vagans **L.**   Cornish heath (ON, E)
Gaultheria L.
    G. hispidula **(L.) Muhl.**   creeping-snowberry, moxie

(***Chiogenes hispidula*** L.)
G. procumbens L.   wintergreen, checkerberry
Gaylussacia Kunth
    G. baccata (Wangenh.) K. Koch   black huckleberry
    G. brachycera (Michx.) A. Gray   box huckleberry
    G. dumosa (Andr.) Torr. & A. Gray   dwarf huckleberry
    G. frondosa (L.) Torr. & A. Gray   dangleberry
    G. ursina (M. A. Curtis) Torr. & A. Gray ex A. Gray (s)
Harrimanella Coville
    H. hypnoides (L.) Coville   moss-heather
    (***Cassiope hypnoides*** L.)
Kalmia L.
    K. angustifolia L.   sheep-laurel
    K. cuneata Michx. (s)
    K. latifolia L.   mountain-laurel
    K. polifolia Wangenh.   pale-laurel, swamp-laurel
Ledum L.
    L. groenlandicum Oeder   Labrador-tea
Leiophyllum Hedwig f.
    L. buxifolium (Bergius) Elliot   sand-myrtle (s)
Leucothoe D. Don
    L. axillaris (Lam.) D. Don   swamp dog-laurel (s)
    L. fontanesiana (Steud.) Sleumer   dog-hobble, mountain
        dog-laurel (s)
        (***L. editorum* Fernald**)
    L. populifolia (Lam.) Dippel (s)
    L. racemosa (L.) A. Gray   sweetbells, swamp-fetterbush
    L. recurva (Buckl.) A. Gray   redtwig leucothoe, mountain-
        fetterbush (s)
Loiseluria Desv.
    L. procumbens (L.) Desv.   alpine-azalea
Lyonia Nutt.
    L. ferruginea (Walter) Nutt. (s)
    L. ligustrina (L.) DC.   maleberry, he-huckleberry
    L. lucida (Lam.) K. Koch   staggerbush, fetterbush,
        hurrahbush (s)
    L. mariana (L.) D. Don   staggerbush
Menziesia Sm.
    M. pilosa (Michx.) Juss.   minniebush

Oxydendrum DC.

    O. arboreum (L.) DC.   sourwood, sorrel tree

Phyllodoce Salisb.

    P. caerulea (L.) Bab.   mountain-heath

Pieris D. Don

    P. floribunda (Pursh) Bentham & Hooker   mountain-fetterbush (s)

    P. japonica (Thunb.) D. Don ex G. Don   andromeda, lily-of-the-valley bush (A)

Rhododendron L.

    R. arborescens (Pursh) Torr.   smooth azalea

    R. atlanticum (Ashe) Rehder   coastal azalea

    R. bakeri (W. P. Lemmon & McKay) Hume   Cumberland azalea

        (R. cumberlandense E. L. Braun.)

    R. calendulaceum (Michx.) Torr.   flame azalea

    R. canadense (L.) Torr.   rhodora

    R. canescens (Michx.) Sweet   piedmont azalea

    R. catawbiense Michx.   mountain rosebay, red-laurel

    R. indicum (L.) Sweet   macranthum azalea (A)

    R. japonicum (A. Gray) Suringar   Japanese azalea (A)

    R. lapponicum (L.) Wahlenb.   Lapland rosebay, Lapland rhododendron

    R. maximum L.   rosebay, white-laurel

    R. minus Michx. (s)

        (R. carolinianum Rehder)

    R. mucronulatum Turcz. (A)

    R. periclymenoides (Michx.) Shinners   pinxterflower

        (R. nudiflorum [L.] Torr.)

    R. prinophyllum (Small) Millais   rosebud azalea

    R. serrulatum (Small) Millais   hammock sweet azalea (s)

    R. vaseyi A. Gray (s)

    R. viscosum (L.) Torr.   swamp azalea, white azalea

Vaccinium L.

    V. angustifolium Aiton   low sweet blueberry, dwarf blueberry

    V. arboreum Marshall   farkleberry, tree-huckleberry (s)

    V. boreale I. V. Hall & Halders   sweet hurts

    V. caespitosum Michx.   dwarf bilberry

    V. corymbosum L.   highbush blueberry

(*V. atrococcum* [A. Gray] A. Heller)
(*V. corymbosum* L. var. *glabrum* [A. Gray] Camp.)
(*V. constablaei* A. Gray)
(*V. fuscatum* Aiton)
V. crassifolium Andrews
V. elliottii Chapm. (s)
V. erythocarpum Michx.   southern mountain cranberry,
    dingleberry (s)
V. hirsutum Buckley (s)
V. macrocarpon Aiton   large cranberry
V. membranaceum Torr.   mountain bilberry
V. myrtilloides Michx.   Canada blueberry
V. ovalifolium Sm.   tall bilberry
V. oxycoccos L.   small cranberry
V. pallidum Aiton   late low blueberry
    (*V. vacillans* Kalm)
V. stamineum L.   deerberry
V. tenellum Aiton (s)
V. uliginosum L.   great bilberry, bog bilberry
V. vitis-idaea L.   lingonberry, mountain cranberry, foxberry,
    partridgeberry
Zenobia
    Z. pulverulenta (Bartram) Pollard (s)

Pyrolaceae

Chimaphila Pursh
    C. maculata (L.) Pursh   spotted wintergreen
    C. umbellata (L.) Barton   prince's-pine, pipsissewa

Diapensiaceae

Diapensia L.
    D. lapponica L.   diapensia
Pyxidanthera Michx.
    P. barbulata Michx.   pyxie, flowering-moss
    (*P. brevifolia* Wells)

Sapotaceae

Bumelia Swartz
    B. lanuginosa (Michx.) Pers. (s)

B. lycioides (L.) **Pers.** (s)
B. tenax (L.) **Willd.** (s)

Ebenaceae

Diospyros L.
   D. lotus L.   date-plum (OC, A)
   D. virginiana L.   persimmon

Styracaceae

Halesia Ellis
   H. diptera **Ellis** (s)
   H. monticola (Rehder) **Sarg.** (s)
   H. tetraptera **Ellis**   silverbell, wild-olive, shittimwood,
      oppossumwood
      (*H. carolina* L.)
      (*H. parviflora* Michx.)
Pterostyrax Siebold & Zucc.
   P. corymbosus Siebold & Zucc.   epaulette tree, asagara (OC, A)
   P. hispidus Siebold & Zucc.   epaulette tree (OC, A)
Styrax L.
   S. americanus **Lam.**   American snowbell
   S. grandifolius **Aiton**   bigleaf snowbell (s)
   S. japonicus Siebold & Zucc.   Japanese snowbell (A)
   S. obassia Siebold & Zucc.   fragrant snowbell (OC, A)

Symplocaceae

Symplocos Jacq.
   S. paniculata (Thunb.) Miq.   sapphire berry (A)
   S. tinctoria (L.) **L'Her.**   sweetleaf, horse-sugar (s)

Myrsinaceae

Ardisia Swartz
   A. crenata Sims.   coralberry, spiceberry (CS, A)

ROSIDAE

Pittosporaceae

Pittosporum L.
   P. tobira (Thunb.) Aiton   Japanese pittosporum, Australian-
      laurel (CS, A)

Hydrangeaceae

Decumaria L.
   D. barbara L.   climbing hydrangea (s)
Deutzia Thunb.
   D. gracilis Siebold & Zucc.   deutzia (A)
   D. scabra Thunb.   deutzia (ON, A)
Hydrangea L.
   H. anomala D. Don (A)
      (*H. petiolaris* Siebold & Zucc.)
   H. arborescens L.   American hydrangea, smooth hydrangea
      (**H. cinerea Small**)
      (**H. radiata Walter**)
   H. aspera D. Don (A)
   H. heteromala D. Don (A)
   H. macrophylla (Thunb.) Ser.   French hydrangea, hortensia (A)
   H. paniculata Siebold   panicle hydrangea, peegee hydrangea
      (ON, A)
   H. quercifolia Bartram   oakleaf hydrangea, sevenbark (s)
Philadelphus L.
   P. coronarius L.   European mock-orange (N, E)
   P. floridus Beadle (CS)
   P. hirsutus Nutt.   Cumberland mock-orange
   P. inodorus L.   Appalachian mock-orange
      (**P. grandiflorus Willd.**)
   P. ×*lemoinei* Hort. Lemoine
      (**P. coronarius L.** × **P. microphyllus A. Gray**)
   P. lewisii Pursh (W)
   P. microphyllus A. Gray (s)
   P. pubescens Loisel.   Ozark mock-orange (s)
   P. tomentosus Wallich. (ON, A)
   P. ×**virginalis Rehder**
Schizophragma Siebold & Zucc.
   S. hydrangeoides Siebold & Zucc. (OC, A)
   S. integrifolium (Franco) Oliv. (CS, A)

Grossulariaceae

Itea L.
   I. virginica L.   Virginia sweetspire

Ribes L.

R. americanum Mill.   Eastern black currant

R. cynosbati L.   dogberry

R. glandulosum Grauer   skunk currant

R. hirtellum Michx.

R. hudsonianum Richardson   western black currant

R. lacustre (Pers.) Poiret   bristly black currant, swamp currant

R. missouriense Nutt.   Missouri gooseberry
  (**R. gracile Pursh**)

R. nigrum L.   garden black currant (ON, A, E)

R. odoratum H. Wendl.   buffalo currant

R. oxyacanthoides L.   northern gooseberry

R. rotundifolium Michx.   Appalachian gooseberry

R. rubrum L.   European northern red currant, garden red
  currant (ON, E)
  (**R. sativum Syme**)

R. triste Pall.   swamp red currant

R. uva-crispa L.   garden gooseberry (ON, A, E)
  (**R. grossularia L.**)

Rosaceae

Amelanchier Medikus

A. alnifolia Nutt.   saskatoon, western serviceberry

A. arborea (F. Michx.) Fernald   downy serviceberry

A. bartramiana (Tausch) Roemer   mountain serviceberry

A. canadensis (L.) Medikus   shadbush
  (**A. intermedia Spach.**)

A. fernaldii Wiegand   St. Lawrence serviceberry

A. humilis Wiegand   bush serviceberry

A. ×interior Nielsen (**A. arborea [F. Michx.] Fernald** ×
  A. sanguinea [Pursh] DC.)

A. laevis Wiegand   smooth serviceberry

A. obovalis (Michx.) Ashe   coastal plain serviceberry

A. sanguinea (Pursh) DC.   New England serviceberry
  (**A. amabilis Wiegand**)

A. spicata (Lam.) K. Koch   dwarf serviceberry
  (**A. stolonifera Wiegand**)

Aronia Medikus

A. arbutifolia (L.) Elliott   red chokeberry

A. melanocarpa (Michx.) Elliott   black chokeberry

A. ✕prunifolia (Marshall) Rehder (*A. arbutifolia* [L.] **Elliott** ✕
    A. melanocarpa [Michx.] **Elliott**)
    (*A. arbutifolia* **Spach var.** *atropurpurea* **Britton**)
    (*A. floribunda* [Lindl.] **Spach**)

Chaenomeles Lindl.

C. japonica (Thunb.) Lindl. ex Spach   lesser flowering-quince
    (N, A)

C. speciosa (Sweet) Nakai   Japanese flowering-quince (N, A)
    (*C. legeneria* [Loisel.] **Koidz.**)

Cotoneaster Medikus

C. horizontalis Decne.   rock cotoneaster (A)

Crataegus L.

C. aestivalis (Walter) Torr. & A. Gray (s)

C. berberifolia Torr. & A. Gray   barberry hawthorn (s)

C. brainerdii Sarg.   Brainerd hawthorn

C. calpodendron (Ehrh.) Medikus   pear hawthorn

C. chrysocarpa Ashe   fireberry hawthorn
    (*C. dodgei* **Ashe**)
    (*C. margaretta* **Ashe**)

C. coccinea L.   scarlet hawthorn
    (*C. holmesiana* **Ashe**)

C. coccinioides Ashe   Kansas hawthorn (s)

C. collina Chapm. (s)

C. crus-galli L.   cockspur hawthorn, cockspur thorn

C. dilatata Sarg.   broadleaf hawthorn

C. douglasii Lindl.   black hawthorn

C. flabellata (Bosc) K. Koch   fanleaf hawthorn
    (*C. beata* **Sarg.**)
    (*C. filipes* **Ashe**)
    (*C. macrosperma* **Ashe**)

C. flava Aiton (s)

C. harbisonii Beadle (s)

C. intricata Lange   Biltmore hawthorn

C. marshallii Eggl.   parsley hawthorn (s)

C. michauxii Pers.   yellow hawthorn (s)

C. mollis (Torr. & A. Gray) Scheele   downy hawthorn
    (*C. submollis* **Sarg.**)

C. monogyna Jacq.   oneseed hawthorn (N, E)

(*C. oxycantha* **L.**)

C. phaenopyrum **(L.f.) Medik.**　　Washington thorn,
　Washington hawthorn

C. pruinosa **(J. D. Wendl.) K. Koch**　　frosted hawthorn

C. punctata **Jacq.**　　dotted hawthorn

C. spathulata **Michx.**　　littlehip hawthorn (s)

C. schuettei **Ashe** (s)

C. succulenta **Schrader**　　fleshy hawthorn
　(*C. macracantha* **Lodd.**)

C. triflora **Chapm.** (s)

C. uniflora **Muenchh.**　　oneflower hawthorn

C. viridis **L.**　　green hawthorn

Cydonia Mill.

　C. oblonga Mill.　　common quince (A)

Eriobotrya Lindl.

　E. japonica (Thunb.) Lindl.　　loquat (cs, A)

Exochorda Lindl.

　E. giraldii **Hesse**　　pearlbush (ON, A)

　E. racemosa **(Lindl.) Rehder**　　pearlbush (ON, A)

Kerria DC.

　K. japonica **DC.**　　kerria (ON, A)

Licania Aublet

　L. michauxii Prance　　gopher-apple (cs)
　(*Chrysobalanus oblongifolius* Michx.)

Malus Mill.

　M. angustifolia **(Aiton) Michx.**　　southern wild crabapple (s)

　M. baccata **(L.) Borkh.**　　Siberian crabapple (N, A)

　M. coronaria **(L.) Mill.**　　American crabapple, wild crab, sweet
　　crab
　(*M. glaucescens* **Rehder**)
　(*Pyrus coronaria* **L.**)

　M. floribunda **Siebold ex Van Houtte**　　Japanese flowering
　　crabapple (N, A)

　M. ioensis **(A. Wood) Britton**　　prairie crabapple

　M. prunifolia **(Willd.) Borkh.**　　Chinese crabapple, plumleaf
　　apple (N, A)

　M. pumila **Mill.**　　common apple (N, A, E)
　(*Pyrus malus* **L.**) (as used by Muenscher in earlier editions)

　M. sargentii Rehder (A)

M. sieboldii (Regel) Rehder   toringo crabapple (ON, A)

Mespilus L.

M. germanica L.   medlar (A, E)

Neillia D. Don

N. sinensis Oliv.   neillia (OC, A)

Neviusia A. Gray

N. alabamensis A. Gray   Alabama snow wreath (S)

Oemleria Reichb.

O. cerasiformis (Hooker & Arn.) Landon   Indian-plum,
    osoberry (CS, W)

    (*Osmaronia cerasiformis* Hooker & Arn.)

Photinia Lindl.

P. ×*fraseri* Dress (A)

    (*P. glabra* (Thunb.) Maxim. × *P. serratifolia* (Desf.) Kalkman)

P. glabra (Thunb.) Maxim. (A)

P. parvifolia (E. Pritzel) C. K. Schneid. (OC, A)

P. serratifolia (Desf.) Kalkman   Chinese photinia (A)

    (*P. serrulata* Lindl.)

P. villosa (Thunb.) DC.   Asiatic photinia (A)

Physocarpus Maxim.

P. opulifolius (L.) Maxim.   ninebark

Potentilla L.

P. fruticosa L.   shrubby cinquefoil, golden hardhack

P. tridentata Aiton   three-toothed cinquefoil

Prinsepia Royle

P. sinensis (Oliv.) Oliv. ex Bean (OC, A)

P. uniflora Batalin (OC, A)

Prunus L.

P. alleghaniensis Porter   Allegheny plum

P. americana Marshall   wild plum

P. angustifolia Marshall   chickasaw plum (S)

P. armeniaca L.   apricot (A)

P. avium L.   sweet cherry (N, A, E)

P. cerasifera Ehrh.   cherry plum (A, E)

P. cerasus L.   sour cherry (ON, E)

P. domestica L.   common plum (ON, E)

P. fruticosa Pall.   ground cherry (N, A, E)

P. glandulosa Thunb.   flowering almond (A)

P. hortulana L. H. Bailey   Hortulan plum (S)

P. laurocerasus L.   cherry-laurel, English-laurel (A, E)

P. mahaleb L.   Mahaleb cherry (ON, E)
(**P. caroliniana Aiton**)

P. maritima Marshall   beach plum

P. mexicana S. Watson

P. mume Siebold & Zucc.   Japanese flowering apricot (CS, A)

P. munsoniana Wight & Hedrick   wildgoose plum (S)

P. nigra Aiton   Canada plum

P. padus L.   European bird cherry (N, E)

P. pensylvanica L.f.   pin cherry, fire cherry

P. persica (L.) Batsch.   peach (N, A)

P. pumila L.   sand cherry

P. serotina Ehrh.   wild black cherry

P. serrulata Lindl.   Japanese flowering cherry (A)

P. spinosa L.   blackthorn, sloe (ON, A, E)

P. subhirtella Miq.   rosebud cherry, higan cherry (A)

P. tomentosa Thunb.   Nanking cherry (N, A)

P. triloba Lindl.   flowering almond (A)

P. umbellata Ellis   hog plum, flatwoods plum (S)

P. virginiana L.   chokecherry

P. yedoensis Matsum.   Yoshino cherry (OC, A)

Pyracantha M. Roem.

P. angustifolia (Franch.) C. K. Schneid.   pyracantha, firethorn
(OC, A)

P. atalantioides (Hance) Stapf   pyracantha, firethorn (OC, A)

P. coccinea M. Roem.   pyracantha, firethorn (ON, A)

P. crenulata (D. Don) M. Roem.   pyracantha, firethorn (OC)

P. koidzumii (Hayata) Rehder   pyracantha, firethorn (A)

Pyrus L.

P. calleryana Decne.   Bradford pear (A)

P. communis L.   pear (ON, A, E)

P. ×*lecontei* Rehder (**P. communis L.** × *P. pyrifolia* (Burm. f.)
Nakai)

P. pyrifolia (Burm. f.) Nakai   Chinese pear (A)

Raphiolepis Lindl.

R. indica (L.) Lindl.   Indian-hawthorn (CS, A)

R. umbellata (Thunb.) Makino   yedda-hawthorn, Indian-
hawthorn (CS, A)

Rhodotypos Siebold & Zucc.

R. scandens (Thunb.) Makino   jetbead (ON, A)
(**R. kerrioides Siebold & Zucc.**)
(**R. tetrapetala [Siebold] Makino**)

Rosa L.

R. acicularis Lindl.   bristly rose
R. arkansana Porter   dwarf prairie rose
R. banksiae W. T. Aiton   Banks rose (A)
R. blanda Aiton   smooth rose (N, A, E)
R. bracteata Wendl.   McCartney rose (N, CS, A)
R. canina L.   dog rose (N, E)
R. carolina L.   pasture rose
R. centifolia L.   cabbage rose (E)
R. cinnamomea L.   cinnamon rose (N, A)
R. damascena Mill.   damask rose (S)
R. eglanteria L.   eglantine, sweetbrier (N, E)
R. gallica L.   French rose (ON, E)
R. laevigata Michx.   Cherokee rose (S)
R. majalis Herrm.   cinnamon rose (N, A, E)
R. micrantha Sm.   sweetbrier, eglantine (N, E)
R. multiflora Thunb.   multiflora rose (N, A)
R. nitida Willd.   New England rose
R. palustris Marshall   swamp rose
R. rugosa Thunb.   Japanese rose (N, E)
R. setigera Michx.   climbing prairie rose
R. spinosissima L.   Scotch rose
R. tomentosa Sm. (ON, A, E)
R. villosa L. (E)
R. virginiana Mill.   Virginia rose
R. wichuraiana Crepin   memorial rose (A)
R. woodsii Lindl.   western rose

Rubus L.

R. acaulis Michx.
R. allegheniensis Porter   common blackberry
R. argutus Link   southern blackberry
R. betulifolius Small (S)
R. bifrons Tratt. (ON, E)
R. caesius L.   European dewberry (N, A, E)
R. canadensis L.   smooth blackberry
R. chamaemorus L.   cloudberry

R. cuneifolius **Pursh.**   sand blackberry

R. discolor **Weihe & Nees.**   Himalayan blackberry (ON, A)
   (*R. procereus* **P. J. Muell.**)

R. enslenii **Tratt.**   southern dewberry

R. flagellaris **Willd.**   northern dewberry

R. hispidus **L.**   swamp dewberry

R. idaeus **L.**   European red raspberry (N, E)

R. illecebrosus **Focke** (ON, A)

R. laciniatus **Willd.**   evergreen blackberry (N, E)

R. occidentalis **L.**   black raspberry

R. odoratus **L.**   flowering raspberry

R. parviflorus **Nutt.**   thimbleberry

R. pensilvanicus **Poiret.**   Pennsylvania blackberry

R. phoenicolasius **Maxim.**   wineberry (N, A)

R. pubescens **Raf.**   dwarf raspberry

R. recurvicaulis **Blanch.**   Blanchard's dewberry

R. setosus **Bigelow**   bristly blackberry

R. strigosus **Michx.**   American red raspberry

R. trivialis **Michx.**   coastal plain dewberry (S)

Sorbaria (Ser.) A. Braun

S. sorbifolia (L.) A. Braun   Ural false-spiraea (ON, A)

×*Sorbaronia* C. K. Schneid.

   S. ×*hybrida* (Moench.) C. K. Schneid. (***Aronia arbutifolia* (L.)**
      Elliot × *Sorbus aucuparia* **L.**)

Sorbus L.

   S. americana Marshall   American mountain-ash
      (***Pyrus americana* DC.**)

   S. aucuparia L.   European mountain-ash (N, A, E)
      (***Pyrus aucuparia* Ehrh.**)

   S. decora (Sarg.) C. K. Schneid.   showy mountain-ash
      (***Pyrus decora* [Sarg.] Hyland**)

Spiraea L.

   S. alba Du Roi   meadowsweet
      (**S. latifolia (Ait.) Borkh.**)

   S. *betulifolia* Pall.   birch-leaved spiraea (OC, A)

   S. ×*billiardii* Herincq (*S. douglasii* Hooker × **S. salicifolia** L.)

   S. ×*bumalda* Burv. (*S. albiflora* [Miq.] Zabel × **S. japonica** L.f.)

   S. *cantoniensis* Lour. (A)
      (*S. corymbosa* Raf.)

S. chamaedryfolia L. (ON, A, E)
S. japonica L.f.   Japanese spiraea (N, A)
S. nipponica Maxim. (OC, A)
S. prunifolia Siebold & Zucc.   bridal wreath (N, A)
S. salicifolia L. (ON, A, E)
S. septentrionalis (Fernald) A. & D. Love (OC, A)
S. thunbergii Siebold (ON, A)
S. tomentosa L.   hardhack, steeplebush
S. trilobata L. (A)
S. ×vanhouttei (C. Briot) Zabel (*S. cantoniensis* Lour. × *S. trilobata* L.)   bridal wreath (ON, A)
S. virginiana Britton (S)
Stephanandra Siebold & Zucc.
    S. incisa (Thunb.) Zabel   lace shrub, cutleaf stephanandra (A)
Stranvaesia Lindl.
    S. davidiana Decne.   stranvaesia (OC, A)

Mimosaceae

Albizia Durazz.
    A. julibrissin Durazz.   silk tree, mimosa (N, A)
Mimosa L.
    M. pudica L.   mimosa, sensitive brier (N, extreme south, rare)

Caesalpinaceae

Cercis L.
    C. canadensis L.   redbud
    C. chinensis Bunge   Chinese redbud (OC, A)
    C. siliquastrum L.   Judas tree, love tree (OC, A, E)
Chamaecrista Moench
    C. fasciculata (Michx.) Greene   partridge-pea, locust weed
    (*Cassia fasciculata* Michx.)
Gleditsia L.
    G. aquatica Marshall   water-locust, swamp-locust (S)
    G. triacanthos L.   honey-locust, sweet-locust
Gymnocladus Lam.
    G. dioica (L.) K. Koch   Kentucky coffeetree

Leguminosae (Fabaceae)

Amorpha L.
    A. canescens Pursh   lead plant

A. fruticosa L.   false-indigo, bastard-indigo
   (*A. nitens* **Boynton**)
A. georgiana **Wilbur** (s)
A. glabra **Desf. ex Poiret** (s)
A. herbacea **Walter** (s)
A. nana **Nutt.**   fragrant-indigo, smooth lead plant
A. schwerinii **C. K. Schneid.** (s)
Caragana Lam.
   C. arborescens Lam.   Siberian pea tree (A)
   C. frutex (L.) K. Koch   Russian peashrub (A)
Cladrastis Raf.
   C. kentuckea (**Dum. Cours.**) **Rudd**   yellowwood
      (*C. lutea* [**F. Michx.**] **K. Koch**)
Colutea L.
   C. arborescens **L.**   bladder-senna (ON, E, F)
Cytisus L.
   C. scoparius (**L.**) **Link**   Scotch-broom (N, E)
Genista L.
   G. tinctoria **L.**   Dyer's greenwood, woodwaxen, woadwaxen
      (N, E)
Indigofera L.
   I. kirilowii Maxim.   indigo (A)
Laburnum Medikus
   L. alpinum (Mill.) Bercht. & J. Presl   golden chain tree (E)
   L. anagyroides Medikus   golden chain tree (E)
   L. ×*watereri* (Kirchn.) Dippel (*L. alpinum* (Mill.) Bercht & J. Presl
      × *L. anagyroides* Medikus)   golden chain tree (OC)
Lespedeza Michx.
   L. bicolor **Turcz** (ON, CS, A)
   L. thunbergii (**DC.**) **Nakai** (ON, CS, A)
Maackia Rupr. & Maxim.
   M. amurensis Rupr. & Maxim. (OC, A)
Pueraria DC.
   P. lobata (**Willd.**) **Ohwi**   kudzu, kudzu vine (N, A)
Robinia L.
   R. boyntonii **Ashe** (s)
   R. elliottii (**Chapm.**) **Ashe ex Small** (s)
   R. hartwigii **Koehne** (s)
   R. hispida **L.**   bristly locust, rose-acacia
   R. kelseyi **Hutchinson** (s)

R. nana **Elliot** (s)
R. pseudoacacia **L.**   black locust
R. viscosa **Vent.**   clammy locust
Sesbania Scop.
   S. punicea (**Cav.**) **Bentham**   rattle-box, purple sesban (N, CS)
     (*Daubentonia punicea* [**Cav.**] **DC.**)
Sophora L.
   S. japonica L.   Japanese pagoda tree, Chinese scholar tree (A)
Ulex L.
   U. europaeus **L.**   gorse (N, E)
Wisteria Nutt.
   W. floribunda (**Willd.**) **DC.**   Japanese wisteria (N, A)
   W. frutescens (**L.**) **Poiret.**   Atlantic wisteria (s)
   W. macrostachya **Nutt.**   Mississippi wisteria (s)
   W. sinensis (**Sims**) **Sweet**   Chinese wisteria (N, A)

Elaeagnaceae

Elaeagnus L.
   E. angustifolius **L.**   Russian-olive (N, A, E)
     (*E. argentea* **Moench, not Pursh**)
   E. commutata **Bernh.**   silverberry
     (*E. argentea* **Pursh, not Moench**)
   E. multiflora **Thunb.**   cherry elaeagnus (ON, A)
   E. pungens **Thunb.**   thorny elaeagnus (ON, A)
   E. umbellata **Thunb.**   autumn elaeagnus (N, A)
Hippophae L.
   H. rhamnoides **L.**   sea-buckthorn (ON, A, E)
Shepherdia Nutt.
   S. argentea **Nutt.**   buffalo berry
   S. canadensis (**L.**) **Nutt.**   buffalo berry, soapberry

Lythraceae

Lagerstroemia L.
   L. fauriei Koehne   Japanese crape-myrtle (CS, A)
   L. indica L.   crape-myrtle (A)

Thymeleaceae

Daphne L.
Many Asian species and hybrids of *Daphne* are cultivated to some

extent in the southern United States. The species listed here are found throughout the range covered in this book.

D. ×*burkwoodii* Turrill (*D. caucasica* Pallas × *D. cneorum* L.) Burkwood daphne

D. cneorum L.   garland flower (E)

D. mezereum L.   mezereum, February daphne (N, A, E)

D. odora Thunb.   winter daphne (A)

Dirca L.

D. palustris L.   leatherwood, wicopy

Edgeworthia Meissn.

E. papyrifera Siebold & Zucc.   paperbush, mitsumata (OC, A)

Punicaceae

Punica L.

P. granatum L.   pomegranate (CS, E)

Cornaceae

Aucuba Thunb.

A. japonica Thunb.   Japanese aucuba, Japanese-laurel (A)

Cornus L.

C. alba L.   Tartarian dogwood, Siberian dogwood (A)

C. alternifolia L.f.   pagoda dogwood, alternate-leaved dogwood, green osier

C. amomum Mill.   silky cornel, kinnikinnik

C. asperifolia Michx. (S)

C. controversa Hemsl.   giant dogwood (OC, A)

C. coreana Wangerin (OC, A)

C. drummondii C. A. Mey.   roughleaf dogwood (**C. asperifolia of authors, not Michx.**)

C. florida L.   flowering dogwood

C. foemina Mill.   gray dogwood (**C. racemosa Lam.**) (**C. paniculata L'Her.**) (**C. stricta Lam.**)

C. kousa Hance   Japanese dogwood, kousa (A)

C. macrophylla Wallich. (CS, A)

C. mas L.   cornelian-cherry (A, E)

C. officinalis Siebold & Zucc.   Japanese cornelian-cherry, Japanese cornel (A)

C. rugosa **Lam.**   round-leaf dogwood
C. sanguinea **L.**   bloodtwig dogwood, pegwood (ON, E)
C. stolonifera **Michx.**   red-osier dogwood, American dogwood
  (*C. sericea* **L.**)
Davidia Baill.
  D. involucrata Baill.   dovetree, handkerchief tree (OC, A)
Nyssa L.
  N. aquatica **L.**   water tupelo
  N. sylvatica **Marshall**   blackgum, black tupelo, tupelo

Santalaceae

Buckleya Torr.
  B. distichophylla **Torr.** (S)
Nestronia Raf.
  N. umbellulata **Raf.** (S)
Pyrularia Michx.
  P. pubera **Michx.**   buffalo nut, oil nut (S)

Viscaceae

Arceuthobium M. Bieb.
  A. pusillum **Peck**   Eastern dwarf-mistletoe
Phoradendron Nutt.
  P. leucarpum **(Raf.) Reveal & M. C. Johnston**   American
    mistletoe, Christmas mistletoe
    (*P. flavescens* **Nutt. ex Engelm.**)
    (*P. serotinum* **[Raf.] M. C. Johnston**)

Celastraceae

Celastrus L.
  C. orbiculatus **Thunb.**   Oriental bittersweet (N, A)
  C. scandens **L.**   American bittersweet
Euonymus L.
  E. alata **(Thunb.) Siebold**   winged spindle tree, winged
    burning bush (N, A)
  E. americana **L.**   strawberry bush, American burning bush
  E. atropurpurea **Jacq.**   wahoo, burning bush
  E. europaea **L.**   European spindle tree (N, E)
  E. fortunei **(Turcz.) Hand.-Mazz.**   spindle tree (ON, A)
  E. japonica Thunb.   Japanese spindle tree (A)

E. kiautschovica Loes.   spreading euonymus (cs, A)

E. obovata Nutt.   running strawberry bush

Paxistima Raf.

P. canbyi A. Gray   mountain-lover, cliff-green
(***Pachistima canbyi* A. Gray**)

Aquifoliaceae

Ilex L.

I. ×*altaclarensis* (Dallim.) Rehder (*I. aquifolium* L. × *I. perado*
Aiton)   highclere holly (OC)

I. aquifolium L.   English holly (A, E, F)

I. ×*aquipernyi* Gable ex W. B. Clarke (*I. aquifolium* L. × *I. pernyi*
Franch.) (OC)

I. ×*attenuata* Ashe (**I. cassine L.** × **I. opaca Sol. ex Aiton**)
topec holly (OC)

I. cassine L.   dahoon, cassine (s)

I. chinensis Sims   kashi holly (A)

I. ciliospinosa Loes. (A)
(*I. purpurea* Hassk.)

I. coriacea (Pursh) Chapm.   gallberry (s, ON NORTH)

I. cornuta Lindl. & Paxton (A)

I. crenata Thunb.   Japanese holly, box-leaved holly (ON, A)

I. decidua Walter   deciduous holly, possum-haw
(**I. longipes Chapm.**)

I. glabra (L.) A. Gray   inkberry

I. laevigata (Pursh) A. Gray   smooth winterberry

I. ×*meserveae* S. Y. Hu. (*I. rugosa* Schmidt × *I. aquifolium* L.)

I. montana (Torr. & A. Gray) A. Gray   mountain winterberry,
bigleaf holly
(**I. ambigua [Michx.] Torrey**)
(**I. beadlei [Ashe] Fern.**)
(**I. mollis A. Gray**)
(**I. monticola A. Gray**)

I. opaca ex Aiton   American holly

I. pedunculosa Miq. (A)

I. pernyi Franch. (A)

I. rotunda Thunb.   kurogane holly (CS, A)

I. rugosa Schmidt (OC, A)

I. serrata Thunb.   Japanese holly (OC, A)

I. verticillata (L.) **Gray**   winterberry, black-alder
I. vomitoria **Aiton**   yaupon (s)
Nemopanthus Raf.
    N. collinus (Alexander) **R. C. Clark**   Appalachian mountain-
        holly
    N. mucronata (L.) **Trelease**   mountain-holly

Buxaceae

Buxus L.
    B. microphylla Siebold & Zucc.   Asian boxwood (A)
    B. sempervirens **L.**   common boxwood (ON, A, E, F)
Sarcococca Lindl.
    S. hookerana Baill.   sarcococca (A)
    S. ruscifolia Stapf   fragrant sarcococca, sweet-box (A)
    S. saligna (D. Don) Muell. Arg.   willow-leaf sarcococca (A)

Euphorbiaceae

Sapium (L.) Roxb.
    S. sebiferum (L.) **Roxb.**   Chinese tallow tree (N, CS, A)
Sebastiania Spreng.
    S. ligustrina (Michx.) **Muell. Arg.** (s)
Securinega Comm. ex Juss.
    S. suffruticosa (Pall.) Rehder (CS, A)

Rhamnaceae

Berchemia Neck.
    B. scandens (Hill) **K. Koch**   supplejack (s)
Ceanothus L.
    C. americanus **L.**   New Jersey–tea, redroot
    C. herbaceus **Raf.**   prairie-redroot
        (*C. ovatus* **Desf.**)
    C. sanguineus **Pursh**   wild-lilac
Hovenia Thunb.
    H. dulcis **Thunb.**   Japanese raisin tree (ON, A)
Paliuris Mill.
    P. spina-christi Mill.   Christ thorn (A, E)
Rhamnus L.
    R. alnifolia **L'Her.**   American alder-leaved buckthorn
    R. caroliniana **Walter**   Carolina buckthorn, Indian-cherry (s)

R. cathartica **L.**   common buckthorn (N, A, E)
R. frangula **L.**   glossy buckthorn (N, A, E)
R. lanceolata **Pursh**   lance-leaved buckthorn
Ziziphus Mill.
  Z. jujuba **Mill.**   common jujube, Chinese-date (OC, A, E)

Vitaceae

Ampelopsis Michx.
  A. arborea **(L.) Koehne**   pepper vine (S)
  A. brevipedunculata (Maxim.) Trautv.   porcelain berry (A)
  A. cordata **Michx.**   raccoon-grape
Parthenocissus Planch.
  P. quinquefolia **(L.) Planch.**   Virginia creeper
  P. tricuspidata **(Siebold & Zucc.) Planch.**   Boston-ivy (N, A)
  P. vitacea **(Knerr) Hitchc.**   grape-woodbine
Vitis L.
  V. aestivalis **Michx. var.** *aestivalis*   summer grape, pigeon
      grape, bunch grape
  V. aestivalis **Michx. var.** *argentifolia* **(Munson) Fernald**
  V. baileyana **Munson**   possum grape
  V. cinerea **Engelm.**   graybark grape, pigeon grape
  V. labrusca **L.**   fox grape, skunk grape
  V. ×novae-angeliae **Fernald** (*V. labrusca* **L.** × *V. riparia* **Michx.**)
      New England grape
  V. palmata **M. Vahl**   red grape, cat grape, catbird grape (S)
  V. riparia **Michx.**   river-bank grape, frost grape
  V. rotundifolia **Michx.**   muscadine grape, scuppernong
      grape
  V. rupestris **Scheele**   sand grape, sugar grape
  V. vinifera **L.**   wine grape, European grape (ON, A, E)
  V. vulpina **L.**   winter grape, frost grape, chicken grape

Staphyleaceae

Staphylea L.
  S. bumalda DC.   Japanese bladdernut (OC, A)
  S. colchica Steven   colchis bladdernut (OC, A, E)
  S. pinnata L.   European bladdernut (E)
  S. trifolia **L.**   American bladdernut

Sapindaceae

Koelreuteria

   K. paniculata **Laxm.**   golden-rain tree (ON, A)

Hippocastinaceae

Aesculus L.

   A. ×*carnea* Hayne (**A. hippocastanum L.** × **A. pavia L.**)   red horse-chestnut

   A. flava **Aiton**   yellow buckeye, sweet buckeye (**A. octandra Marshall**)

   A. glabra **Willd.**   Ohio buckeye

   A. hippocastanum **L.**   horse-chestnut (N, A, E)

   A. parviflora **Walter**   bottlebrush buckeye, dwarf horse-chestnut (S)

   A. pavia **L.**   red buckeye (S)

   A. sylvatica **Bartram**   painted buckeye, Georgia buckeye (S)

Aceraceae

Acer L.

   A. buergeranum Miq.   trident maple (OC, A) (*A. trifidum* Hooker & Arn.)

   A. campestre **L.**   hedge maple, English maple (ON, A, E)

   A. cappadocicum Gled. (OC, A)

   A. carpinifolium Siebold & Zucc.   hornbeam maple (OC, A)

   A. ×*freemanii* E. Murr. (**A. rubrum L.** × **A. saccharinum L.**) Freeman maple

   A. ginnala **Maxim.**   amur maple, Siberian maple (ON, A)

   A. griseum (Franch.) Pax   paperbark maple (A)

   A. japonicum Thunb.   Japanese maple (A)

   A. maximowiczianum Miq.   nikko maple (OC, A) (*A. nikoense* Maxim.)

   A. negundo **L.**   boxelder

   A. nigrum **F. Michx.**   black maple, black sugar maple (**A. saccharum Marshall var. nigrum [F. Michx.] Britton**)

   A. palmatum **Thunb.**   Japanese maple (ON, A)

   A. pensylvanicum **L.**   striped maple, moosewood

   A. platanoides **L.**   Norway maple (N, A, E)

   A. pseudoplatanus **L.**   sycamore maple (ON, A, E)

***A. rubrum* L.**   red maple, soft maple

A. rufinerve Siebold & Zucc. (OC, A)

***A. saccharinum* L.**   silver maple

***A. saccharum* Marshall**   sugar maple, hard maple, rock maple

(***A. barbatum* Michx.**)

***A. spicatum* Lam.**   mountain maple

A. tataricum L.   tatarian maple (OC, A, E)

A. truncatum Bunge   shantung maple (OC, A)

Anacardiaceae

Cotinus Mill.

***C. coggygria* Scop.**   smoketree (ON, A, E)

***C. obovatus* Raf.**   American smoketree, chittamwood (S)

Pistacia L.

P. chinensis Bunge   pistachio (CS, A)

Rhus L.

***R. aromatica* Aiton**   fragrant sumac, aromatic sumac

***R. copallina* L.**   dwarf sumac, shining sumac

***R. glabra* L.**   smooth sumac

***R. michauxii* Sarg.** (S)

***R. typhina* L.**   staghorn sumac

Toxicodendron L.

***T. pubescens* Mill.**   poison-oak

***T. radicans* (L.) Kuntze**   poison-ivy

(***Rhus radicans* L.**)

***T. rydbergii* (Small) Greene**   western poison-ivy

***T. vernix* (L.) Kuntze**   poison-sumac, swamp sumac

Simaroubaceae

Ailanthus Desf.

***A. altissima* (Mill.) Swingle**   ailanthus, tree-of-heaven (N, A)

Meliaceae

Melia L.

***M. azedarach* L.**   Chinaberry (N, CS, A)

Toona (Endl.) M. Roem.

T. sinensis (Juss.) M. Roem.   cedrela (A)

(*Cedrela sinensis* Juss.)

Rutaceae

Phellodendron Rupr.
   P. amurense Rupr.   amur cork tree (A)
   P. chinense C. K. Schneid.   Chinese cork tree (OC, A)
   P. japonicum Maxim. (ON, A)
   P. sachalinense (Friedr. Schmidt) Sarg.   Sakhalin cork tree
       (OC, A)
Poncirus Raf.
   P. trifoliata (L.) Raf.   trifoliate-orange (A)
       (*Citrus trifoliata* Raf.)
Ptelea L.
   P. trifoliata L.   hop tree, wafer-ash
Ruta L.
   R. graveolens L.   common rue (N, E, F)
Skimmia Thunb.
   S. japonica Thunb.   skimmia (A)
Tetradium Lour.
   T. danielii (Benn.) T. G. Hartley   evodia (OC, A)
       (*Evodia danielii* [Benn.] Hemsl.)
       (*E. henryi* Dode)
       (*E. hupehensis* Dode)
Zanthoxylum L.
   Z. americanum Mill.   prickly-ash
   Z. clava-herculis L.   Hercules'-club

Araliaceae

Aralia L.
   A. elata (Miq.) Seem.   Japanese-angelica (CS, A)
   A. hispida Vent.   bristly sarsaparilla
   A. spinosa L.   Hercules'-club, devil's walking stick
Eleutherococcus Maxim.
   E. sieboldianus (Makino) Maxim.   spiny-panax, five-leaf-aralia
       (OC, A)
       (*Acanthopanax sieboldianus* Makino)
       (*Aralia pentaphylla* Thunb.)
×*Fatshedera* Guillaumin (*Fatsia* Decne. & Planch. × *Hedera* L.)
   F. lizei (Cochet) Guillaumin   aralia-ivy, tree-ivy (CS)
Fatsia Decne. & Planch.

F. japonica (Thunb.) Decne. & Planch.   Japanese fatsia,
      Formosa rice tree, paper plant (cs, A)
Hedera L.
  H. helix L.   English ivy (ON, E)
Kalopanax Miq.
  K. pictus (Thunb.) Nakai.   sennoki, kalopanax, castor-aralia
      (OC, A)
Oplopanax (Tor. & A. Gray) Miq.
  O. horridus (Sm.) Miq.   devil's club
    (***Echinopanax horridus* Hort.**)

ASTERIDAE

Loganiaceae

Gelsemium Juss.
  G. rankinii Sm. (s)
  G. sempervirens Juss.   yellow jessamine

Apocynaceae

Nerium L.
  N. oleander L.   common oleander (cs, A, E)
Trachelospermum Lemaire
  T. asiaticum (Siebold & Zucc.) Nakai (A)   star-jasmine
    (*T. divaricatum* [Thunb.] Kanitz)
    (*Rhynchospermum asiaticum* Hort.)
    (*R. divaricatum* Hort.)
  T. difforme (Walter) A. Gray (s)
Vinca L.
  V. major L.   greater periwinkle (N, E)
  V. minor L.   periwinkle (N, E)

Asclepiadaceae

Periploca L.
  P. gracea L.   silk vine (ON, cs, E)

Solanaceae

Lycium L.
  L. barbarum L.   matrimony vine (ON, A, E)
    (**L. halimifolium Mill.**)
Solanum L.

S. dulcamara L.   European-bittersweet, bitter-nightshade
(N, E)

Boraginaceae
Ehretia R. Br.
  E. acuminata R. Br. (OC, A)

Verbenaceae
Callicarpa L.
  C. americana L.   American beautyberry
  C. dichotoma (Lour.) K. Koch   Chinese beautyberry (ON, A)
Clerodendrum L.
  C. indicum (L.) Kuntze   tubeflower, Turk's turban (CS, A)
  C. japonicum (Thunb.) Sweet   glory-bower (ON, A)
    (*C. fragrans* Vent)
  C. trichotomum Thunb.   glory-bower (A)
Vitex L.
  V. agnus-castus L.   chaste tree (ON, A, E)
  V. negundo L.   negundo chaste tree (ON, A)

Labiatae (Lamiaceae)
Lavandula L.
  L. angustifolia Mill.   English lavender (ON, E)
Rosmarinus L.
  R. officinalis L.   rosemary (ON, E)

Buddleiaceae
Buddleia L.
  B. davidii Franch.   butterfly bush (N, A)
  B. lindleyana Fortune (S)

Oleaceae
Abeliophyllum Nakai
  A. distichum Nakai (OC, A)   abelialeaf, abeliophyllum
Chionanthus L.
  C. retusus Lindl. & Paxton (ON, A)
  C. virginicus L.   fringetree, old man's beard
Fontanesia Labill.
  F. fortunei Carriere   Fortune fontanesia (CS, A)

F. phillyreoides Labill. (CS, A)
Forestiera Poiret
   F. acuminata (Michx.) Poiret   swamp-privet (S)
   F. ligustrina (Michx.) Poiret   swamp-privet (S)
Forsythia Vahl
   F. ×*intermedia* Zabel. (**F. suspensa Vahl** × *F. viridissima* **Lindl.**)
     border forsythia (A)
   F. suspensa Vahl   weeping forsythia (ON, A)
   F. viridissima viridissima Lindl.   golden bells, greenstem
     forsythia (ON, A)
Fraxinus L.
   F. americana L.   white ash
   F. caroliniana Mill.   water ash (S)
   F. excelsior L.   European ash (E)
   F. excelsior L. 'Diversifolia'   simple-leaved European ash
   F. excelsior L. 'Hessei'   Hesse simple-leaved European ash
   F. nigra Marshall   black ash, hoop ash
   F. ornus L.   flowering ash (OC, A, E)
   F. pennsylvanica Marshall var. lanceolata (Borkh.) Sarg.
     green ash
   F. pennsylvanica Marshall var. pennsylvanica   red ash
   F. profunda (Bush) Bush   pumpkin ash
     (**F. tomentosa F. Michx.**)
   F. quadrangulata Michx.   blue ash
Jasminum L.
   J. humile L.   jasmine (CS, A)
   J. nudiflorum Lindl.   jasmine (A)
   J. officinale L.   poet's jasmine (CS, A)
Ligustrum L.
   L. amurense Carriere (A)
   L. japonicum Thunb.   waxleaf privet, Japanese privet (A)
   L. lucidum Aiton   Chinese privet (A)
   L. obtusifolium Siebold & Zucc. var. obtusifolium   (ON, A)
   L. obtusifolium Siebold & Zucc. var. *regelianum* (Koehne) Rehder
     Regel's privet (A)
   L. ovalifolium Hassk.   California privet (ON, A)
   L. sinense Carriere (A)
   L. vulgare L.   common privet (N, A, E)
Osmanthus Lour.

O. americanus (L.) A. Gray (s)

O. ×fortunei Carriere (*O. fragrans* (Thunb.) Lour. × *O. heterophyllus* [G. Don] P. S. Green) (cs)

O. fragrans (Thunb.) Lour.   fragrant-olive, sweet-olive (A)

O. heterophyllus (G. Don) P. S. Green   holly-olive, Chinese-holly, false-holly (A)

Phillyrea L.

P. decora Boisson & Balansa   phillyrea (oc, A)

P. latifolia L.   tree phillyrea (oc, A, E)

Syringa L.

S. ×chinensis Willd. (*S. persica* L. × **S. vulgaris L.**)   Chinese lilac

S. laciniata Mill.   cutleaf lilac (oc, A)

S. meyeri C. K. Schneid. (A)

S. microphylla Diels (oc, A)

S. oblata Lindl. (oc, A)

S. pekinensis Rupr. (on, A)

S. ×persica L. (*S. afghanica* C.K. Scneid. × *S. laciniata* Mill.) Persian lilac (A)

S. pinnatifolia Hemsl. (A)

S. pubescens Turcz. (oc, A)

S. reticulata (Blume) Hara   Japanese tree lilac (A)
(*S. amurensis* Rupr. var. *japonica* [Maxim.] Franch. & Sav.)
(*S. japonica* Maxim.)

S. villosa Vahl   late lilac (A)

S. vulgaris L.   common lilac (on, E)

Bignoniaceae

Bignonia L.

B. capreolata L.   trumpet flower, cross-vine
(***Anisostichus capreolata* [L.] Bureau**)

Campsis Lour.

C. grandiflora (Thunb.) K. Schum.   Chinese trumpet creeper (A)

C. radicans (L.) Seem.   trumpet creeper, trumpet vine

C. ×tagliabuana (Vis.) Rehder *(C. grandiflora* [Thunb.] K. Schum. × **C. radicans** [L.] **Seem.**)   trumpet creeper

Catalpa Scop.

C. bignonioides Walter   northern catalpa, Indian-bean

C. ×*hybrida* Hort. ex Spath (*C. bignonioides* Walter × *C. ovata*
   G. Don)   hybrid catalpa
C. ovata **G. Don.**   Chinese catalpa (ON, A)
C. speciosa **Warder ex Engelm.**   western catalpa
Paulownia Siebold & Zucc.
   P. tomentosa **(Thunb.) Steud.**   empress tree, princess tree
   (N, A)

Rubiaceae

Cephalanthus L.
   C. occidentalis **L.**   buttonbush
Gardenia Ellis
   G. jasminoides Ellis   common gardenia, cape-jasmine (CS, A)
   (*G. radicans* Thunb.)
Mitchella L.
   M. repens **L.**   partridgeberry
Pinckneya Michx.
   P. pubens Michx.   fever tree, pinckneya (CS)
Serissa Comm. ex Juss.
   S. japonica Lam.   yellow rim (CS, A)
   (*S. foetida* [L.f.] Lam.)

Caprifoliaceae

Abelia R. Br.
   A. ×*grandiflora* (Andre) Rehder (*A. chinensis* R. Br. × *A. uniflora*
      R. Br. ex Wallich)   glossy abelia (A)
Diervilla Mill.
   D. lonicera **Mill.**   bush-honeysuckle
   D. sessilifolia **Buckley**   bush-honeysuckle
Heptacodium Rehder
   H. miconioides Rehder   seven son flower (OC, A)
Kolkwitzia Graebn.
   K. amabilis Graebn.   beautybush (A)
Leycesteria Wallich
   L. formosa Wallich (CS, A)
Linnaea L.
   L. borealis **L.**   twinflower
Lonicera L.

L. ×*bella* Zabel. (*L. tatarica* L. × *L. morrowii* A. Gray)   belle honeysuckle (N)

L. canadensis Marshall   American fly honeysuckle

L. caprifolium L.   sweet honeysuckle (ON, E)

L. dioica L.   climbing honeysuckle

L. flava Sims   yellow honeysuckle (S)

L. fragrantissima Lindl. & Paxton   fragrant honeysuckle (ON, A)

L. hirsuta Eaton   hairy honeysuckle

L. involucrata Spreng.   bearberry honeysuckle

L. japonica Thunb.   Japanese honeysuckle (N, A)

L. maackii (Rupr.) Herder   amur honeysuckle (N, A)

L. morrowii A. Gray   Morrow honeysuckle (ON, E)

L. nitida E. H. Wilson   Kokonor honeysuckle (CS, A)

L. oblongifolia (Goldie) Hooker   swamp fly honeysuckle

L. periclymenum L.   woodbine (N, A, E, F)

L. prolifera (Kirchn.) Rehder   grape honeysuckle

L. ruprechtiana Regel (OC, A)

L. sempervirens L.   trumpet honeysuckle

L. standishii Jacques   standish honeysuckle (A)

L. tatarica L.   Tatarian honeysuckle (N, A, E)

L. villosa (Michx.) Schultes   mountain fly honeysuckle (N, E)
   (*L. caerulea* L. var. *villosa* Michx.)

L. xylosteum L.   European fly honeysuckle (N, E)

Sambucus L.

S. canadensis L.   black elderberry, common elder

S. nigra L.   European elder (ON, E)

S. racemosa L.   red elderberry
   (*S. pubens* Michx.)

Symphoricarpos Duhamel

S. albus (L.) S. F. Blake   snowberry, waxberry

S. occidentalis Hooker   wolfberry

S. orbiculatus Moench   coralberry

Viburnum L.

V. acerifolium L.   maple-leaved viburnum, dockmackie, flowering-maple

V. alnifolium Marshall   hobblebush
   (*V. lanatanoides* Michx.)

V. ×*burkwoodii* Hort. (*V. carlesii* Hemsl. × *V. utile* Hemsl.)
   Burkwood viburnum
V. ×*carlcephalum* Hort. (*V. carlesii* Hemsl. × *V. macrocephalum*
   Fortune)
V. carlesii Hemsl.   Korean spice viburnum (A)
V. cassinoides L.   withe-rod, wild-raisin
   (**V. nudum L. var. cassinoides [L.] Torr. and A. Gray**)
V. davidii Franch. (CS, A)
V. dentatum L.   arrowwood
   (**V. scabrellum [Torr. & A. Gray] Chapm.**)
   (**V. pubescens [Aiton] Pursh**)
V. dilatatum Thunb.   Linden viburnum (N, A)
V. edule (Michx.) Raf.   squashberry
   (**V. pauciflorum Raf.**)
V. farreri Stearn   fragrant viburnum (A)
   (*V. fragrans* Bunge)
V. lantana L.   wayfaring tree (ON, A)
V. lentago L.   nannyberry, sheepberry
V. macrocephalum Fortune   Chinese snowball (OC, A)
V. molle Michx.   soft-leaved arrowwood, poison-haw,
   black-alder
V. nudum L.   possum-haw, swamp-haw, withe-rod,
   Appalachian-tea, teaberry
V. odoratissimum Ker Gawl. var. *awabuki* K. Koch   awabuke
   viburnum (CS, A)
V. odoratissimum Ker Gawl. var. *odoratissimum*   sweet viburnum
   (CS, A)
V. opulus L. var. americanum (Mill.) Aiton   highbush-
   cranberry
   (**V. trilobum Marshall**)
V. opulus L. var. opulus   European cranberry bush, Guelder-
   rose, snowball (N, E)
V. plicatum Thunb. var. plicatum K. Koch (ON, A)
V. plicatum Thunb. var. *tomentosum* (Thunb.) Rehder   Japanese
   snowball, snowball viburnum, doublefile viburnum (A)
   (*V. tomentosum* Thunb.)
V. prunifolium L.   black-haw
V. rafinesquianum Schultes   downy arrowwood
   (**V. affine Bush ex C. K. Schneid.**)

(*V. pubescens* of authors, not Pursh)

V. recognitum Fernald

   (*V. dentatum* L. var. *lucidum* Aiton)

V. rhytidophyllum Hemsl.   leatherleaf viburnum (A)

V. rufidulum Raf.   southern black-haw, alligator bar (s)

V. setigerum Hance   tea viburnum (ON, A)

V. sieboldii Miq.   Japanese viburnum (ON, A)

V. suspensum Lindl. (CS, A)

V. tinus L.   laurustinus viburnum (A)

Weigela Thunb.

   W. coraeensis Thunb. (OC, A)

   W. floribunda (Siebold & Zucc.) K. Koch (A)

   W. florida (Bunge) A. DC. (A)

      (*W. rosea* Lindl.)

   W. hortensis (Siebold & Zucc.) K. Koch (OC, A)

   W. japonica Thunb. (OC, A)

   W. praecox (Hort. Lemoine) L. H. Bailey (A)

Compositae (Asteraceae)

Artemisia L.

   A. abrotanum L.   wormwood, southernwood, old man (N, E)

Baccharis L.

   B. angustifolia Michx. (s)

   B. glomerulifolia Pers. (s)

   B. halimifolia L.   groundsel tree

Borrichia Adans.

   B. frutescens (L.) DC.   sea ox-eye (s)

Iva L.

   I. frutescens L.   marsh-elder, highwater shrub

   I. imbricata Walter (s)

Solidago L.

   S. pauciflosculosa Michx. (s)   woody goldenrod

LILIOPSIDA (MONOCOTYLEDONS)

ARECIDAE

Arecaceae (Palmae)

Sabal Adans.

   S. minor (Jacq.) Pers.   palmetto (s)

S. palmetto Lodd. ex Schultes   cabbage palmetto (s)
Serenoa Hook f.
  S. repens (Bartram) Small   saw-palmetto (s)

COMMELINIDAE

Poaceae (Gramineae)

Arundinaria Michx.
  A. gigantea (Walter) Muhl.   giant cane
Cortaderia Stapf
  C. selloana (Schult. & Schult.f.) Asch. & Graebn.   pampas grass
    (cs)
Miscanthus Andersson
  M. sinensis Andersson   eulalia (N, A)
Phragmites Trin.
  P. australis (Cav.) Trin. ex Steud.   common reed
    (**P. communis Trin.**)
Phyllostachys Siebold & Zucc.
  P. aurea Riv.   fishpole bamboo, golden bamboo (cs, A)
  P. aureosulcata McClure   stake bamboo, yellow-groove bamboo
    (cs, A)
  P. bambusoides Siebold & Zucc.   timber bamboo (cs, A)
  P. nigra (Lodd. ex Lindl.) Munro   black bamboo (cs, A)
Sasa Makino & Shibata
  S. palmata E. G. Camus   sasa-bamboo (N, A)

LILIIDAE

Liliaceae

Ruscus L.
  R. aculeatus L. (A, E)   butcher's broom

Agavaceae

Yucca L.
  Y. aloifolia L.   Spanish bayonet, dagger plant (w)
  Y. glauca Nutt. ex J. Fraser   soapweed, soapwell (ON, W)
  Y. filamentosa L.   Adam's needle, needle-palm (s)
    (**Y. smalliana Fernald**)
  Y. flaccida Haw. (s)
  Y. gloriosa L.   Spanish dagger, palm-lily, Roman candle, Lord's
    candlestick (s)

Smilacaceae

Smilax L.

S. *bona-nox* L.   halberd-leaved smilax
S. *glauca* Walter   saw brier, greenbrier, wild-sarsaparilla
S. *laurifolia* L.   laurel greenbrier, laurel-leaf catbrier
S. *rotundifolia* L.   common greenbrier, horsebrier
S. *smallii* Morong   Jackson brier, Jackson vine
  (**S. *lanceolata* of authors, not L.**)
S. *tamnoides* L.   bristly greenbrier
  (**S. *hispida* Muhl.**)
S. *walteri* Pursh   coral greenbrier

GLOSSARY

Accessory buds Additional buds when more than one occurs in or near the axil: of two kinds, collateral or superposed.

Achene Small, dry, hard, one-seeded indehiscent fruit.

Acicular Needle-shaped.

Acuminate Gradually tapering to a long point.

Acute Sharp-pointed.

Adnate Formed of the congenital union of two different organs or structures.

Aggregate fruit One formed by the coherence of several pistils that were distinct in the flower.

Alternate One (leaf or bud) at a node; placed singly at different heights on the stem.

Anastomosing Rejoining after branching, forming a network.

Angiosperms Plants with seeds borne in an ovary.

Anterior The site away from the axis or stem.

Apiculate Ending abruptly in a sharp, fine point.

Appressed Lying close and flat against.

Arborescent Tree-like in size or form.

Ascending Rising obliquely upwards.

Auriculate Having ear-like appendages.

Axil Upper angle formed where the leaf joins the stem.

Axillary buds Buds in or from an axil.

Berry Fleshy fruit, soft throughout.

Biennial Of two-year duration.

Blade The expanded part of a leaf; the leaf excluding the petiole

Bract A reduced or modified leaf, usually below a flower.

Bracteole A small bract.

Branchlet An ultimate division of a branch, not including the last season's growth.

Bundle scars Scars formed in a leaf scar by the breaking of the vascular bundles of the petiole.

Capsule Dry, dehiscent fruit of a compound pistil.

Carinate Having a keel or projecting a longitudinal medial line on the lower or outer surface.

Chambered pith Pith in transverse plates with air cavities between them.

Ciliate Fringed with hairs on the margin.

Collateral buds Accessory buds at the side of the main axillary bud.

Compound leaves Those in which the blade consists of two or more separate parts (leaflets).

Conduplicate Having two parts folded lengthwise along the midrib.

Connate Having a congenital union of like structures or organs.

Convolute Rolled up lengthwise.

Cordate Heart-shaped; notched at the base.

Coriaceous Of the texture of leather.

Corymb A flat or convex flower cluster with the outer flower opening first.

Crenate Dentate with the teeth greatly rounded.

Crenulate Finely crenate.

Cuneate Wedge-shaped.

Cuspidate Tipped with an abrupt, sharp, sometimes rigid, point.

Cyme A broad, more or less flat-topped flower cluster with central flowers opening first.

Deciduous Falling off in autumn or before.

Decompound Several times compound or divided.

Decumbent Reclining at the base but with the summit ascending.

Decurrent Extending down the stem below the insertion.

Deliquescent Having the main stem branching off into numerous smaller ones.

Deltoid Triangular.

Dentate Toothed, with teeth spreading or pointing outward.

Denticulate Having minute, spreading teeth; diminutive of dentate.

Depressed Somewhat flattened from the end.

Dichotomous Forking regularly by pairs.

Divaricate Widely divergent or spreading.

Drupe Fleshy stone fruit with inner part of wall bony and outer part soft.

Elliptical Oval or oblong with the ends rounded; widest at the middle.

Emarginate With a shallow notch at the apex.

Entire Having an even margin, not toothed, notched, or divided.

Ephemeral Short-lived.

Excurrent Having the stem or trunk continuing to the top of the tree.

Exfoliating Cleaving off in thin layers.

Exserted Projecting beyond, sticking out.

Falcate Scythe-shaped.

Fastigiate (branches) Erect, close together and more or less parallel, in dense masses.

Floccose With clusters or bunches of soft or woolly hair.

Foliate Suffix describing the number of leaflets.

Foliolate Suffix describing further divisions of the leaflet.

Follicle A fruit consisting of a single carpel splitting along the inner or upper suture only.

Fusiform Spindle-shaped, swollen in the middle and tapering toward each end.

Glabrate Nearly glabrous or becoming glabrous.

Glabrous Without hairs.

Gland A secreting part or appendage; term is often used for small swellings or projections on various organs.

Glandular Furnished with glands, or gland-like.

Glaucous Covered or whitened with a bloom.

Globose Spherical in form or nearly so.

Globular Nearly globose.

Glutinous Sticky.

Granulose Composed of or appearing as if covered by minute grains.

Gymnosperms Plants with seeds borne naked.

Hastate Shaped like an arrowhead but with the basal lobes spreading.

Head A short, compact flower cluster of more or less sessile flowers.

Herbaceous Not woody; of the texture of an herb.

Hirsute With rather coarse hairs.

Hispid With rigid hairs or bristles.

Imbedded buds Buds completely or partially sunken in the bark.

Imbricate Overlapping like the shingles on a roof.

Inflorescence Mode of flower-bearing; a flower cluster.

Involucre A whorl of small leaves or bracts subtending or positioned just below a flower or flower cluster.

Lanceolate Several times longer than wide, broadest near the base and narrowed to the apex.

Leaflet One part of a compound leaf.

Leaf scar A scar left on the twig when a leaf falls.

Legume Dehiscent dry fruit of a simple pistil normally splitting along two sides.

Lenticel A small corky area or speck serving as a breathing pore.

Ligulate Strap-shaped.

Linear Long and narrow with parallel margins; line-shaped.

Lobed Divided into segments about halfway to the middle; segments are larger than teeth.

Lunate Crescent-shaped.

Membranous Thin, rather soft, and somewhat translucent.

Mucronate Having an abrupt, minute point.

Multiple fruit A fruit formed by the coherence of pistils and associated parts of the several flowers of an inflorescence.

Netted venation Pattern in which the principal veins of a leaf form a network.

Node A joint or place where leaves are attached to a stem.

Nut A hard, indehiscent, one-celled and one-seeded fruit, usually resulting from a compound ovary.

Nutlet A small nut.

Oblanceolate Like lanceolate but with the narrow end toward the base.

Oblique (leaves) Having unequal sides or a base with sides of unequal lengths.

Oblong Longer than broad, and with the sides nearly parallel most of their length.

Obovate Inverted ovate; the widest part toward the apex.

Obtuse Blunt or rounded at the end.

Opposite Two (leaves or buds) at a node.

Orbicular Circular.

Ovate Of the shape of a longitudinal section through a chicken egg, with the broad end toward the base.

Ovoid Of the shape of a chicken egg, with the broad end toward the base.

Palmate Radiating fan-like from approximately one point.

Panicle An elongated irregularly branched inflorescence.

Pappus The modified calyx-limb in Compositae, forming a crown of various characters at the summit of the achene.

Parallel venation Pattern in which the principal veins of a leaf run parallel or nearly so.

Pedicel Stem of an individual flower of a flower cluster.

Peduncle Stem of a solitary flower or of a flower cluster.

Pellucid Clear, nearly transparent.

Peltate Attached to its stalk inside the margin; shield-shaped.

Pendulous More or less hanging, drooping, or declined.

Perennial Of three or more seasons' duration.

Perfoliate Of a leaf in which the stem appears to pass through the blade.

Persistent Remaining attached; leaves not all falling off at the same time.

Petiole The stalk of a leaf.

Petiolule The stalk of a leaflet.

Pinnate Arranged feather-like on each side of a common axis.

Plicate Folded, usually lengthwise, like a closed fan.

Pome Fleshy fruit with a bony or leathery several-celled core and a soft outer part.

Prickle A small, sharp outgrowth from the bark or rind.

Prostrate Lying flat upon the ground.

Pseudo-terminal bud The uppermost lateral bud on a twig lacking a terminal bud, appearing as a terminal bud The bud scar with its ring of vascular tissue and the axillary position of the bud will enable one to determine this condition.

Puberulent   Minutely pubescent.
Pubescent   Covered with soft, short
hairs.

Raceme   A simple flower cluster of
pedicelled flowers upon a common
elongated axis.
Rachis   The main axis of an inflorescence
or compound leaf.
Reniform   Kidney-shaped.
Retuse   With a shallow notch at a
rounded apex.
Revolute   Rolled backwards from the
margins or apex.
Rugose   Wrinkled; generally due to the
depression of the veins in the upper surface
of the leaf.
Rugulose   Slightly wrinkled.

Sagittate   Shaped like an arrowhead;
triangular, with the basal lobes pointing
downward.
Samara   An indehiscent winged fruit.
Scabrous   Rough to the touch.
Scurfy   Covered with small bran-like
scales.
Serrate   Having sharp teeth pointing
forward.
Serrulate   Finely serrate; the teeth shorter
and closer together.
Sessile   Without a stalk.
Simple leaves   Those in which the blade
is all in one piece; may be lobed or cleft
but not divided all the way to the
midrib.
Sinus   The space or recess between two
lobes of a leaf.
Spatulate   Gradually narrowed
downward from a rounded summit.
Spike   A flower cluster like a raceme, but
with sessile flowers.
Spine   A sharp, rather slender, rigid
outgrowth, representing a modified leaf or
stipule or, loosely, a modified tooth.
Spur   A short, slowly grown branchlet.
Stalked bud   A bud in which the outer
scales are attached above the base.

Stellate   Star-shaped; several or many
branches from the base.
Stipel   Stipule of a leaflet or basal
appendage of a petiolule.
Stipule   A basal appendage of a
petiole.
Stoloniferous   Bearing runners or shoots
that take root.
Striate   Marked with fine longitudinal
lines or ridges.
Strigose   With appressed, sharp, straight,
and stiff hairs.
Sub-   A prefix meaning somewhat or
nearly.
Subcordate   Slightly cordate.
Subtending   Postioned just below
another structure on the same axis.
Subulate   Awl-shaped; broad at base,
narrow and tapering from the base to a
sharp, rigid point, the sides generally
concave.
Suffrutescent   Slightly or obscurely
shrubby.
Superposed buds   Accessory buds above
the axillary bud.
Supra-axillary   Located above an
axil.

Tendril   A thread-like leaf or stem
part by which a plant may cling to a
support.
Tepal   Sepal (outer set of floral leaves)
or petal (inner set of floral leaves) that
cannot be distinguished from each
other.
Terete   Round in cross section.
Terminal bud   The bud formed at the
tip of a twig.
Thorn   A modified stem or branch with a
sharp point.
Tomentose   Densely hairy with matted
or tangled woolly hairs.
Tomentulose   Tomentose but with fewer
or shorter hairs.
Trifoliate   Of three leaflets.
Truncate   Ending abruptly as if cut off
transversely.

Twig A young shoot; the growth of the past season only.

Umbel An umbrella-like flower cluster.
Undulate With a wavy margin or surface.

Valvate With scales of the bud meeting at the edges and not overlapping.

Whorled Having three or more leaves or buds at a node.

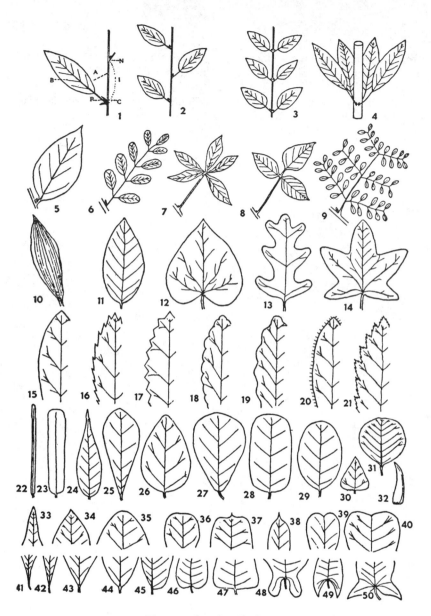

Terms used to describe leaves

Generalized simple leaf: 1, B, blade; C, stipule; P, petiole; A, leaf axil; N, node; I, internode.

Arrangement: 2, alternate (spiral); 3, opposite; 4, whorled.

Composition: 5, simple; 6, pinnately compound; 7, palmately compound; 8, trifoliate; 9, decompound.

Venation: 10, parallel; 11, pinnately netted; 12, palmately netted.

Lobing: 13, pinnately lobed; 14, palmately lobed.

Margin: 15, entire; 16, serrate; 17, dentate; 18, crenate; 19, undulate; 20, ciliate; 21, doubly-serrate.

Outline: 22, acicular; 23, linear; 24, lanceolate; 25, oblanceolate; 26, ovate; 27, obovate; 28, oblong; 29, elliptical; 30, deltoid; 31, orbicular; 32, subulate.

Apex: 33, acuminate; 34, acute; 35, obtuse; 36, truncate; 37, mucronate; 38, cuspidate; 39, emarginate; 40, retuse.

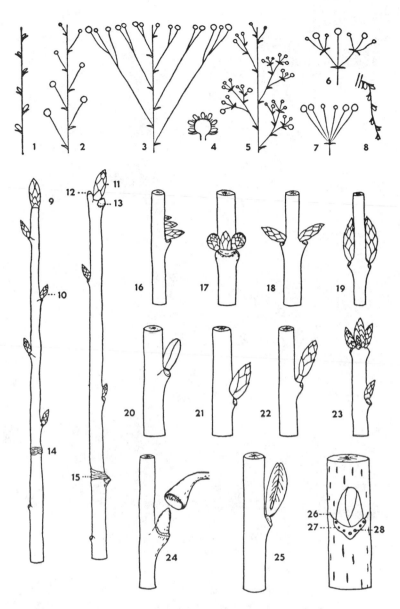

Terms used to describe inflorescences and winter buds

Inflorescences: 1, spike; 2, raceme; 3, corymb; 4, head; 5, panicle; 6, cyme; 7, umbel; 8, ament (catkin).

Buds: 9, terminal; 10, lateral; 11, pseudo-terminal; 23, clustered terminal.

Attitude of buds: 18, divaricate; 19, appressed.

Position of buds: 21, axillary; 16, superposed; 17, collateral.

Insertion of buds: 21, sessile; 22, stalked; 24, sub-petiolar.

Bud scales: 20, valvate; 21, imbricate; 25, absent (naked bud).

Scars: 12, abscission scar of growing point; 14, terminal bud scar; 15, pseudo-terminal bud scar; 13, upper leaf scar; 27, leaf scar; 26, stipule scar; 28, bundle scar.

ANNOTATED BIBLIOGRAPHY

The references listed below are useful companion volumes for complete descriptions, ranges, illustrations, or additional identification aids to the plants treated in this book. Many of these have been consulted in the shaping of this revision.

Bailey, L. H. 1949. *Manual of Cultivated Plants, Most Commonly Grown in the United States and Canada*. New York: MacMillan. This book is the only one of its kind, as it offers keys and descriptions to many of the plants in cultivation fifty years ago in the United States. It is still very useful for identification and comparison of many plants in cultivation today.

Bailey Hortorium staff. 1976. *Hortus Third: A Concise Dictionary of Plants Cultivated in the United States and Canada*. New York: MacMillan. A comprehensive alphabetical listing of common and uncommon plants cultivated in North America, north of Mexico (as well as most native plants). Excellent descriptions, common names, cultivated varieties, and other horticultural information has made *Hortus* the standard reference for the last twenty years.

Blackburn, Benjamin. 1952. *Trees and Shrubs in Eastern North America, Keys to the Wild and Cultivated Woody Plants Grown in the Temperate Regions Exclusive of Conifers*. New York: Oxford University Press. The only comprehensive key to cultivated and native plants exclusive of conifers of eastern temperate North America to this point, this book includes some cultivated varieties.

Campbell, C. S., F. Hyland, and M. L. F. Campbell. 1978. *Winter Keys to Woody Plants of Maine*. Orno, Maine: University of Maine at Orono Press. Keys and sixty-three plates of excellent illustrations of twigs, buds, and fruits of woody plants.

Collingwood, G. H. and W. D. Brush. 1974. *Knowing Your Trees*. Washington, D.C.: The American Forestry Association. Descriptions, range maps, and photographs, including close-ups of the bark, branches, and fruits of each North American native tree.

Cope, E. A. 1986. *Native and Cultivated Conifers of Northeastern North America*. Ithaca, NY: Cornell University Press. Keys and detailed line drawings for conifers that grow in the northeastern North America.

Dirr, M. A. 1990. *Manual of Woody Landscape Plants*. Champaign, Ill.: Stipes. Descriptions and some illustrations of woody plants in cultivation. An exhaustive and valuable resource, particularly for horticulturists.

Duncan, W. H. and M. B. Duncan. 1988. *Trees of the Southeastern United States*. Athens, Ga.: University of Georgia Press. Keys, descriptions, and often color photographs of trees of the Southeast.

Fernald, M. L. 1950. *Gray's Manual of Botany, a Handbook of the Flowering Plants and Ferns of the Central and Northeastern United States and Adjacent Canada*, 8th edition. New York: American Book Co. This book is one of the bibles of the American botanist. It has complete descriptions, keys, derivations of the scientific names, and distributional and habitat information, and it contains a suprising amount of information on cultivated plants. I have often consulted this volume in preparing these keys.

Flint, H. L. 1997. *Landscape Plants of Eastern North America, Exclusive of Florida and the Immediate Gulf Coast*, 2nd edition. New York: Wiley. This book is extensive in its coverage of cultivated plants north of Florida. It contains photographs; size and habit drawings; information about optimal growing conditions, maintenance, and cultivated varieties; and limited descriptions. This volume was useful in my decisions about which cultivated species to include.

Flora of North America Editorial Committee. 1993. *Flora of North America*, vol. 1. New York: Oxford University Press.

———. 1997. *Flora of North America*, vol. 3. New York: Oxford University Press. Written by experts in each plant group, this book provides descriptions, keys, illustrations, range maps, and current taxonomic judgments on the flora of North America.

Gleason, H. A. 1952. *The New Britton and Brown Illustrated Flora of the Northeastern United States and Adjacent Canada*. New York: The New York Botanical Garden. This three-volume set contains keys, descriptions, and line drawings of native and many naturalized plants of northeastern North America. It can be used as a companion to Gleason and Cronquist 1991. A recent revison is a better companion in terms of the nomenclature, but the reduced size of the drawings makes it less useful.

Gleason, H. A. and Arthur Cronquist. 1991. *Manual of Vascular Plants of Northeastern United States and Adjacent Canada*. New York: The New York Botanical Garden. This book is a complete treatment, with keys and descriptions, of the native and naturalized plants of northeastern North America. I have consulted the descriptions extensively, especially for range, rarity, and habitat information.

Godfrey, R. K. 1988. *Trees, Shrubs, and Woody Vines of Northern Florida and Adjacent Georgia and Alabama*. Athens, Ga.: University of Georgia Press. This work is a valuable and exhaustive reference for much of the Southeast. I have frequently consulted the descriptions, keys, and the range, habitat, and taxonomic information in this thorough book.

Graves, A. H. 1952. *Illustrated Guide to Trees and Shrubs*. Wallingford, CT: Arthur Graves. An out-of-date book with short descriptions, keys, and useful illustrations of selected woody plants.

Harlow, W. M. 1959. *Fruit Key and Twig Key to Trees and Shrubs*. New York: Dover. Keys and photographs of twigs and fruits of woody plants of northeastern North America.

Little Jr., E. L. 1979. *Checklist of United States Trees Native and Naturalized*. Washington, D.C.: USDA handbook. An excellent source of synonyms and the reference for standardized common names for United States trees.

Mitchell, R. S. and G. C. Tucker. 1997. *Revised Checklist of New York State Plants*. Albany, NY: New York State Museum Bulletin No. 490. A source for synonyms, common names, and rarity status in New York.

Muenscher, W. C. 1922. *Keys to Woody Plants*. Ithaca, NY: Cornell University Press. This and subsequent editions have an extensive bibliography of woody plant books listed by geographical area.

———. 1950. *Keys to Woody Plants*, 6th edition, revised. Ithaca, NY: Cornell University Press.

Radford, A. E., H. E. Ahles, and C. R. Bell. 1968. *Manual of the Vascular Flora of the Carolinas*. Chapel Hill, N.C.: The University of North Carolina Press. Descriptions, keys, and range and habitat information for Carolina plants. I have consulted this volume frequently.

Sargeant, C. S. 1922. *Manual of Trees of North America Exclusive of Mexico*, 2nd edition. Boston: Houghton-Mifflin. Extensive descriptions and illustrations of North American trees.

Scoggan, H. J. 1979. *The Flora of Canada*. Ottawa: National Museum of Natural Sciences. Descriptions and keys to plants of Canada.

Trelease, William. 1918. *Winter Botany*. Urbana, Ill.: William Trelease. A comprehensive (for its time) manual of cultivated and native woody plants, including many tropical genera. The book includes short keys (using mainly one character), excellent descriptions, and very useful illustrations of winter characteristics.

Tutin, T. G., V. H. Heywood, N. A. Burges, D. M. Moore, D. H. Valentine, S. M. Walters, and D. A. Webb. 1968. *Flora Europaea*. London: Cambridge University Press. Keys, descriptions, and ranges of native plants of Europe. This five-volume set provides information on many of the naturalized plants in North America.

Viertel, A. T. 1970. *Trees, Shrubs and Vines: A Pictorial Guide to the Ornamental Woody Plants of the Northern United States Exclusive of Conifers*. Syracuse, NY: Syracuse University Press. A handy, inexpensive, quick reference for leaf shape for a large number of woody plants of the Northeast.

Voss, E. G. 1972, 1985, 1996. *Michigan Flora, A Guide to the Identification and Occurrence of the Native and Naturalized Seed-Plants of the State*, 3 volumes. Ann Arbor, Mich.: Cranbrook Institute of Science Bulletin 59 and University of Michigan Herbarium. This three-volume set contains excellent keys and much range and distributional information. I have consulted these volumes extensively for their taxonomic information, especially for difficult groups with unclear species limits.

INDEX

Boldface type indicates the page on which a genus is listed in the Keys to Species.

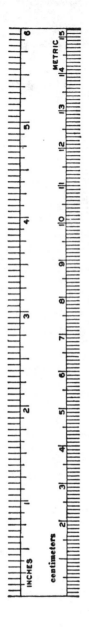